Omics in Soil Science

Edited by

Paolo Nannipieri

Giacomo Pietramellara

and

Giancarlo Renella

Department of Agrifood Production and Environmental Sciences
University of Florence
Italy

Caister Academic Press

Caister Academic Press
Norfolk, UK

www.caister.com

British Library Cataloguing-in-Publication Data
A catalogue record for this book is available from the British Library

ISBN: 978-1-908230-32-4 (hardback)
ISBN: 978-1-908230-94-2 (ebook)

Description or mention of instrumentation, software, or other products in this book does not imply endorsement by the author or publisher. The author and publisher do not assume responsibility for the validity of any products or procedures mentioned or described in this book or for the consequences of their use.

Cover design adapted from Figure 6.1.

Printed and bound in Great Britain

Contents

Contributors

Mariarita Arenella
Department of Agrifood Production and Environmental Sciences
University of Firenze
Italy

mariarita.arenella@unifi.it

Jean Armengaud
CEA, DSV, IBEB, Lab Biochim System Perturb
Bagnols-sur-Cèze
France

jean.armengaud@cea.fr

Matthieu Barret
BIOMERIT Research Centre
Department of Microbiology
University College Cork
Cork
Ireland

matthieu.barret@angers.inra.fr

Céline Bland
CEA, DSV, IBEB, Lab Biochim System Perturb
Bagnols-sur-Cèze
France

celine.bland@cea.fr

PierSandro Cocconcelli
Istituto di Microbiologia
Università Cattolica del Sacro Cuore
Piacenza
Italy

pier.cocconcelli@unicatt.it

Mariana Silvia Cretoiu
University of Groningen
Groningen
The Netherlands

m.s.cretoiu@rug.nl

Francisco Dini-Andreote
Linneausborg
Groningen
The Netherlands

f.dini.andreote@rug.nl

Jan Dirk van Elsas
University of Groningen
Groningen
The Netherlands

j.d.van.elsas@rug.nl

Laura Giagnoni
Department of Agrifood Production and Environmental Sciences
University of Firenze
Italy

laura.giagnoni@unifi.it

Daniel H. Huson
Center for Bioinformatics
University of Tübingen
Tübingen
Germany

daniel.huson@uni-tuebingen.de

Heribert Insam
University of Innsbruck
Institute of Microbiology
Innsbruck
Austria

heribert.insam@uibk.ac.at

Anna Maria Kielak
Linneausborg
Groningen
The Netherlands

a.m.kielak@rug.nl

Yongkyu Kim
Max Planck Institute for Terrestrial Microbiology
Marburg
Germany

yongkyu.kim@mpi-marburg.mpg.de

Werner Liesack
Max Planck Institute for Terrestrial Microbiology
Marburg
Germany

liesack@mpi-marburg.mpg.de

John P. Morrissey
Department of Microbiology
University College Cork
Cork
Ireland

j.morrissey@ucc.ie

David D. Myrold
Department of Crop and Soil Science
Oregon State University
Corvallis, OR
USA

david.myrold@oregonstate.edu

Paolo Nannipieri
Department of Agrifood Production and Environmental Sciences
University of Florence
Florence
Italy

paolo.nannipieri@unifi.it

Fergal O'Gara
BIOMERIT Research Centre
Department of Microbiology
University College Cork
Cork
Ireland

f.ogara@ucc.ie

Giacomo Pietramellara
Department of Agrifood Production and Environmental Sciences
University of Florence
Florence
Italy

giacomo.pietramellara@unifi.it

Edoardo Puglisi
Istituto di Microbiologia
Univesità Cattolica del Sacro Cuore
Piacenza
Italy

edoardo.puglisi@unicatt.it

Giancarlo Renella
Department of Agrifood Production and Environmental Sciences
University of Florence
Florence
Italy

giancarlo.renella@unifi.it

Marco Trevisan
Istituto di Chimica Agraria ed Ambientale
Università Cattolica del Sacro Cuore
Piacenza
Italy

marco.trevisan@unicatt.it

Sotirios Vasileiadis
Istituto di Chimica Agraria ed Ambientale
Università Cattolica del Sacro Cuore
Piacenza
Italy

sotirios.vasileiadis@unicatt.it

Nico Weber
Center for Bioinformatics
University of Tübingen
Tübingen
Germany

weber@informatik.uni-tuebingen.de

Carl-Eric Wegner
Max Planck Institute for Terrestrial Microbiology
Marburg
Germany

carl-eric.wegner@mpi-marburg.mpg.de

Current books of interest

Preface

In 2013 the 'genomic' era gets to the age of 60 after the seminal papers by F. Crick and J. Watson where they showed the DNA molecular structure and formalized the central dogma in biology, i.e. the genetic information is encoded by DNA by the mean of its transcription into messenger RNA and by its further translation to proteins. Formalization of the universal biological programming language has indisputably revolutionized our cultural and scientific approaches in all biological sciences.

Soil is a peculiar biological system with several distinctive properties including a huge microbial diversity (Chapter 1). Only a small percentage of microorganisms inhabiting soil is culturable and thus the majority of soil microbial species are unknown. Their characterization can give further insights in soil as a biological system. In addition the expression of genes of these unknown microbial species can synthesize unknown molecules, such as novel enzymes and antibiotics, whose use can have important implications for human activities and health. Chapters 3, 5 and 6 discuss advantages and disadvantages, including methodological problems, of studies on collective genomes, transcripts and proteins of soil. Bibliography on soil metagenomics is more extensive than those on soil metatranscriptomics and soil proteomics also due to the development of second generation sequencing technologies. However, the development of the latter is needed because the integration of metagenomics, metatranscriptomics and proteomics can give a more complete picture of living organisms. Protogenomics, involving both proteomics and metagenomics, is important because the detection of a specific protein can confirm the presence of protein encoding genes hypothesized by a metagenomic screening (Chapter 8). Bioinformatics is rapidly developing being important in all omics studies since both sequencing or mass spectrometry reads have to be compared against a reference database and the subsequent steps are based on the obtained alignments. In chapter 9 it is discussed the use of MEGAN (Metagenome Analyzer), a program, designed to work with very large datasets, which can support various input formats for loading data and export analysis results in different text-based and graphical formats. An overview of methodologies and concepts related to marker gene screening from soil is discussed in chapter 4.

Plant-microbes interactions are important for plant growth. The analysis of the rhizospheric communities incorporating both established techniques, and recently developed 'omic technologies' can now facilitate investigations into the molecular basis underpinning the establishment of plant-microbes interactions in the rhizosphere.

The collective analysis of genomes, transcripts and proteins are not the only meta approaches to characterize microbial communities; for example, collective soil volatile organic compounds (VOCs), mainly produced by microorganisms, can be characterized to evaluate composition and activities of soil microbial communities (Chapter 7).

As discussed in the last chapter the various omics approaches hold much promise but further refinement is needed for their widespread adaptation; their comparison with rapid and inexpensive methods, that have become standards for soil microbiology research, can evaluate their readiness to be used. The very exciting aspect of soil 'omics' is that we do not currently understand where most of DNA is located, belongs to or codes for, and what environmental signals control its expression within the genetic and metabolic networks which are being slowly identified in the pedosphere. Integrative omic and hypothesis driven research answer these difficult questions. In this sense, development of soil metabolomics, lipidomic and glycomics, missing in this volume, is urgent to fill the gaps in the current knowledge of the complex genotype-phenotype relationships in the soil microbial communities, and how species/environment equilibria are perturbed by changes in the environment. The next integrative soil 'omic' studies undertaken by environmental biologists in the coming years will surely reveal what role the physico-chemical environment plays for microbial communities complexity and functions far more than what we can understand by continuing to analyse soils with traditional soil microbial and biochemical techniques.

Soil as a Biological System

1

Paolo Nannipieri, Giacomo Pietramellara and Giancarlo Renella

Abstract

Soil is an unique biological system with an abundant microflora and a very high microbial diversity. The space occupied by microorganisms is very low because only few microsites have the right set of conditions suitable for microbial life. Surface-reactive particles can adsorb important biological molecules, such as DNA and enzymes, which become resistant to microbial degradation and thus genes are preserved and extracellular enzymes can be reactive when conditions are not suitable for microbial activity. Most soil functions mainly depend on microbial activity but soil fauna can accelerate microbial processes and complete food webs in soil. Omics techniques, such as metagenomics, metatranscriptomics and proteomics, have several problems when applied to soil. However, if used in a complementary way these techniques are promising for providing an integrated picture of the relationship between composition and activity of soil microflora.

Introduction

Soil is considered a biological system with the most abundant living organisms on the earth (Paul and Nannipieri, 2012). The importance of soil organisms in completing important cycles of nutrients such as carbon, nitrogen and sulphur has been recognized in the 1890–1910 period and due to the important advances in the knowledge of soil, this period has been termed 'golden age' of soil microbiology (Waksman, 1932). Then in the mid-twentieth century the use of tracers has allowed quantifying rates of important nutrient transformations (Paul and Nannipieri, 2012). Nowadays the use of molecular techniques, based on the extraction of nucleic acids from soil followed by their characterization, can determine unculturable microorganisms and their activities with profound progresses in the knowledge of soil as a biological system. Therefore, new reactions and microorganisms are being discovered. For this reason Paul and Nannipieri (2012) have proposed to denominate this period as the 'second golden age' of soil biology. The use of molecular techniques in soil has allowed not only a better understanding of soil functionality but also the study of unknown microorganisms with discovery of novel antibiotics and novel enzymes with important consequences on human health and development of biotechnology.

The aim of this introductory chapter is to discuss the particular aspects of soil as a biological system and to introduce the reader to the omics approaches which has the potential to carry out in-depth analysis of gene detection and expression in soil. Soil metagenomics has decreased the threshold of microbial species detection with the possibility to detect rare species escaping the common used fingerprinting methods separating DNA amplified by PCR. Both metatranscriptomics and proteomics can determine gene expression in soil.

Main characteristics of soil as a biological system

Soil is a structured, heterogeneous and discontinuous system with the solid phase prevailing on the liquid and gaseous phases, and is a peculiar living system (Nannipieri *et al.*, 2003) because:

1 Soil contains a huge biological biomass with a high microbial diversity. The number of microbial cells can range from 10^8 to 10^{10} cells per g of soil and that of microbial species from thousand to millions per g of soil (Chapter 3). The amount of genetic information stored in 1 g of soil is supposed to be 4000 human genomes (Schloter, 2012) and the amount of carbon and nitrogen stored in microbial biomass is greater than in plant biomass (Whitman *et al.*, 1998).

2 The biological space, that is the space occupied by living organisms, is very small, accounting for 0.1–5% of the total available volume. Indeed only a few microsites have the right set of conditions for soil microbial activity and microorganisms are generally concentrated in hotspots such as rhizosphere and soil surrounding particulate organic matter, animal manures, liquid flow paths, etc. (Nannipieri *et al.*, 2003).

3 Surface-reactive soil particles can absorb important biological molecules such as proteins and nucleic acids. Proteins can also be entrapped by humic molecules (Nannipieri *et al.*, 2012). In this way active extracellular enzymes can maintain their activities being protected against proteolytic degradation and the role of these stabilized extracellular enzymes in microbial ecology has been also hypothesized (Burn, 1982; Burns *et al.*, 2013). In a pioneering review McLaren (1975) suggested that soil is a system of humus and minerals containing both immobilized enzymes and occluded microbial cells. Notably, stabilized extracellular enzymes can be active even under unfavourable conditions for soil microflora.

   Humic molecules, clay and sand particles can also adsorb or bound DNA, which is then protected against degradation by nucleases but still able to transform competent bacterial cells. Extracellular DNA can move through soil transported by water (Ceccherini *et al.*, 2007) and then be adsorbed and stabilized at lower soil depth (Agnelli *et al.*, 2004). Therefore, soil can preserve genes outside microbial cells and this, with the presence of active and stabilized extracellular enzymes, is an unique characteristic of the system.

   Microbial degradation of organic substrates can be retarded if these substrates are adsorbed by surface-reactive particles or localized inside pores or microaggregates since the accessibility of microorganisms to these substrates is reduced (Ladd *et al.*, 1996). The degradation of polymeric components of plant, microbial and animal residues reaching soil necessitates the activities of microbial consortia, with the release of arrays of extracellular enzymes acting in a synergic way (Burns *et al.*, 2913).

4 Enzyme-like reactions, such as oxidation, deamination, polymerization, polycondensation and ring cleavage, can occur in soil due to catalysts such as clay minerals and several oxides (Huang *et al.*, 1990; Ruggiero *et al.*, 1996). There are no methods to distinguish between these abiotic reactions and those catalysed by enzymes. Probably the latter are

prevailing over the former under natural conditions being abiotic reactions important under adverse conditions for biological activity (Nannipieri *et al.*, 2003). Other abiotic reactions, such as photochemical breakdown of litter, can occur in arid and semiarid environments (Pancotto *et al.*, 2005; Gallo *et al.*, 2006, 2009).

5 The role of the smallest soil biota, virus and prions has been neglected despite their importance on microbial ecology. The study of virus has generally concerned their effects as pathogens of plants, animals and insects (Williamson *et al.*, 2012). However, most of soil viruses are bacteriophages and need to infect a bacterial cell to replicate. The ratio between viral abundance to bacterial abundance increases in agricultural soils (330–470) compared to waterlogged soils (10–60) and aquatic systems (0.5–50), because viral production is much faster than viral decay in agricultural soils. The very high ratio of Antarctic soils (170–8100) is due to low bacterial abundances of these soils; however, the low bacterial biomass can support the relatively high number of virus of these soils. Virus adsorption to surface-reactive particles can increase their persistence in soil (Lipson and Stotzky, 1986; Vettori *et al.*, 1999)

6 Complex trophic interactions exist in soil. For example, in the rhizosphere soil bacteria are more active than fungi due to the release of organic exudates by roots. However, nutrients blocked in the bacterial biomass are released and made available to roots by microfaunal grazers, such as nematodes and protozoa (Griffiths *et al.*, 2012). In addition, protozoa show feeding preferences and in the rhizosphere this preference can stimulate root growth by promoting specific rhizobacteria.

Soil biota and their functions in soil

Biological habitats in soil differ in both time and space for their biological, chemical and physical properties and scales of these habitats mainly depend on the organism size. Virus and prions have molecular size. Indeed some virus can consist of a nucleic acid surrounded by a protein coat (Williamson *et al.*, 2012). The other soil biota can range from microns for bacteria and archaea ($\geq 0.5\,\mu m$) to less than $100\,\mu m$ for fungi, and between $20\,\mu m$ and less than 100 mm for fauna (Nannipieri *et al.*, 2003). According to their size faunal organisms are divided in microfauna (protozoa), mesofauna (Nematoda, Enchytraeidae, Mollusca, Rotifera, Collembola, Acari, etc.) and macrofauna (Chilopoda, Lipotyphla and Lumbricidae) (St John *et al.*, 2012).

Prokaryotes (Eubacteria and Archaea) exclusively perform reactions such as nitrogen fixation and biotic methane synthesis and with fungi carry out most of the reactions in soil (Schloter, 2012). Prokaryotes can survive to extreme conditions, such as high salinity and drought, degrade pollutants and affect either positively (plant growth-promoting rhizobacteria) or negatively (phytopathogens) plant growth (Schloter, 2012). In addition to genomic DNA, they often contain plasmid DNA, which is easily transferable to other bacterial cells and encode for resistance genes making these bacteria resistant against antibiotics and heavy metals. Prokaryotes with fungi are also important in the formation of soil structure through various mechanisms, including the release of organic compounds acting as glue.

Fungi are involved in the decomposition of polymeric substances, such as lignin, cellulose chitin, etc., of plant and animal residues, in the decomposition of organic pollutants and in both beneficial (mycorrhizae) and detrimental (fungal pathogens) symbiosis with plant (Thorn, 2012). Generally their biomass is higher than that of other soils biota.

Protozoa are active in water-filled pores and water films of large pores and are important links in the soil food web contributing to the carbon and nitrogen cycle in soil. Indeed they are mainly bacterial feeders but they can also feed on fungi and other faunal organisms, such as nematodes (Griffiths *et al.*, 2012). On the other hand they are consumed individually by nematodes, collembola and mites, and together soil and organic matter by enchytraeids and earthworms.

Other important positions in soil food webs are occupied by nematodes, which can be herbivores consuming roots, feeders of bacteria and fungi, predators of other nematodes and other small invertebrates (McSorley, 2012). They can increase decomposition rates in soil and thus recycling carbon and other nutrients by dispersing microflora to new sites. They can be consumed by mites and other invertebrate predators. Nematode herbivores can also be detrimental to plants causing crop yield losses if endoparasites.

Microarthropods (collembolan and mites, which include acari), macroarthropods (include as classes Arachnida, Malacostraca, Diploda, Chilopoda and Hexapoda) and Enchytraeidae (family of the subclass Oligochaeta) are other components of soil fauna participating to soil food webs. Some can markedly modify soil structure as the cases of ants and termites (both macroarthropods) and earthworms (Oligochaeta). The bibliography of earthworms is extensive due to their beneficial effects on agriculture, waste management and land reclamation (Van Vliet *et al.*, 2012).

Microbial diversity, soil functions and the holistic approach

Here we consider microbial and not biological diversity because soil functions mainly depends on the activity of both protokaryotes and fungi. Microbial diversity is a general term including genetic diversity (amount and distribution of genes within microbial species), diversity of bacterial and fungal communities and ecological diversity (number of trophic levels, number of guilds and changes in microbial composition) (Nannipieri *et al.*, 2003). Generally, microbial diversity is simply considered to include the number of different microbial species (richness) and their relative abundance (eveness). In terrestrial ecosystems the plant yield increases by increasing plant diversity until a certain value and then for an further increase in diversity there is a decrease in the yield (Loreau *et al.*, 2001). In soil, in contrast to what is seen for higher organisms, there is no consistent relationships between microbial diversity and soil processes such as C and N mineralization because soil microbial communities are characterized by high functional redundancy (Nannipieri *et al.*, 2003; Nielsen *et al.*, 2011). For example, Chander *et al.* (2002) observed that $CHCl_3$ fumigated soil with a lower microbial biomass than the corresponding non-fumigated soil respired the same amount of ^{14}C-CO_2 from the added labelled straw. It is not clear, however, if there is a threshold level under which these processes are affected and if this threshold value depend on soil type. Therefore, the links between microbial diversity and soil functions as well as those between microbial diversity and stability (resilience or resistance) require an accurate determination of microbial diversity; it is needed to determine rare species which escape the fingerprinting methods. However, the situation may be different for processes such as nitrification which is carried out by specific microbial communities (Nannipieri *et al.*, 2003).

Other approaches than molecular techniques and omics approaches are important to quantify important soil functions. Indeed to quantify nutrient rates in soil is not needed to determine microbial diversity, gene expression, the concentration of each metabolite and the rate of each reaction involved in the target nutrient transformation (Nannipieri *et al.*, 2003). The use of labelled compounds (for example ^{14}C-, ^{13}C-labelled or ^{15}N-labelled) combined with the holistic approach can trace the fate of the main nutrients in the soil–plant system (Chapter 10).

The omics approaches in soil

The accurate determination of the presence and expression of taxonomical and functional genes is important to understand better the links between microbial diversity and soil functions, the interactions among microbial species and interactions between microbial species and fauna or plants, with consequent better understanding of processes such pathgogenicity, beneficial effects on plants, etc. Soil metagenomics aims to detect all species in soil since it is the study of the collective microbial genome of soil. As reviewed by Van Elsas *et al.* in Chapter 3 there are (1) methodological problems, mainly related to the quantity and quality of DNA extracted from soil; and (2) computational problems related to the huge amount of information. Nowadays it is impossible to combine millions of single reads to reconstruct the complete microbial genome of soil since we are unable to properly assemble the huge amount of short reads generated by soil metagenomics. Mining genetic novelty is another potential use of soil metagenomics and it is generally based on clone libraries prepared by using soil DNA. Also here the length of extracted DNA is important since large DNA fragments (even up to 200 kb) are required for detecting expression of gene sequences including operons responsible of antibiotic synthesis; this is only possible by preparing bacterial artificial chromosomes (BAC). The exploring for novel genes has generally given poor results due to the low level of gene expression in the host with expression of genes of known functions. However, as discussed by Van Elsas *et al.* in Chapter 3, future research is needed to overcome the present problems in soil metagenomics also by using the old culture techniques to study the behaviour of unculturable species discovered by soil metagenomics.

Studies on gene expression by soil metatranscriptomics and soil proteomics are still in their infancy if compared with similar studies in other environments. As reviewed by Kim *et al.* (Chapter 5), there are (i) methodological problems in soil metatranscriptomics, due to the lack of a method to extract total RNA and enrich mRNA which can be applied to all soils; and (ii) computational problems due to the lack of a representative database of representative genomes and metagenomes for transcript mapping. Further research should aim in setting up methods for extracting total RNA and enrich mRNA from groups of soils since it is problematic setting up an universal method; a method for clay soils can be different from a method for sandy soils due to the RNA adsorption by clays (Wang *et al.*, 2012). Research on soil metagenomics is being developed and database of soil microbial genes is increasing and this could also increase database of transcripts.

Soil proteomics, which aims to study the whole protein expression profile of soil, presents the same problems of soil metatranscriptomics, as discussed by Renella *et al.* (Chapter 6). Indeed soil proteomics is still in its infancy due to low recovery of soil proteins, because of protein adsorption by surface-reactive particles in soil, and poor information on the relative databases. However, soil proteomics presents a further problem compared with soil metagenomics and soil metatranscriptomics because interactions of proteins with soil colloids may affect the conformation of the protein molecule with consequences on the trypsin digestion of the molecule prior to the analyses by mass spectrometry (Arenella *et al.*, personal communication). Model studies based on the use of microbial species with known proteomes are useful for understanding effects of soil colloids on soil proteome. Future research should combine metagenomics, metatranscriptomics and proteomics in soil. Proteogenomics combines proteomics with metagenomics and has improved the annotation of genomes and discovered new genes that were not annotated, as discussed by Bland and Armengaud, in this book. This has been got by refining the annotation of the genome of model organisms with proteomic data.

Soil volatilomics is an approach which avoids the soil extraction step and uses volatile organic compounds released from soil as indicators of composition and activity of soil microflora (Chapter 7). However, other factors such as changes in environmental conditions can markedly affect the composition of volatile organic compounds released from soil.

Despite the many challenges that omics techniques have in soil, including the fact that replicated experiments are impractical, the three omics methods, if used in a complementary way, are still promising approaches for providing an integrated picture of the relationship between composition and activity of soil microflora. However, more hypothesis-driven than technology-driven research is needed in soil omics studies with more attention to imaginative research and experimental plans rather than technological skill as discussed by Myrold and Nannipieri in Chapter 10.

References

Agnelli, A., Ascher, J., Corti, G., Ceccherini, M.T., Nannipieri, P., and Pietramellara, G. (2004). Distribution of microbial communities in a forest soil profile investigated by microbial biomass, soil respiration and DGGE of total and extracellular DNA. Soil Biol. Biochem. *36*, 859–868.

Arenella, M., Giagnoni, L., Masciandaro, G., Ceccanti, B., Nannipieri, P., and Renella, G. (2013). Interaction between proteins and humic substances and protein identification by mass spectrometry. Soil Biol. Biochem. (submitted)

Burns, R.G. (1982). Enzyme activity in soil: location and a possible role in microbial ecology. Soil Biol. Biochem. *14*, 423–427.

Burns, R.G., DeForest J.L., Marxsen, J., Sinsabaugh, R.L., Stromberger, M.E., Wallenstein, M.D., Weintraub, M.N., and Zoppini, A. (2013). Soil enzymes in a changing environment: current knowledge and future directions. Soil Biol. Biochem. *58*, 216–234.

Ceccherini, M.T., Ascher, J., Pietramellara, G., Vogel, T., and Nannipieri, P. (2007). Vertical advection of extracellular DNA by water capillarity in soil columns. Soil Biol. Biochem. *39*, 158–163.

Gallo, M.E., Sinsabaugh, R.L., and Cabaniss, E.A. (2006). The role of ultraviolet radiation in litter decomposition in arid ecosystems. Appl. Soil Ecol. *34*, 82–91.

Gallo, M.E., Porras-Alfaro, A., Odenbach, K.J., and Sinsabaugh, R.L. (2009). Photoacceleration of plant litter decomposition in an arid environment. Soil Biol. Biochem. *41*, 1433–1441.

Griffiths, B., Clarholm, M., and Bonkowski, M. (2012). Protozoa. In Handbook of Soil Sciences: Properties and Processes, 2nd edition, P.M. Huang, Y. Li, and M.E. Summer, eds. (CRC Press, Taylor & Francis Group, Boca Raton, FL), pp. 25.1–25.5.

Ladd, J.N., Foster, R., Nannipieri, P., and Oades, J.M. (1996). Soil structure and biological activity. In Soil Biochemistry, vol. 9, G. Stotzky, and J-.M. Bollag, eds. (Marcel Dekker, New York), pp. 23–78.

Lipson, S.M., and Stotzky, G. (1986). Effect of kaolinite on the specific infectivity or reovirus. FEMS Microbiol. Lett. *37*, 83–88.

Loreau, M., Naeem, S., Inchausti, P., Bengtsson, J., Grime, J.P., Hector, A., Hooper, D.U., Huston, M.A., Rastelli, D., Schmid, B., *et al.* (2001). Biodiversity and ecosystem functioning: current knowledge and future challenge. Science *294*, 804–808.

McLaren, A.D. (1975). Soil as a system of humus and clay immobilized enzymes. Chemical Scripta *8*, 97–99.

McSorley R. (2012). Nematodes. In Handbook of Soil Sciences: Properties and Processes, 2nd edition, P.M. Huang, Y. Li, and M.E. Summer, eds. (CRC Press, Taylor & Francis Group, Boca Raton, FL), pp. 25-5–25-12.

Nannipieri, P., Ascher, J., Ceccherini, M.T., Landi, L., Pietramellara, G., and Renella, G. (2003). Microbial diversity and soil functions. Eur. J. Soil Sci. *54*, 655–670.

Nannipieri, P., Giagnoni, L., Renella, G., Puglisi, E., Ceccanti, B., Masciandaro, G., Fornasier, F., Moscatelli, M.C., and Marinari, S. (2012). Soil enzymology: classical and molecular approaches. Biol. Fertil. Soils *48*, 743–762.

Nielsen, U.N., Ayres, N., Wall, D.H., and Bardgett, R.D. (2011). Soil biodiversity and carbon cycling: a review and synthesis of studies examining diversity–function relationships. Eur. J. Soil Sci. *62*, 105–116.

Pancotto, V.A., Sala, O.E., Robson, T.M., Caldwell, M.M., and Scopel, A.L. (2005). Direct and indirect effects of solar ultraviolet-B radiation on long-term decomposition. Global Change Biol. *11*, 1982–1989.

Paul, E.A., and Nannipieri, P. (2012). Soil Biology and biochemistry: soil biology in its second golden age. In Handbook of Soil Sciences: Properties and Processes, 2nd edition, P.M. Huang, Y. Li, and M.E. Summer, eds. (CRC Press, Taylor & Francis Group, Boca Raton, FL), pp. IV2–IV4.

Schloter, M. (2012). Structure and function of prokaryotes in soil. In Handbook of Soil Sciences: Properties and Processes, 2nd edition, P.M. Huang, Y. Li, and M.E. Summer, eds. (CRC Press, Taylor & Francis Group, Boca Raton, FL), pp. 24.11–24.18.

St John, M.G., Crossley, D.A. jr, and Coleman, D.C. (2012). Microarthropods. In Handbook of Soil Sciences: Properties and Processes, 2nd edition, P.M. Huang, Y. Li, and M.E. Summer, eds. (CRC Press, Taylor & Francis Group, Boca Raton, FL), pp. 25.12–25.18.

Thorn, R.G. (2012). Soil Fungi. In Handbook of Soil Sciences: Properties and Processes, 2nd edition, P.M. Huang, Y. Li, and M.E. Summer, eds. (CRC Press, Taylor & Francis Group, Boca Raton, FL), pp. 24.18–24.29.

Van Vliet, P.C.J., Hendrix, P.F., and Callaham, M.A. Jr (2012). Earthworms. In Handbook of Soil Sciences: Properties and Processes, 2nd edition, P.M. Huang, Y. Li, and M.E. Summer, eds. (CRC Press, Taylor & Francis Group, Boca Raton, FL), pp. 25.35–25.44.

Vettori, C., Stotzky, G., Yoder, M., and Gallori, E. (1999). Interaction between bacteriophage PBS1 and clay minerals and transduction of *Bacillus subtilis* by clay–phage complexes. Environ. Microbiol. *1*, 347–355.

Waksman, S.A. (1932). The Principles of Soil Microbiology (Williams & Wilkins, Baltimore, MD).

Wang, Y., Hayatsu, M., and Fuji, T. (2012). Extraction of bacterial RNA from soil: challenges and solution. Microbes Environ. *27*, 111–121.

Whitman, W.B., Coleman, D.C., and Wiebe, W. (1998). Prokaryotes: the unseen majority. Proc. Natl. Acad. Sci. U.S.A. *95*, 6578–6583.

Williamson, K.E., Srinivasiah, S., and Wommack, K.E. (2012). Viruses in soil ecosystem. In Handbook of Soil Sciences: Properties and Processes, 2nd edition, P.M. Huang, Y. Li, and M.E. Summer, eds. (CRC Press, Taylor & Francis Group, Boca Raton, FL), pp. 24.1–24.10.

Functional Genomics Analysis of Key Bacterial Traits Involved in Rhizosphere Competence During Microbial–Host Interactions

2

Matthieu Barret, John P. Morrissey and Fergal O'Gara

Abstract

The rhizosphere is a nutrient rich environment, where numerous interactions between plant and microorganisms occur, ranging from mutualism to parasitism. The enrichment of specific microbial populations in the rhizosphere is dependent on the capability of these microorganisms to utilize root exudates, to effectively colonize the root surface and to interact or compete with other micro-organisms. Analysis of the rhizospheric communities incorporating both established techniques, and recently developed 'omic technologies' can now facilitate investigations into the molecular basis underpinning the establishment of plant-microbial interactomes in the rhizosphere. Therefore, the aim of this chapter is to present an overview of bacterial functions enriched in the rhizosphere of different plant species using data obtained from several functional genomics analyses.

Introduction

The rhizosphere, first described by Hiltner in 1904, is classically defined as the area of soil directly influenced by the presence and the activities of the root system. A combination of processes, including root respiration (Hanson *et al.*, 2000), uptake of water and nutrients by the root system (Hinsinger, 1998), and exudation of small molecular weight compounds (Bais *et al.*, 2006), is involved in the local modification of the surrounding bulk soil. The increase of carbon availability in the rhizosphere generally results in a less diverse microbial community with an enhanced biomass and activity in comparison with the root-free soil (Morgan *et al.*, 2005; Nannipieri *et al.*, 2008a,b; Dennis *et al.*, 2010). The enrichment of specific microbial populations within the rhizosphere is dependent on the ability of these microorganisms to utilize root exudates, to effectively colonize the root surface and to interact or compete with other microorganisms (Barret *et al.*, 2011a). Therefore, the selection of microbial population within the rhizosphere is thought to result primarily from specific biological pathways, which are encoded by specific suites of genes in resident microbial populations.

One of the major aims of rhizosphere microbial ecology is to improve the management of agricultural systems to produce sufficient food supply in an environmentally sustainable manner. This could be achieved in part by encouraging the proliferation of plant-beneficial soil micro-organisms. Indeed, microbes residing in the rhizosphere can positively affect plant growth and health through processes such as nitrogen fixation, nutrient solubilisation and phytohormone production. On the other hand, plant yield could also be increased through the suppression or reduction of phytopathogenic microorganisms by antagonistic bacteria. A better understanding of the microbial traits involved in colonization and survival of beneficial and detrimental microbes within the rhizospheres of different plant species and different soil types can therefore provide new strategies to improve plant productivity, while helping to protect the environment and maintain global biodiversity (Morrissey *et al.*, 2004). Therefore, microbial ecologists seek to establish key issues such as (i) which bacterial species are specifically selected in the rhizosphere of different plant species and, more importantly (ii) which bacterial functions are enriched by this specific association. Consequently, this book chapter will focus on the bacterial species and functions enriched in the rhizosphere of different plant species. Although other microorganisms such as oomycetes, fungi and protists also play key ecological functions in the rhizosphere environment, these groups of organisms will not be discussed in this chapter.

Bacterial species specifically selected by the rhizosphere

Rhizospheric bacteria play important roles in plant growth and health. Therefore over decades, significant attention has been devoted to the characterization of the structure and these rhizosphere microbial communities. Cultivation-based approaches were initially employed to dissect the bacterial community structure of the rhizosphere and the bulk soil (Barber and Lynch, 1977; Nannipieri *et al.*, 2008a,b). However, the main limitation of these approaches is the narrow spectrum of microorganisms studied due to the small recovery of the soil indigenous population by traditional cultivation techniques. Indeed, based on different analyses, it is believed that the proportion of cultivable rhizosphere micro-organism vary between 0.1% and 10% (Nichols, 2007; Ritz, 2007). Moreover, this proportion is predicted to be even smaller in the bulk soil (Zelenev *et al.*, 2005). To overcome this limitation, fingerprinting techniques based on the direct extraction of nucleic acids from the soil have been employed to unravel the structure of different bacterial rhizosphere communities (Table 2.1). Firstly, the total DNA is extracted from a specific soil environment, then the bacterial diversity is usually inferred by amplification through polymerase chain reaction (PCR) of the small subunit rRNA (16S rRNA for bacteria) and subsequently analysed by different typing methods such as DGGE, T-RFLP or SSCP (Sorensen *et al.*, 2009). These methods have been successfully used to show differences in bacterial community composition between the rhizosphere and the bulk soil (Smalla *et al.*, 2007). Moreover, the large amount of samples analysed through these methods has been used to assess the spatiotemporal dynamics of bacterial populations from the rhizosphere and have highlighted that the rhizosphere microbial community is influenced by the combined influence of the root zone location, the plant species/cultivar and the soil type (Berg and Smalla, 2009).

While the classic fingerprinting methods are useful to study the microbial community structure, the taxonomic information relative to the different groups of microorganisms is generally difficult to obtain and often restricted to groups of major abundance. Therefore other approaches such as microarrays containing 16S rRNA gene probes or deep-sequencing of 16S rRNA have also been developed to extend the repertoire of key bacterial and archaeal community members associated with the rhizosphere (Table 2.1). Although these studies have revealed that the taxonomical structure of soil bacterial communities fluctuates

Table 2.1 Fingerprinting techniques that have been employed to assess factors that influenced the microbial community structure within the rhizosphere

Main findings	References	Gene	Technics
Bacterial community composition is affected by soil types, plant species and root zone location	Marschner *et al.* (2001)	16S	PCR-DGGE
Soil type is the major factor influencing the bacterial communities in the soybean rhizosphere	Xu *et al.* (2009)	16S	PCR-DGGE
Bacterial community structure is influenced by plant ages	Miethling *et al.* (2003)	16S	SSCP
Archael community composition is affected by the rhizosphere	Sliwinski and Goodman (2004)	16S	SSCP
Arable soil type influenced the bacterial community structure	Ulrich and Becker (2006)	16S	T-RFLP
Bacterial community structure is mainly influenced by plant species	Chan *et al.* (2008)	16S	T-RFLP
Identification of bacterial phyla enriched in the rhizosphere of maize	Sanguin *et al.* (2006)	16S	Phylochips
Identification of bacterial phyla enriched in the rhizosphere of grasslands	DeAngelis *et al.* (2009)	16S	Phylochips, T-RFLP
Identification of bacterial phyla enriched in different potato cultivars	Weinert *et al.* (2011)	16S	Phylochips
Diversity of bacterial phyla is enriched in forest soil compared to agricultural soils	Roesch *et al.* (2007)	16S	Pyrosequencing
Identification of bacterial phyla enriched in the oak rhizosphere	Uroz *et al.* (2010)	16S	Pyrosequencing
Identification of bacterial phyla affected by potato cultivar and plant ages	İnceoğlu *et al.* (2011)	16S	Pyrosequencing

in response to biotic (e.g. plant age, plant species and plant cultivar) and abiotic factors (e.g. soil pH and climatic variations), some major taxonomic groups are consistently present in the rhizosphere. For example, a recent literature survey has identified seven bacterial phyla (*Acidobacteria, Actinobacteria, Bacteroidetes, Firmicutes, Planctomycetes, Proteobacteria* and *Verrucomicrobia*), which are dominant bacterial groups in different rhizospheres (da Rocha *et al.*, 2009). Whereas the *Actinobacteria, Bacteroidetes, Firmicutes* and *Proteobacteria* phyla have been frequently isolated from the rhizosphere through cultivation approaches, bacteria belonging to the three other phyla have only been obtained by cultivation-independent approaches. Therefore the functional role that these microorganisms play during plant–bacterial interaction remains to be investigated.

Bacterial functions enriched in the rhizosphere

The formation of specific rhizobacterial communities associated with distinct plant species, from the initial recruitment of bacteria to plant roots to the persistence within the rhizosphere, is determined by 'complex epistatic interactions among many different gene products' (Rainey, 1999). Numerous bacterial functions (summarized in Fig. 2.1) such as motility (de Weert *et al.*, 2002; Barahona *et al.*, 2011), attachment (Rodriguez-Navarro *et al.*, 2007),

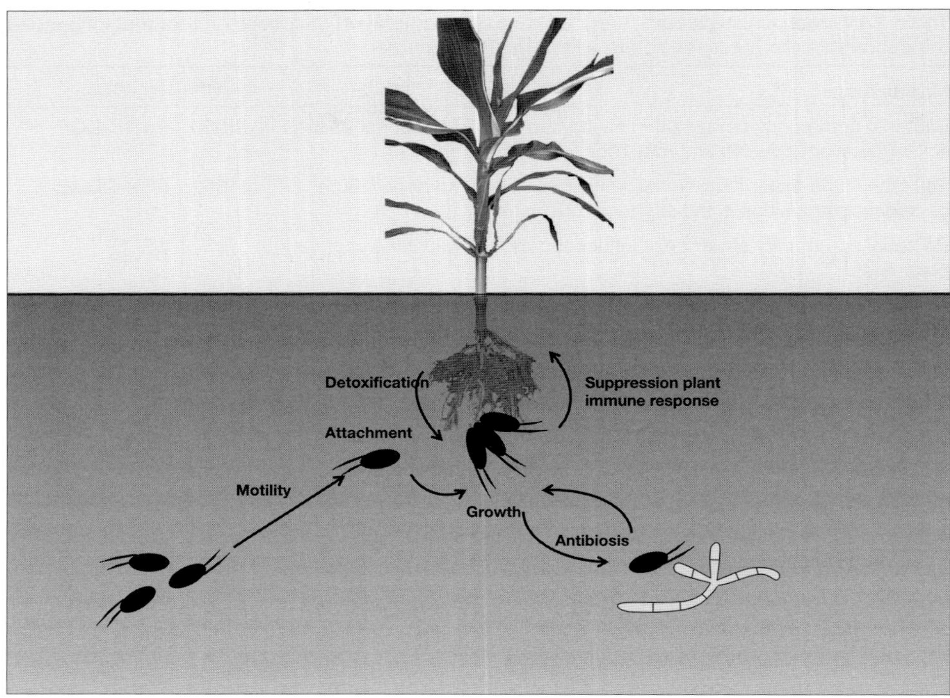

Figure 2.1 Bacterial traits involved in the colonization of the rhizosphere (adapted from Barret *et al.*, 2011). (1) Specific bacterial populations sense and move towards root exudates released in the rhizosphere through chemotaxis and motility. (2) Bacteria adhere on roots or rhizosphere soil surface. (3) Efficient bacterial growth rely on efficient nutrient scavenging systems coupled to synthesis of key components that are not present in the rhizosphere. (4) To survive in the rhizosphere bacteria have to overcome the chemical stress caused by numerous plant-derived toxic compounds through (a) detoxification mechanisms and/or (b) suppression of plant immune response. (5) Finally bacteria have to compete with other microorganisms through antibiosis, competition for nutrient and detoxification mechanisms. A colour version of this figure is available in the plate section at the back of the book.

growth (Lugtenberg and Kamilova, 2009; Browne *et al.*, 2010; Miller *et al.*, 2010), stress resistance (Abbas *et al.*, 2004; Martinez *et al.*, 2009) and production of secondary metabolites (Laue *et al.*, 2000; Delany *et al.*, 2001; Haas and Defago, 2005; Maunsell *et al.*, 2006) have been shown to be involved in rhizosphere colonization on the basis of gene-inactivation or gene expression. The bacterial traits involved in rhizosphere colonization and survival (henceforth defined as rhizosphere competence) have been initially ascribed to bacteria that are easy to grow and amenable to genetic analysis. For example, rhizosphere competence of plant-beneficial bacteria has been mostly assessed in genera of the *Proteobacteria* phylum such as: *Azospirillum* (Pothier *et al.*, 2007), *Enterobacter* (English *et al.*, 2010), *Pseudomonas* (Barret *et al.*, 2011a), *Serratia* (Mueller *et al.*, 2009) and *Stenotrophomonas* (Ryan *et al.*, 2009), and to a lesser extend in members of the *Firmicutes* such as *Bacillus* (Choudhary *et al.*, 2007). One key question is whether the rhizocompetence traits of these model bacterial strains are also shared by the unculturable bacteria? Two complementary approaches could be used to answer this question (Fig. 2.2).

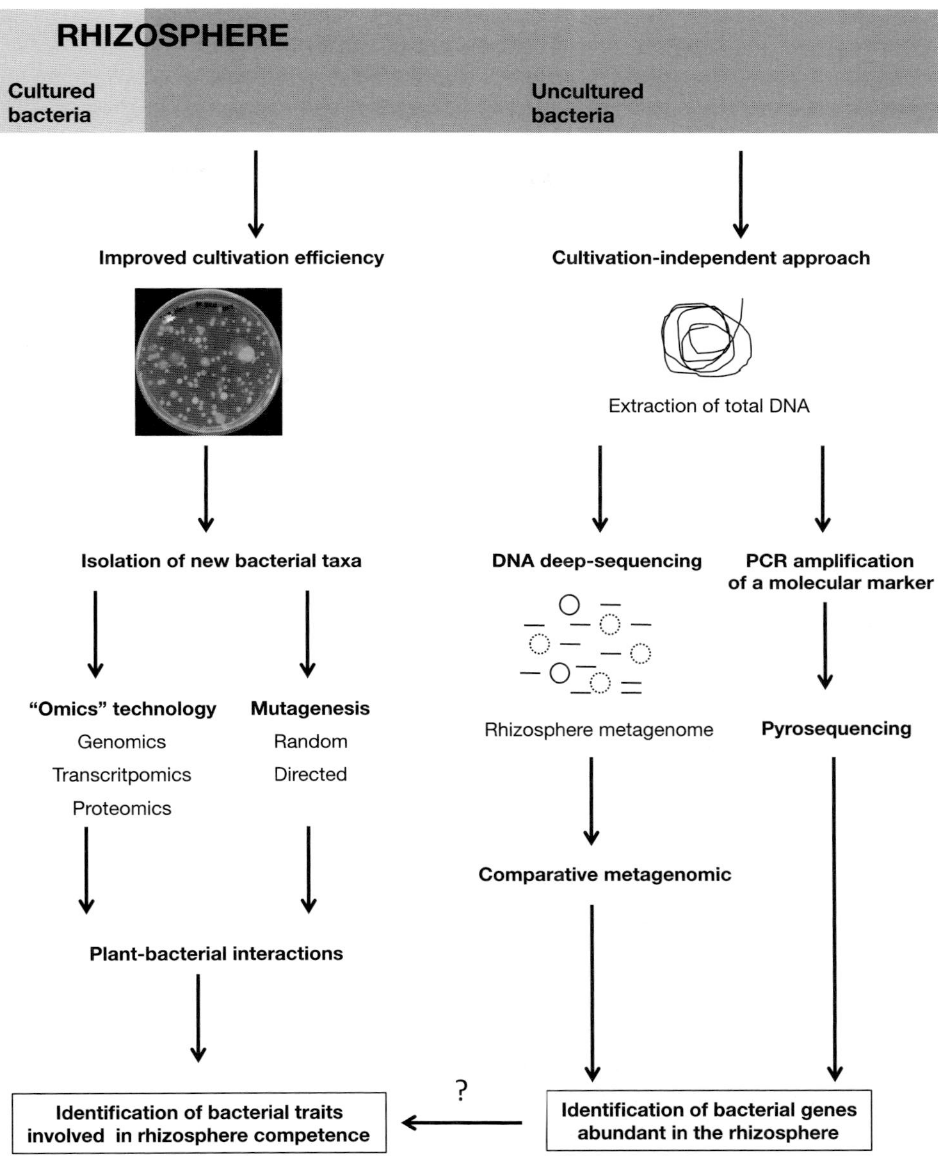

Figure 2.2 Approaches used to decipher the functions involved in the rhizosphere competence of 'uncultured-bacteria'. A colour version of this figure is available in the plate section at the back of the book.

The first option is to developed alternative cultivation methods using, for example, different growth media, temperature, longer incubation time and different concentration of oxygen. Improvement of cultivation methods results in an increase of the recovery of rhizosphere micro-organisms, which can lead to the isolation of new bacterial taxa. Once cultivated, these isolates could be investigated with respect to their metabolic potential

Table 2.2 Genomes sequences publicly available in the IMG database (Markowitz *et al.*, 2008)

Phylum	Class	Genome Name	Status	Gene Count	Genome Size	Reference
Acidobacteria	Division 1	*Acidobacterium capsulatum* ATCC 51196	Finished	3425	4127356	Ward *et al.* (2009)
Acidobacteria	Division 1	*Granulicella mallensis* MP5ACTX8	Draft	4947	6211694	None
Acidobacteria	Division 1	*Granulicella tundricola* MP5ACTX9	Finished	4757	5503984	None
Acidobacteria	Division 1	*Korebacter versatilis* Ellin345	Finished	4837	5650368	Ward *et al.* (2009)
Acidobacteria	Division 3	*Solibacter usitatus* Ellin6076	Finished	8003	9965640	Ward *et al.* (2009)
Acidobacteria	Division 1	*Terriglobus saanensis* SP1PR4	Finished	4333	5095226	None
Planctomycetes	Planctomycetacia	*Blastopirellula marina* SH 106T, DSM 3645	Draft	6090	6653746	Fuchsman *et al.* (2006)
Planctomycetes	Planctomycetacia	*Candidatus* Kuenenia stuttgartiensis	Finished	4710	4218325	Strous *et al.* (2006)
Planctomycetes	Planctomycetacia	*Gemmata obscuriglobus* UQM 2246	Draft	8086	9161841	Fuchsman *et al.* (2006)
Planctomycetes	Planctomycetacia	*Isosphaera pallida* IS1B, ATCC 43644	Finished	3823	5529304	Goeker *et al.* (2011)
Planctomycetes	Planctomycetacia	*Pirellula staleyi* DSM 6068	Finished	4825	6196199	Clum *et al.* (2009)
Planctomycetes	Planctomycetacia	*Planctomyces brasiliensis* IFAM 1448, DSM 5305	Finished	4865	6006602	None
Planctomycetes	Planctomycetacia	*Planctomyces limnophilus* Mu 290, DSM 3776	Finished	4372	5460085	LaButti *et al.* (2010)
Planctomycetes	Planctomycetacia	*Planctomyces maris* DSM 8797	Draft	6542	7777997	None
Planctomycetes	Planctomycetacia	*Rhodopirellula baltica* SH 1	Finished	7414	7145576	Glockner *et al.* (2003)
Verrucomicrobia	Verrucomicrobiae	*Akkermansia muciniphila* ATCC BAA-835	Finished	2238	2664102	van Passel *et al.* (2011)
Verrucomicrobia	Spartobacteria	*Chthoniobacter flavus* Ellin428	Draft	6778	7848700	Kant *et al.* (2011)
Verrucomicrobia	Opitutae	*Coraliomargarita akajimensis* DSM 45221	Finished	3191	3750771	Mavromatis *et al.* (2010)
Verrucomicrobia	unclassified	*Methylacidiphilum infernorum* V4	Finished	2521	2287145	Hou *et al.* (2008)
Verrucomicrobia	Opitutae	*Opitutaceae* sp. TAV2	Draft	4105	4954527	Isanapong *et al.* (2012)
Verrucomicrobia	Opitutae	*Opitutus terrae* PB90–1	Finished	4701	5957605	van Passel *et al.* (2011)
Verrucomicrobia	Verrucomicrobiae	*Pedosphaera parvula* Ellin514	Permanent Draft	6573	7414222	Kant *et al.* (2011)
Verrucomicrobia	Verrucomicrobiae	*Verrucomicrobiales* sp. DG1235	Draft	4957	5775745	None
Verrucomicrobia	Verrucomicrobiae	*Verrucomicrobium spinosum* DSM 4136	Draft	6584	8220857	Sait *et al.* (2011)

and subsequent rhizocompetence traits using a range of techniques such as mutagenesis (Domman *et al.*, 2011), genomics (Table 2.2) (Ward *et al.*, 2009; Santarella-Mellwig *et al.*, 2010), transcriptomics (Khadem *et al.*, 2011) and proteomics (for further information concerning cultivability of hitherto-uncultured bacteria, the reader is referred to Chapter 11).

The second approach is to use again cultivation-independent methods to identify processes that are important for the selection of specific population of rhizospheric bacteria by determining whether particular classes of genes are enriched in the rhizosphere. This could be achieved through the random sequencing of an environmental microbial community (also known as the microbiome) to obtain all the individual genomes of this community (also known as the metagenome). For example, the complete or near-complete metagenome of acid mine drainage microbial communities have been successfully sequenced (Tyson *et al.*, 2004). However, most soil microbial communities are extremely complex, with more than 1000 species per gram of soil (Schloss and Handelsman, 2006), and therefore the complete coverage of a rhizosphere metagenome is currently impractical (for further information concerning metagenomes, the reader is referred to Chapter 3). An alternative strategy to address the functionality of the total bacterial community is to analyse unassembled sequences through a gene- or process-centric approach (Tringe *et al.*, 2005). The process-centric approach considers a community from the point of view of its functions rather than its organisms. With such an approach, functions over-represented in one environmental sample are expected to be selected by the local environment and therefore could possibly confer an important function on that ecological niche. Abundance comparison of the proteins encoded by the rhizosphere microbial community to those encoded in the bulk soil could then reveal insights into traits involved in the establishment of microbes in the rhizosphere environment (see Chapter 5 for further information). Finally, a last approach is to assess the abundance of specific genes encoding well-characterized proteins, involved in rhizosphere competence of model bacterial strains, within the total microbial community. The amplification of such genes (henceforth defined as molecular markers) among a pool of nucleic acids could provide useful information on the abundance of key functions that play habitat-specific roles (Table 2.3).

In this chapter, the distribution of bacterial traits involved in rhizosphere competence within the total microbial community will be discussed in light of the key processes involved in bacterial colonization and survival within the rhizosphere (summarized in Fig. 2.1). For each process, a distinction between ubiquity (indicator of essentiality) and abundance (indicator of selective advantage) of the process/gene will be highlighted.

Motility and chemotaxis: early phase traits required for rhizocompetence

The initial step involved in the bacterial colonisation of the rhizosphere is related to the ability to sense and move towards specific molecules released by the plant roots system, a process called chemotaxis. Chemotaxis has been proved to be a general bacterial trait involved in the rhizosphere competence of numerous beneficial or deleterious bacteria belonging to different species such as *Bacillus subtilis* (Rudrappa *et al.*, 2008), *Pseudomonas fluorescens* (de Weert *et al.*, 2002), *Pseudomonas aeruginosa* (Mark *et al.*, 2005), *Rhizobium leguminosarum* (Miller *et al.*, 2007) or *Ralstonia solanacearum* (Yao and Allen, 2006). Mechanistically, changes in attractants or repellents concentration are sensed by specific

Table 2.3 Examples of molecular markers that could be applied to the rhizosphere microbial communities

Genes	Functions	Processes	Primers specificity	References
amoA	Ammonia monooxygenase	N cycle	Bacteria	Hornek *et al*. (2006)
nirS	Nitrite reductase	N cycle	Bacteria	Throback *et al*. (2004)
nirK	Nitrite reductase	N cycle	Bacteria	Throback *et al*. (2004)
nosZ	Nitrous oxide reductase	N cycle	Bacteria	Throback *et al*. (2004)
norB	Nitric oxide reductase	N cycle	Bacteria	Braker *et al*. (2003)
nifH	Nitrogenase reductase	N cycle	Bacteria	Diallo *et al*. (2008)
cbbL	Large subunit of the form I RubisCO	C cycle, CO_2 fixation	Bacteria	Selesi *et al*. (2005)
phoD	Alkaline phosphatase	P cycle	Bacteria	Sakurai *et al*. (2008)
pqqC	Pyrroloquinoline synthase C	P cycle, secondary metabolism	*Pseudomonas* spp.	Meyer *et al*. (2011)
frc	Formyl-CoA-transferase	Central metabolism	Bacteria	Khammar *et al*. (2009)
cheA	Histidine kinase	Chemotaxis	Bacteria	Buchan *et al*. (2010)
copA	Periplasmic multi-copper oxidase	Copper resistance	*Proteobacteria*	Lejon *et al*. (2007)
PKS	KS domain of type I PKS	Secondary metabolism	Bacteria	Parsley *et al*. (2011)
phzE	Phenazine biosynthetic pathway	Secondary metabolism	Bacteria	Schneemann *et al*. (2011)
hcnAB	Hydrogen cyanide biosynthesis	Secondary metabolism	*Pseudomonas* spp.	Svercel *et al*. (2007)
gacA	Response regulator	Secondary metabolism	*Pseudomonas* spp.	Costa *et al*. (2007)
phl	DAPG biosynthetic cluster, type III PKS	Secondary metabolism	*P. fluorescens*	Moyhinan *et al*. (2009)
hrcRST	Type III secretion system apparatus	Protein secretion	Hrp1 phylogenetic cluster	Mazurier *et al*. (2004)

chemoreceptor, which activates through a linker protein CheW the histidine kinase protein CheA. CheA communicates in turn the signal to the flagellar motor through the response regulator CheY (for more details on chemotaxis the readers is referred to Porter *et al.*, 2011). The relative simplicity of this transduction system coupled to its widespread presence in the genome of diverse *Bacteria* and *Archaea* (Wuichet and Zhulin, 2010) might offers a good opportunity to assess its contribution to the rhizosphere competence of uncultured bacteria. Recently, the genetic diversity of chemotaxis systems from natural bacterial populations associated with the rhizosphere or the surrounding bulk soil has been examined (Buchan *et al.*, 2010). In this study, the authors have monitored the temporal and dynamic changes in chemotactic-competent bacterial populations associated to the rhizosphere of cow pea and wheat through the genetic diversity of *cheA*. Results from the *cheA*-based analysis indicate that distinct genetics subgroups of chemotactic-competent microbes are present in natural soil populations and that some of these genetic groups belonging to the *Acidobacteria* and *Verrucomicrobia* phyla are specifically enriched in the rhizosphere (Buchan *et al.*, 2010). However, it remains to be determined whether the chemotaxis function itself is responsible for this selection and if so, what chemicals (attractants or repellents) are sensed by these bacteria.

A recent comparative metagenomic analysis performed on six publicly available soil microbiomes present in the IMG/M database (Markowitz *et al.*, 2008) has revealed an over-abundance of genes linked to flagella biogenesis in the rhizosphere of different C4 grasses (Barret *et al.*, 2011a). According to the process-centric postulate (see section *Bacterial functions enriched in the rhizosphere*), this abundance is indicative of a habitat-specific functionality, which implies that flagella-driven motility is probably also an important trait involved in rhizosphere competence of the total bacterial community. This is in accordance with previous findings showing that mutants deficient in flagella biogenesis of different bacterial strains are severely affected in root colonization (De Weger *et al.*, 1987; Capdevila *et al.*, 2004; Gorski *et al.*, 2009).

Microbial growth in the rhizosphere: the contribution and relevance of central metabolism

Rhizodeposition accounts for approximately 20% of the photosynthetically fixed carbon in the rhizosphere, thereby providing nutrients for the microbial communities. Thus, the ability of soil microorganisms to use a diverse array of nutrients coupled with efficient nutrient scavenging systems, are decisive traits for a successful colonization of the rhizosphere. Given the wide variety of substances present in different rhizodeposits (Dennis *et al.*, 2010), it is somewhat challenging to summarize findings on the bacterial protein-coding genes involved in nutrient acquisition within the rhizosphere. In addition, the extent to which results obtained with model bacterial strains could be extrapolated to the entire microbial community is hampered by the fact that each bacterial species or genus has its own nutrient preference. For instance, bacteria belonging to the *Pseudomonas* genus use preferentially tricarboxylic acid cycle intermediates and amino acids in contrast to carbohydrates, a regulatory process called catabolite repression control (Browne *et al.*, 2010). Consequently, protein-coding genes involved in the catabolism of specific amino acids, such as histidine catabolism, have been frequently reported to be induced in the rhizosphere of different plant species (Mark *et al.*, 2005; Matilla *et al.*, 2007; Silby *et al.*, 2009). However, for members of the *Enterobacteriaceae* family (e.g. *Erwinia* spp. or *Serratia* spp.), the catabolism of amino acids is probably less important than the catabolism of carbohydrates in rhizosphere competence, as the preferred carbon source of these bacteria is glucose (Rojo, 2010). Moreover in certain rhizosphere micro-environments (e.g. at the proximity of root nodules), CO_2 fixation by diverse autotrophic microorganisms could be larger than heterotrophic soil respiration (Dong and Layzell, 2001), which clearly indicates that a range of microorganisms with opposite metabolic abilities are able to grow in the rhizosphere.

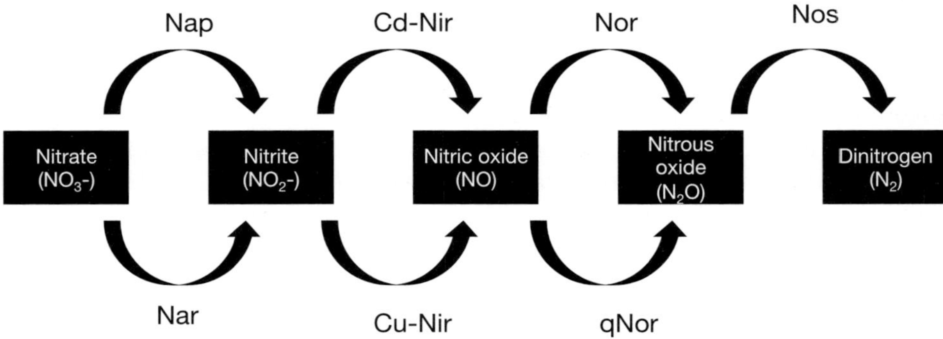

Figure 2.3 Denitrification pathway (adapted from Philippot and Hallin, 2005). Nitrate reduction to nitrite is either performed by a periplasmic (Nap) or membrane (Nar) nitrate reductase. Nitrite is further reduce to nitric oxide gas by a copper (Cu-Nir) or a cytochrome cd1 nitrite reductase (Cd-Nir). Reduction of NO to nitrous oxide is performed by a single (qNor) or two-component type (Nor) nitric oxide reductase. Finally, (iv) N_2O is reduced to dinitrogen by the nitrous oxide reductase (Nos).

Denitrification: a promising model linking microbial metabolic flexibility and community structure

A number of plant-associated bacteria, including *Pseudomonas* spp. (Ghiglione *et al.*, 2000; Rediers *et al.*, 2009) and *Azospirillum brasilense* (Pothier *et al.*, 2008), are able to use nitrogenous compounds as alternative electrons acceptors under oxygen limiting conditions. This process, called denitrification, converts nitrate, nitrite and nitric oxide to nitrous oxide or dinitrogen (Fig. 2.3). The denitrification pathway may constitute a significant advantage for root colonization due to the low oxygen level present in the rhizosphere (Hojberg *et al.*, 1999). Indeed mutants of *P. fluorescens* impaired in nitrate or nitrite reductases are deficient in the colonization of the rhizosphere (Philippot *et al.*, 1995; Ghiglione *et al.*, 2000), although some fluctuations occur between experiments (Rediers *et al.*, 2009). Moreover, the dissimilatory nitrite reductase gene *nirK* has been shown to be induced on wheat roots (Pothier *et al.*, 2008). Although bacterial denitrification could decrease the level of nitrate available for the plant, this process might have a number of positive effect on root development, such as root growth, nodulation and inhibition of pathogens via the production of nitric oxide (Richardson *et al.*, 2009).

Soil denitrifying bacteria have been intensively used as a model microbial community for understanding the relationship between community structure and activity (Philippot and Hallin, 2005). Indeed the fact that denitrifying pathways are well-conserved in a wide range of bacteria, including *Acidobacteria* (Ward *et al.*, 2009), *Proteobacteria* (Philippot and Hallin, 2005) and *Firmicutes* (Verbaendert *et al.*, 2011) has provided a good opportunity to use protein-coding genes as molecular markers (Table 2.3). For example, using *nirS* and *nirK* as molecular markers, Graham *et al.* have linked denitrification efficiency with genes abundances (Graham *et al.*, 2010). The gene *nirS* has also been employed to show that the denitrifiers were dominant in the rice rhizosphere in comparison to the ammonia oxidizer (using *amoA* as a molecular marker), which suggest a better adaptability of denitrifier in this environment (Hussain *et al.*, 2011).

Surviving in the rhizosphere: the relevance of stress and detoxification traits

While regards as a nutrient-rich niche, the rhizosphere environment generates a bacterial stress response through the release of plant-derived toxic molecules such as flavonoids and terpenoids (Bais *et al.*, 2006; Cesco *et al.*, 2012). Indeed, numerous genes coding for stress response and detoxification proteins are up-regulated in the rhizosphere of different plant species (Matilla *et al.*, 2007; Barr *et al.*, 2008, 2009). In addition, bacteria must compete for space and nutrients with other members of the rhizosphere microbial community. Therefore, the selection of specific microbial populations in the rhizosphere may be explained in part by means of different mechanisms to cope with these chemical stresses, such as active extrusion and/or degradation of the toxic compounds. For example, export of toxic molecules from the bacterial cells is generally driven by multidrug resistance pumps (Martinez *et al.*, 2009). Remarkably, bacterial genomes with the largest repertoire of multidrug resistance efflux pumps are derived from microorganisms living in the soil or in association with plants (Konstantinidis and Tiedje, 2004). In accordance with this finding, a recent comparative metagenomic analysis has revealed that genes encoding ABC-type antimicrobial peptide transport system are overabundant in the rhizosphere of maize, miscanthus and switchgrass in comparison to the bulk soil (Barret *et al.*, 2011a).

Genes encoding antibiotic resistance determinants have also been found in uncultured bacteria via metagenomic approaches (reviewed in Monier *et al.*, 2011). For example, two new proteins involved in cephalosporin (ceftazidime) and aminoglycoside (kanamycin) resistances have been identified by screening a metagenomic library constructed from an apple orchard soil (Donato *et al.*, 2010). Using the same experimental strategy, another research group has identified eleven new antibiotic resistance genes from three different soil metagenomic libraries (Torres-Cortes *et al.*, 2011). Altogether, these studies demonstrated that both cultured and uncultured soil display a high level of diversity of gene encoding proteins involved in stress resistance and detoxification.

Secretion systems: important functional traits involved in rhizosphere competence

Bacterial protein secretion plays a key role in the modulation of different biotic interactions, ranging from symbiosis to pathogenesis. In every case, the secretion of protein(s) is dependent on specific secretion systems, which have been divided into an array of functional classes. For example, secretion systems in Gram-negative bacteria have been classified into at least six types: from type I through type VI (T1SS–T6SS) (Filloux, 2011). Initially these secretion systems were studied in plant, animal or human pathogenic bacteria, but more recently, studies on the influence of proteins secreted by T1SS (Fauvart and Michiels, 2008), T3SS (Mavrodi *et al.*, 2011) and T6SS (Records, 2011) in the rhizosphere colonisation of plant beneficial bacteria have been initiated.

Type I secretion systems (T1SSs)

T1SSs are a class of ATP-binding cassette (ABC) transporter, which is linked to an outer-membrane protein by a connector (Delepelaire, 2004). Proteins secreted by the T1SS of plant-beneficial bacteria have been implicated in different processes of plant–bacterial interactions. For example the large surface proteins LapA and LapF secreted by the T1SS of *Pseudomonas putida* KT2440 have been involved in the bacterial attachment to seeds (Espinosa-Urgel *et al.*, 2000) and roots (Yousef-Coronado *et al.*, 2008; Martinez-Gil *et al.*, 2010). Alternatively, the type I effector NodO of *Rhizobium leguminosarum* bv. *viciae* confers the ability to nodulate *Leucaena leucocephala* to (reviewed in Fauvart and Michiels, 2008). Based on genomic analysis it seems that T1SS are not encoded in the genome of bacteria belonging to the rare phyla *Acidobacteria*, *Planctomycetes* and *Verrucomicrobia*.

Type III secretion systems (T3SSs)

T3SS is a complex secretion machinery made up of approximately 25 proteins (reviewed in Cornelis, 2010). Phylogenetic studies have divided T3SS into seven different families: *Chlamydia*, Hrp1, Hrp2, SPI-1, SPI-2, *Rhizobiaceae* and Ysc (Pallen *et al.*, 2005; Troisfontaines and Cornelis, 2005). The role of type 3 effectors (T3E) during plant–bacterial interactions has initially been studied in phytopathogenic bacteria such as *Pseudomonas syringae*, *Xanthomonas* spp. and *Ralstonia solanacearum* (Büttner and He, 2009), or in symbiotic bacteria such as *Rhizobium* spp (Fauvart and Michiels, 2008). However, T3SS is also widely encoded in the genome of other bacteria selected within the rhizosphere. For example, T3SSs belonging to the Hrp1 family have been identified in approximately 50% of culturable *P. fluorescens* strains examined using highly conserved genes such as *hrcRST* (Mazurier *et al.*, 2004) and *hrcN* (Rezzonico *et al.*, 2004). The induction of Hrp1 in different rhizosphere environments (Rainey, 1999; Jackson *et al.*, 2005; Mavrodi *et al.*, 2011) begs the question as to its ecological significance. Recent data seem to indicate that this secretion system might be involved in the persistence of bacterial population in the rhizosphere, through inhibition of the plant immune response (Mavrodi *et al.*, 2011) or through growth inhibition of other microorganisms (Rezzonico *et al.*, 2005). However other Hrp1-derived functions have been proposed such as the promotion of the symbiosis between mycorrhizal fungus and plants by mycorrhiza helper bacteria strain (Cusano *et al.*, 2010), which is consistent with the fact that T3SSs are enriched in the mycorrhizosphere (Warmink and van Elsas, 2008; Viollet *et al.*, 2011).

The distribution of T3SSs in the rhizosphere or in the associated bulk soil has mainly been assessed for particular T3SS phylogenetic clusters such as Hrp1. Consequently, the prevalence of the other T3SS phylogenetic clusters in the rhizosphere has been largely ignored, which hampers further conclusion on the prevalence of T3SSs in non-cultured soil bacteria. However, comparative genomic analysis has highlighted the presence of a T3SS belonging to the Ysc family in the genome of *Verrucomicrobium spinosum* (Pallen *et al.*, 2005). Moreover, T3SS transcripts of *Verrucomicrobium spinosum* are induced *in vitro* and one putative T3E is involved in growth inhibition of yeast when expressed into *Saccharomyces cerevisiae* (Sait *et al.*, 2011), although the production of the structural apparatus has yet to be demonstrated. Collectively, these results suggest that some bacteria belonging to the *Verrucomicrobia* phylum might contain T3SSs and that these secretion systems could possibly be involved in rhizosphere competence.

Type VI secretion systems (T6SSs)

T6SSs are macromolecular machineries composed of approximately 15 conserved proteins, which are involved in virulence towards different eukaryotes and/or in bacterial killing (Schwarz *et al.*, 2010). T6SSs are widely encoded in the genomes of phytopathogenic (i.e. *P. syringae*, *Agrobacterium tumefaciens*, *Pectobacterium atrosepticum*), symbiotic (i.e. *Azoarcus* sp., *Rhizobacterium leguminosarum*), or plant-associated proteobacteria (i.e. *P. fluorescens*, *Lysobacter enzymogenes*) (Barret *et al.*, 2011b; Records, 2011), but are also present within in the genomes of planctomycetes (Persson *et al.*, 2009; Tseng *et al.*, 2009), acidobacteria (Boyer *et al.*, 2009; Tseng *et al.*, 2009) and verrucomicrobia (M. Barret, unpublished observation). Except from specific involvement during plant pathogenesis (Lesic *et al.*, 2009) or symbiosis (Bladergroen *et al.*, 2003), data on the role of T6SS during bacterial rhizosphere colonization are limited. Nevertheless, mounting evidence indicates that T6SS of different bacterial strains is involved in interbacterial interactions (Schwarz *et al.*, 2010). Moreover, two reports have highlighted a possible involvement of T6SS during bacterial–fungal interactions (Barret *et al.*, 2009; Patel *et al.*, 2009). Based on these results, it is tempting to speculate that T6SS could therefore be an important trait for PGPR to compete with other rhizosphere micro-organisms.

Secondary metabolism: specialized functions involved in competitive rhizosphere fitness

Roots are frequently colonized with filamentous fungi, which could be either plant-beneficial or phyopathogenic (de Boer *et al.*, 2005). The combined influence of roots and fungi generates an environmental niche that diverges from the rhizosphere *sensu stricto* (i.e. area of soil directly influenced by the presence and the activities of the root system), which results ultimately in modification of the bacterial community structure (Frey-Klett *et al.*, 2005; Leveau and Preston, 2008). This well-established phenomenon happens in disease suppressive soils, in which disease severity of plants is decreased as a result of activities of the soil microbial community (Haas and Defago, 2005; Mendes *et al.*, 2011). For instance, a reduction of take-all disease severity after a severe outbreak caused by the pathogenic fungus *Gaeumannomyces graminis* var. *tritici* (*Ggt*) is observed after several years of wheat monoculture and is related to the selection of specific bacterial population, including *Pseudomonas* spp. (Cook and Rovira, 1976; Sanguin *et al.*, 2008). The effectiveness of these so-called biocontrol bacteria depends primarily on their rhizosphere competence but also on the inhibition of plant-pathogenic fungi through production of a variety of secondary metabolites (Fenton *et al.*, 1992; Haas and Defago, 2005; Gross and Loper, 2009).

Non-ribosomal peptide synthase (NRPS) and polyketide synthase (PKS) enzymes are involved in the synthesis of diverse antimicrobial compounds (e.g. 2,4-diacetylphloroglucinol) and siderophores (e.g. pyoverdine) (reviewed in Gross and Loper, 2009). The molecule 2,4-diacetylphloroglucinol (DAPG), is a key determinant that contribute to the disease suppressiveness of soils (Raaijmakers and Weller, 1998) and has antifungal, antibacterial and nematicidal activities (Fenton *et al.*, 1992; Cronin *et al.*, 1997a,b; Gleeson *et al.*, 2010). The presence of DAPG gene cluster in bacterial genomes is unusual and restricted to some *Pseudomonas fluorescens* strains (Moynihan *et al.*, 2009). However, other NRPS/PKS gene clusters have been detected in the genome of soil bacteria such as *Bacillus* spp. (Deng *et al.*, 2011), *Burkholderia* spp. (Partida-Martinez and Hertweck, 2005) or *Sorangium* spp. (Julien *et al.*, 2000). More recently, a study has identified 29 type I modular polyketide synthase genes in a soil metagenomic library using degenerate primers and probes, which indicate that approximately 30% of the genomes contained a KS domain (Parsley *et al.*, 2011). Interestingly, numerous KS domains were attributed to *Acidobacteria* genomes (Parsley *et al.*, 2011) confirming the presence of PKS cluster in this bacterial phylum (Ward *et al.*, 2009).

Phenazines are a large family of flavine coenzymes analogues with antibiotic activity (Haas and Defago, 2005; Gross and Loper, 2009). Phenazines are involved in the biocontrol abilities of *Pseudomonas* spp. and *Pantoea* spp. towards a range of phytopathogens such as *Fusarium* or *Erwinia* (Giddens *et al.*, 2003; *Mazurier et al.*, 2009). Phenazine clusters are abundant within soil bacteria or plant-associated bacteria (Mavrodi *et al.*, 2010), suggesting that this determinant is required for the successful colonisation and persistence of these environments. Using degenerate primers targeting *phzE,* whose product is involved in the conversion of chorismate to 2-amino-2-deoxyisochorismic acid, phenazine clusters have been identified in isolates of the *Proteobacteria* and *Actinobacteria* phyla (Schneemann *et al.*, 2011). Although the presence of such clusters in *Verrucomicrobia, Acidobacteria* and *Planctomycetes* has not been assessed in this study, the design of *phzE* universal primers could provide a good opportunity to screen these new bacterial phyla.

Conclusions and future directions

Over the last decades, extensive research attempts have focused on bacterial traits contributing to the rhizosphere competence of specific bacterial strains. A number of protein-coding genes involved in chemotaxis, stress resistance and detoxification, secondary metabolism and more recently secretion systems have been identified as important bacterial determinants for successful colonisation and survival within the rhizosphere of different plant species. The isolation of these genetic determinants from fast-growing bacterial strains amenable to genetic manipulations has been subsequently used to assess the distribution of the corresponding genes in the total soil bacterial community. Combination of genetic molecular markers and metagenome datasets (Chapter 3) has highlighted that some of these functional genes are abundant in soil and plant-associated bacteria. However, it remains to be determined how these traits specifically contribute to the ecological fitness of uncultured rhizobacterial populations. Indeed, the identification of a functional gene in the rhizosphere environment does not necessarily mean that its corresponding activity is directly involved in rhizosphere competence. Therefore the development of specific molecular marker targeting the transcript (Chapter 4), protein (Chapters 5 and 7) or metabolite (Chapters 6 and 8) abundances are clearly interesting alternative strategy to monitor the dynamic behaviour of microbial populations within the rhizosphere. The ultimate discovery of the complete repertoire of genetic determinants involved in effective bacterial colonization and survival within the rhizosphere will, in the future, lead to the creation of a minimal bacterial genome required for efficient rhizosphere competence. Additionally, advances in new genetic platforms linked to synthetic biology will, in the near future, facilitate the assembly of novel microbial genomes using different genetic elements including natural or artificial genes (Gibson *et al.*, 2010).

Acknowledgements

This research was supported in parts by grants awarded to FOG by the Science Foundation of Ireland (07IN.1/B948, 08/RFP/GEN1295, 08/RFP/GEN1319, SFI09/RFP/BMT2350); the Department of Agriculture, Fisheries and Food (RSF grants 06-321 and 06-377; FIRM grants 06RDC459 06RDC506 and 08RDC629); the European Commission (MTKD-CT-2006-042062, Marie Curie TOK:TRAMWAYS, EU256596, MicroB3–287589-OCEAN2012, MACUMBA-CP-TP 311975; PharmaSea-CP-TP 312184);

IRCSET (05/EDIV/FP107/INTERPAM, EMBARK), the Marine Institute Beaufort award (C&CRA 2007/082), the Environmental Protection Agency (EPA 2006-PhD-S-21, EPA 2008-PhD-S-2) and the HRB (RP/2006/271, RP/2007/290, HRA/2009/146).

References

Abbas, A., McGuire, J.E., Crowley, D., Baysse, C., Dow, M., and O'Gara, F. (2004). The putative permease PhlE of *Pseudomonas fluorescens* F113 has a role in 2,4-diacetylphloroglucinol resistance and in general stress tolerance. Microbiology *150*, 2443–2450.

Bais, H.P., Weir, T.L., Perry, L.G., Gilroy, S., and Vivanco, J.M. (2006). The role of root exudates in rhizosphere interactions with plants and other organisms. Ann. Rev. Plant Biol. *57*, 233–266.

Barahona, E., Navazo, A., Martinez-Granero, F., Zea-Bonilla, T., Perez-Jimenez, R.M., Martin, M., and Rivilla, R. (2011). *Pseudomonas fluorescens* F113 mutant with enhanced competitive colonization ability and improved biocontrol activity against fungal root pathogens. Appl. Environ. Microbiol. *77*, 5412–5419.

Barber, D.A., and Lynch, J.M. (1977). Microbial growth in the rhizosphere. Soil. Biol. Biochem. *9*, 305–308.

Barr, M., East, A.K., Leonard, M., Mauchline, T.H., and Poole, P.S. (2008). *In vivo* expression technology (IVET) selection of genes of *Rhizobium leguminosarum* biovar *viciae* A34 expressed in the rhizosphere. FEMS Microbiol. Lett. *282*, 219–227.

Barret, M., Frey-Klett, P., Guillerm-Erckelboudt, A.Y., Boutin, M., Guernec, G., and Sarniguet, A. (2009). Effect of wheat roots infected with the pathogenic fungus *Gaeumannomyces graminis* var. *tritici* on gene expression of the biocontrol bacterium *Pseudomonas fluorescens* Pf29Arp. Mol. Plant Microbe In. *22*, 1611–1623.

Barret, M., Morrissey, J., and O'Gara, F. (2011a). Functional genomics analysis of plant growth-promoting rhizobacterial traits involved in rhizosphere competence. Biol. Fertil. Soils *47*, 729–743.

Barret, M., Egan, F., Fargier, E., Morrissey, J.P., and O'Gara, F. (2011b). Genomic analysis of the type VI secretion systems in *Pseudomonas* spp: novel clusters and putative effectors uncovered. Microbiology *157*, 1726–1739.

Braker, G., and Tiedje, J.M. (2003). Nitric oxide reductase (*norB*) genes from pure cultures and environmental samples. Appl. Environ. Microbiol. *69*, 3476–3483.

Berg, G., and Smalla, K. (2009). Plant species and soil type cooperatively shape the structure and function of microbial communities in the rhizosphere. FEMS Microbiol. Ecol. *68*, 1–13.

Bladergroen, M.R., Badelt, K., and Spaink, H.P. (2003). Infection-blocking genes of a symbiotic *Rhizobium leguminosarum* strain that are involved in temperature-dependent protein secretion. Mol. Plant Microbe In. *16*, 53–64.

de Boer, W., Folman, L.B., Summerbell, R.C., and Boddy, L. (2005). Living in a fungal world: impact of fungi on soil bacterial niche development. FEMS Microbiol. Rev. *29*, 795–811.

Boyer, F., Fichant, G., Berthod, J., Vandenbrouck, Y., and Attree, I. (2009). Dissecting the bacterial type VI secretion system by a genome wide *in silico* analysis: what can be learned from available microbial genomic resources? BMC Genomics *10*, 104.

Browne, P., Barret, M., O'Gara, F., and Morrissey, J.P. (2010). Computational prediction of the Crc regulon identifies genus-wide and species-specific targets of catabolite repression control in *Pseudomonas* bacteria. BMC Microbiol. *10*, 300.

Buchan, A., Crombie, B., and Alexandre, G.M. (2010). Temporal dynamics and genetic diversity of chemotactic-competent microbial populations in the rhizosphere. Environ. Microbiol. *12*, 3171–3184.

Büttner, D., and He, S.Y. (2009). Type III protein secretion in plant pathogenic bacteria. Plant Physiol. *150*, 1656–1664.

Capdevila, S., Martinez-Granero, F.M., Sanchez-Contreras, M., Rivilla, R., and Martin, M. (2004). Analysis of *Pseudomonas fluorescens* F113 genes implicated in flagellar filament synthesis and their role in competitive root colonization. Microbiology *150*, 3889–3897.

Cesco, S., Mimmo, T., Tonon, G., Tomasi, N., Pinton, R., Terzano, R., Neumann, G., Weisskopf, L., Renella, G., Landi, L., *et al.* (2012). Plant-borne flavonoids released into the rhizosphere: impact on soil bio-activities related to plant nutrition. A review. Biol. Fertil. Soils *48*, 123–149.

Chan, O.C., Casper, P., Sha, L.Q., Feng, Z.L., Fu, Y., Yang, X.D., Ulrich, A., and Zou, X.M. (2008). Vegetation cover of forest, shrub and pasture strongly influences soil bacterial community structure as revealed by 16S rRNA gene T-RFLP analysis. FEMS Microbiol. Ecol. *64*, 449–458.

Choudhary, D.K., Prakash, A., and Johri, B.N. (2007). Induced systemic resistance (ISR) in plants: mechanism of action. Indian J. Microbiol. *47,* 289–297.

Clum, A., Nolan, M., Lang, E., Del Rio, T.G., Tice, H., Copeland, A., Cheng, J.-F., Lucas, S., Chen, F., Bruce, D., *et al.* (2009). Complete genome sequence of *Acidimicrobium ferrooxidans* type strain (ICPT). Stand. Gen. Sci. *1,* 38–45.

Cook, R.J., and Rovira, A.D. (1976). Role of bacteria in biological control of *Gaeumannomyces graminis* by suppressive soils. Soil. Biol. Biochem. *8,* 269–273.

Cornelis, G.R. (2010). The type III secretion injectisome, a complex nanomachine for intracellular 'toxin' delivery. Biol. Chem. *391,* 745–751.

Costa, R., Gomes, N.C.M., Krogerrecklenfort, E., Opelt, K., Berg, G., and Smalla, K. (2007). *Pseudomonas* community structure and antagonistic potential in the rhizosphere: insights gained by combining phylogenetic and functional gene-based analyses. Environ. Microbiol. *9,* 2260–2273.

Cronin, D., Moenne-Loccoz, Y., Fenton, A., Dunne, C., Dowling, D.N., and Ogara, F. (1997a). Role of 2,4-diacetylphloroglucinol in the interactions of the biocontrol pseudomonad strain F113 with the potato cyst nematode *Globodera rostochiensis.* Appl. Environ. Microbiol. *63,* 1357–1361.

Cronin, D., Moenne-Loccoz, T., Fenton, A., Dunne, C., Dowling, D.N., and Ogara, F. (1997b). Ecological interaction of a biocontrol *Pseudomonas fluorescens* strain producing 2,4-diacetylphloroglucinol with the soft rot potato pathogen *Erwinia carotovora* subsp. *atroseptica.* FEMS Microbiol. Ecol. *23,* 95–106.

Cusano, A.M., Burlinson, P., Deveau, A., Vion, P., Uroz, S., Preston, G.M., and Frey-Klett, P. (2011). *Pseudomonas fluorescens* BBc6R8 type III secretion mutants no longer promote ectomycorrhizal symbiosis. Environ. Microbiol. Rep. *3,* 203–210.

DeAngelis, K.M., Brodie, E.L., DeSantis, T.Z., Andersen, G.L., Lindow, S.E., and Firestone, M.K. (2009). Selective progressive response of soil microbial community to wild oat roots. ISME J. *3,* 168–178.

De Weger, L.A., Vandervlugt, C.I. M., Wijfjes, A.H. M., Bakker, P., Schippers, B., and Lugtenberg, B. (1987). Flagella of a plant-growth-stimulating *Pseudomonas fluorescens* strain are required for colonization of potato roots. J. Bacteriol. *169,* 2769–2773.

Delany, I.R., Walsh, U.F., Ross, I., Fenton, A.M., Corkery, D.M., and O'Gara, F. (2001). Enhancing the biocontrol efficacy of *Pseudomonas fluorescens* F113 by altering the regulation and production of 2,4-diacetylphloroglucinol – improved *Pseudomonas* biocontrol inoculants. Plant Soil *232,* 195–205.

Delepelaire, P. (2004). Type I secretion in Gram-negative bacteria. Biochim. Biophys. Acta Mol. Cell Res. *1694,* 149–161.

Deng, Y., Zhu, Y., Wang, P., Zhu, L., Zheng, J., Li, R., Ruan, L., Peng, D., and Sun, M. (2011). Complete genome sequence of *Bacillus subtilis* BSn5, an endophytic bacterium of *Amorphophallus konjac* with antimicrobial activity for the plant pathogen *Erwinia carotovora* subsp *carotovora.* J. Bacteriol. *193,* 2070–2071.

Dennis, P.G., Miller, A.J., and Hirsch, P.R. (2010). Are root exudates more important than other sources of rhizodeposits in structuring rhizosphere bacterial communities? FEMS Microbiol. Ecol. *72,* 313–327.

Diallo, M.D., Reinhold-Hurek, B., and Hurek, T. (2008). Evaluation of PCR primers for universal *nifH* gene targeting and for assessment of transcribed *nifH* pools in roots of *Oryza longistaminata* with and without low nitrogen input. FEMS Microbiol. Ecol. *65,* 220–228.

Domman, D.B., Steven, B.T., and Ward, N.L. (2011). Random transposon mutagenesis of *Verrucomicrobium spinosum* DSM 4136(T). Arch. Microbiol. *193,* 307–312.

Donato, J.J., Moe, L.A., Converse, B.J., Smart, K.D., Berklein, F.C., McManus, P.S., and Handelsman, J. (2010). Metagenomic analysis of apple orchard soil reveals antibiotic resistance genes encoding predicted bifunctional proteins. Appl. Environ. Microbiol. *76,* 4396–4401.

Dong, Z., and Layzell, D.B. (2001). H-2 oxidation, O-2 uptake and CO_2 fixation in hydrogen treated soils. Plant Soil *229,* 1–12.

English, M.M., Coulson, T.J. D., Horsman, S.R., and Patten, C.L. (2010). Overexpression of *hns* in the plant growth-promoting bacterium *Enterobacter cloacae* UW5 increases root colonization. J. Appl. Microbiol. *108,* 2180–2190.

Espinosa-Urgel, M., Salido, A., and Ramos, J.L. (2000). Genetic analysis of functions involved in adhesion of *Pseudomonas putida* to seeds. J. Bacteriol. *182,* 2363–2369.

Fauvart, M., and Michiels, J. (2008). Rhizobial secreted proteins as determinants of host specificity in the rhizobium–legume symbiosis. FEMS Microbiol. Lett. *285,* 1–9.

Fenton, A.M., Stephens, P.M., Crowley, J., Ocallaghan, M., and Ogara, F. (1992). Exploitation of gene(s) involved in 2.4-diacetylphloroglucinol biosynthesis to confer a new biocontrol capability to a *Pseudomonas* strain. Appl. Environ. Microbiol. *58,* 3873–3878.

Filloux, A. (2011). Protein secretion systems in *Pseudomonas aeruginosa*: an essay on diversity, evolution and function. Front. Microbiol. *2*, 155.

Frey-Klett, P., Chavatte, M., Clausse, M.L., Courrier, S., Le Roux, C., Raaijmakers, J., Martinotti, M.G., Pierrat, J.C., and Garbaye, J. (2005). Ectomycorrhizal symbiosis affects functional diversity of rhizosphere fluorescent pseudomonads. New Phytol. *165*, 317–328.

Fuchsman, C.A., and Rocap, G. (2006). Whole-genome reciprocal BLAST analysis reveals that Planctomycetes do not share an unusually large number of genes with Eukarya and Archaea. Appl. Environ. Microbiol. *72*, 6841–6844.

Ghiglione, J.F., Gourbiere, F., Potier, P., Philippot, L., and Lensi, R. (2000). Role of respiratory nitrate reductase in ability of *Pseudomonas fluorescens* YT101 to colonize the rhizosphere of maize. Appl. Environ. Microbiol. *66*, 4012–4016.

Gibson, D.G., Glass, J.I., Lartigue, C., Noskov, V.N., Chuang, R.Y., Algire, M.A., Benders, G.A., Montague, M.G., Ma, L., Moodie, M.M., *et al.* (2010). Creation of a bacterial cell controlled by a chemically synthesized genome. Science *329*, 52–56.

Giddens, S.R., Houliston, G.J., and Mahanty, H.K. (2003). The influence of antibiotic production and pre-emptive colonization on the population dynamics of *Pantoea agglomerans* (*Erwinia herbicola*) Eh1087 and *Erwinia amylovora* in planta. Environ. Microbiol. *5*, 1016–1021.

Gleeson, O., O'Gara, F., and Morrissey, J.P. (2010). The *Pseudomonas fluorescens* secondary metabolite 2,4 diacetylphloroglucinol impairs mitochondrial function in *Saccharomyces cerevisiae*. Anton. Leeuw. Int. J. G. *97*, 261–273.

Glockner, F.O., Kube, M., Bauer, M., Teeling, H., Lombardot, T., Ludwig, W., Gade, D., Beck, A., Borzym, K., Heitmann, K., *et al.* (2003). Complete genome sequence of the marine planctomycete *Pirellula* sp strain 1. Proc. Natl. Acad. Sci. U.S.A. *100*, 8298–8303.

Goeker, M., Cleland, D., Saunders, E., Lapidus, A., Nolan, M., Lucas, S., Hammon, N., Deshpande, S., Cheng, J.-F., Tapia, R., *et al.* (2011). Complete genome sequence of *Isosphaera pallida* type strain (IS1B(T)). Stand. Gen. Sci. *4*, 63–71.

Gorski, L., Duhe, J.M., and Flaherty, D. (2009). The use of flagella and motility for plant colonization and fitness by different strains of the foodborne pathogen *Listeria monocytogenes*. Plos One 4, e5142.

Graham, D.W., Trippett, C., Dodds, W.K., O'Brien, J.M., Banner, E.B. K., Head, I.M., Smith, M.S., Yang, R.K., and Knapp, C.W. (2010). Correlations between *in situ* denitrification activity and *nir* gene abundances in pristine and impacted prairie streams. Environ. Pollut. *158*, 3225–3229.

Gross, H., and Loper, J.E. (2009). Genomics of secondary metabolite production by *Pseudomonas* spp. Nat. Prod. Rep. *26*, 1408–1446.

Haas, D., and Defago, G. (2005). Biological control of soil-borne pathogens by fluorescent pseudomonads. Nat. Rev. Microbiol. *3*, 307–319.

Hanson, P.J., Edwards, N.T., Garten, C.T., and Andrews, J.A. (2000). Separating root and soil microbial contributions to soil respiration: A review of methods and observations. Biogeochem. *48*, 115–146.

Hinsinger, P. (1998). How do plant roots acquire mineral nutrients? Chemical processes involved in the rhizosphere. Adv. Agro. *64*, 225–265.

Hojberg, O., Schnider, U., Winteler, H.V., Sorensen, J., and Haas, D. (1999). Oxygen-sensing reporter strain of *Pseudomonas fluorescens* for monitoring the distribution of low-oxygen habitats in soil. Appl. Environ. Microbiol. *65*, 4085–4093.

Hornek, R., Pommerening-Röser, A., Koops, H.-P., Farnleitner, A.H., Kreuzinger, N., Kirschner, A., and Mach, R.L. (2006). Primers containing universal bases reduce multiple *amoA* gene specific DGGE band patterns when analysing the diversity of beta-ammonia oxidizers in the environment. J. Microbiol. Methods *66*, 147–155.

Hou, S., Makarova, K.S., Saw, J.H.W., Senin, P., Ly, B.V., Zhou, Z., Ren, Y., Wang, J., Galperin, M.Y., Omelchenko, M.V., *et al.* (2008). Complete genome sequence of the extremely acidophilic methanotroph isolate V4, *Methylacidiphilum infernorum*, a representative of the bacterial phylum *Verrucomicrobia*. Biol. Direct *3*, 26.

Hussain, Q., Liu, Y.Z., Jin, Z.J., Zhang, A.F., Pan, G.X., Li, L.Q., Crowley, D., Zhang, X.H., Song, X.Y., and Cui, L.Q. (2011). Temporal dynamics of ammonia oxidizer (*amoA*) and denitrifier (*nirK*) communities in the rhizosphere of a rice ecosystem from Tai Lake region, China. Appl Soil Ecol. *48*, 210–218.

İnceoğlu, Ö., Al-Soud, W.A., Salles, J.F., Semenov, A.V., and van Elsas, J.D. (2011). Comparative analysis of bacterial communities in a potato field as determined by pyrosequencing. Plos One *6*, e23321.

Isanapong, J., Goodwin, L., Bruce, D., Chen, A., Detter, C., Han, J., Han, C.S., Held, B., Huntemann, M., Ivanova, N., *et al.* (2012). High-quality draft genome sequence of the *Opitutaceae* bacterium strain TAV1, a symbiont of the wood-feeding termite *Reticulitermes flavipes*. J. Bacteriol. *194*, 2744–2745.

Jackson, R.W., Preston, G.M., and Rainey, P.B. (2005). Genetic characterization of *Pseudomonas fluorescens* SBW25 rsp gene expression in the phytosphere and *in vitro*. J. Bacteriol. *187*, 8477–8488.

Julien, B., Shah, S., Ziermann, R., Goldman, R., Katz, L., and Khosla, C. (2000). Isolation and characterization of the epothilone biosynthetic gene cluster from *Sorangium cellulosum*. Gene *249*, 153–160.

Kant, R., van Passel, M.W.J., Palva, A., Lucas, S., Lapidus, A., del Rio, T., Dalin, E., Tice, H., Bruce, D., Goodwin, L., *et al.* (2011). Genome sequence of *Chthoniobacter flavus* Ellin428, an aerobic heterotrophic soil bacterium. J. Bacteriol. *193*, 2902–2903.

Khadem, A.F., Pol, A., Wieczorek, A., Mohammadi, S.S., Francoijs, K.J., Stunnenberg, H.G., Jetten, M.S.M., and Op den Camp, H.J.M. (2011). Autotrophic methanotrophy in *Verrucomicrobia*: *Methylacidiphilum fumariolicum* SolV uses the Calvin-Benson-Bassham cycle for carbon dioxide fixation. J. Bacteriol. *193*, 4438–4446.

Khammar, N., Martin, G., Ferro, K., Job, D., Aragno, M., and Verrecchia, E. (2009). Use of the *frc* gene as a molecular marker to characterize oxalate-oxidizing bacterial abundance and diversity structure in soil. J. Microbiol. Methods *76*, 120–127.

Konstantinidis, K.T., and Tiedje, J.M. (2004). Trends between gene content and genome size in prokaryotic species with larger genomes. Proc. Natl. Acad. Sci. U.S.A. *101*, 3160–3165.

LaButti, K., Sikorski, J., Schneider, S., Nolan, M., Lucas, S., Del Rio, T.G., Tice, H., Cheng, J.-F., Goodwin, L., Pitluck, S., *et al.* (2010). Complete genome sequence of *Planctomyces limnophilus* type strain (Mu 290(T)). Stand. Gen. Sci. *3*, 47–56.

Laue, R.E., Jiang, Y., Chhabra, S.R., Jacob, S., Stewart, G., Hardman, A., Downie, J.A., O'Gara, F., and Williams, P. (2000). The biocontrol strain *Pseudomonas fluorescens* F113 produces the *Rhizobium* small bacteriocin, N-(3-hydroxy-7-cis-tetradecenoyl)homoserine lactone, via HdtS, a putative novel N-acylhomoserine lactone synthase. Microbiology *146*, 2469–2480.

Lejon, D.P.H., Nowak, V., Bouko, S., Pascault, N., Mougel, C., Martins, J.M.F., and Ranjard, L. (2007). Fingerprinting and diversity of bacterial *copA* genes in response to soil types, soil organic status and copper contamination. FEMS Microbiol. Ecol. *61*, 424–437.

Lesic, B., Starkey, M., He, J., Hazan, R., and Rahme, L.G. (2009). Quorum sensing differentially regulates *Pseudomonas aeruginosa* type VI secretion locus I and homologous loci II and III, which are required for pathogenesis. Microbiology *155*, 2845–2855.

Leveau, J.H. J., and Preston, G.M. (2008). Bacterial mycophagy: definition and diagnosis of a unique bacterial–fungal interaction. New Phytol. *177*, 859–876.

Lugtenberg, B., and Kamilova, F. (2009). Plant-growth-promoting rhizobacteria. Annu. Rev. Microbiol. *63*, 541–556.

Mark, G.L., Dow, J.M., Kiely, P.D., Higgins, H., Haynes, J., Baysse, C., Abbas, A., Foley, T., Franks, A., Morrissey, J.P., *et al.* (2005). Transcriptome profiling of bacterial responses to root exudates identifies genes involved in microbe–plant interactions. Proc. Natl. Acad. Sci. U.S.A. *102*, 17454–17459.

Markowitz, V.M., Ivanova, N.N., Szeto, E., Palaniappan, K., Chu, K., Dalevi, D., Chen, I.M. A., Grechkin, Y., Dubchak, I., Anderson, I., *et al.* (2008). IMG/M: a data management and analysis system for metagenomes. Nucleic Acids Res. *36*, D534–D538.

Marschner, P., Yang, C.H., Lieberei, R., and Crowley, D.E. (2001). Soil and plant specific effects on bacterial community composition in the rhizosphere. Soil Biol. Biochem. *33*, 1437–1445.

Martinez-Gil, M., Yousef-Coronado, F., and Espinosa-Urgel, M. (2010). LapF, the second largest *Pseudomonas putida* protein, contributes to plant root colonization and determines biofilm architecture. Mol. Microbiol. *77*, 549–561.

Martinez, J.L., Sanchez, M.B., Martinez-Solano, L., Hernandez, A., Garmendia, L., Fajardo, A., and Alvarez-Ortega, C. (2009). Functional role of bacterial multidrug efflux pumps in microbial natural ecosystems. Fems Microbiol. Rev. *33*, 430–449.

Matilla, M.A., Espinosa-Urgel, M., Rodriguez-Herva, J.J., Ramos, J.L., and Ramos-Gonzalez, M.I. (2007). Genomic analysis reveals the major driving forces of bacterial life in the rhizosphere. Genome Biol. *8*, R179.

Maunsell, B., Adams, C., and O'Gara, F. (2006). Complex regulation of AprA metalloprotease in *Pseudomonas fluorescens* M114: evidence for the involvement of iron, the ECF sigma factor, PbrA and pseudobactin M114 siderophore. Microbiology *152*, 29–42.

Mavrodi, D.V., Joe, A., Mavrodi, O.V., Hassan, K.A., Weller, D.M., Paulsen, I.T., Loper, J.E., Alfano, J.R., and Thomashow, L.S. (2011). Structural and functional analysis of the type III secretion system from *Pseudomonas fluorescens* Q8r1-96. J. Bacteriol. *193*, 177–189.

Mavrodi, D.V., Peever, T.L., Mavrodi, O.V., Parejko, J.A., Raaijmakers, J.M., Lemanceau, P., Mazurier, S., Heide, L., Blankenfeldt, W., Weller, D.M., *et al.* (2010). Diversity and evolution of the phenazine biosynthesis pathway. Appl. Environ. Microbiol. *76*, 866–879.

Mavromatis, K., Abt, B., Brambilla, E., Lapidus, A., Copeland, A., Deshpande, S., Nolan, M., Lucas, S., Tice, H., Cheng, J.-F., *et al.* (2010). Complete genome sequence of *Coraliomargarita akajimensis* type strain (04OKA010-24(T)). Stand. Gen. Sci. *2*, 290–299.

Mazurier, S., Lemunier, M., Siblot, S., Mougel, C., and Lemanceau, P. (2004). Distribution and diversity of type III secretion system-like genes in saprophytic and phytopathogenic fluorescent pseudomonads. FEMS Microbiol. Ecol. *49*, 455–467.

Mazurier, S., Corberand, T., Lemanceau, P., and Raaijmakers, J.M. (2009). Phenazine antibiotics produced by fluorescent pseudomonads contribute to natural soil suppressiveness to Fusarium wilt. ISME J. *3*, 977–991.

Mendes, R., Kruijt, M., de Bruijn, I., Dekkers, E., van der Voort, M., Schneider, J.H. M., Piceno, Y.M., DeSantis, T.Z., Andersen, G.L., Bakker, P.A.H.M., *et al.* (2011). Deciphering the rhizosphere microbiome for disease-suppressive bacteria. Science *332*, 1097–1100.

Meyer, J.B., Frapolli, M., Keel, C., and Maurhofer, M. (2011). Pyrroloquinoline quinone biosynthesis gene *pqqC*, a novel molecular marker for studying the phylogeny and diversity of phosphate-solubilizing pseudomonads. Appl. Environ. Microbiol. *77*, 7345–7354.

Miethling, R., Ahrends, K., and Tebbe, C.C. (2003). Structural differences in the rhizosphere communities of legumes are not equally reflected in community-level physiological profiles. Soil Biol. Biochem. *35*, 1405–1410.

Miller, L.D., Yost, C.K., Hynes, M.F., and Alexandre, G. (2007). The major chemotaxis gene cluster of *Rhizobium leguminosarum* bv. *viciae* is essential for competitive nodulation. Mol. Microbiol. *63*, 348–362.

Miller, S.H., Browne, P., Prigent-Combaret, C., Combes-Meynet, E., Morrissey, J.P., and O'Gara, F. (2010). Biochemical and genomic comparison of inorganic phosphate solubilization in *Pseudomonas* species. Environ. Microbiol. Rep. *2*, 403–411.

Monier, J.M., Demaneche, S., Delmont, T.O., Mathieu, A., Vogel, T.M., and Simonet, P. (2011). Metagenomic exploration of antibiotic resistance in soil. Curr. Opin. Microbiol. *14*, 229–235.

Morgan, J.A. W., Bending, G.D., and White, P.J. (2005). Biological costs and benefits to plant–microbe interactions in the rhizosphere. J. Exp. Bot. *56*, 1729–1739.

Morrissey, J.P., Dow, J.M., Mark, G.L., and O'Gara, F. (2004). Are microbes at the root of a solution to world food production? Rational exploitation of interactions between microbes and plants can help to transform agriculture. EMBO Rep. *5*, 922–926.

Moynihan, J.A., Morrissey, J.P., Coppoolse, E.R., Stiekema, W.J., O'Gara, F., and Boyd, E.F. (2009). Evolutionary history of the *phl* gene cluster in the plant-associated bacterium *Pseudomonas fluorescens*. Appl. Environ. Microbiol. *75*, 2122–2131.

Mueller, H., Westendorf, C., Leitner, E., Chernin, L., Riedel, K., Schmidt, S., Eberl, L., and Berg, G. (2009). Quorum-sensing effects in the antagonistic rhizosphere bacterium *Serratia plymuthica* HRO-C48. FEMS Microbiol. Ecol. *67*, 468–478.

Nannipieri, P., Ascher, J., Ceccherini, M.T., Guerri, G., Renella, G., and Pietramellara, G. (2008a). Recent advances in functional genomics and proteomics of plant associated microbes. In Soil Biology, Nautiyal, C., and Dion, P., ed. (Springer, Berlin, Heidelberg). pp. 215–241.

Nannipieri, P., Ascher, J., Ceccherini, M.T., Landi, L., Pietramellara, G., Renella, G., and Valori, F. (2008b). Effects of root exudates in microbial diversity and activity in rhizosphere soils. In Soil Biology, Nautiyal, C., and Dion, P., ed. (Springer, Berlin, Heidelberg). pp. 339–365.

Nichols, D. (2007). Cultivation gives context to the microbial ecologist. FEMS Microbiol. Ecol. *60*, 351–357.

Pallen, M.J., Beatson, S.A., and Bailey, C.M. (2005). Bioinformatics, genomics and evolution of non-flagellar type-III secretion systems: a Darwinian perpective. FEMS Microbiol. Rev. *29*, 201–229.

Parsley, L.C., Linneman, J., Goode, A.M., Becklund, K., George, I., Goodman, R.M., Lopanik, N.B., and Liles, M.R. (2011). Polyketide synthase pathways identified from a metagenomic library are derived from soil Acidobacteria. FEMS Microbiol. Ecol. *78*, 176–187.

Partida-Martinez, L.P., and Hertweck, C. (2005). Pathogenic fungus harbours endosymbiotic bacteria for toxin production. Nature *437*, 884–888.

van Passel, M.W.J., Kant, R., Zoetendal, E.G., Plugge, C.M., Derrien, M., Malfatti, S.A., Chain, P.S.G., Woyke, T., Palva, A., de Vos, W.M., *et al.* (2011). The genome of *Akkermansia muciniphila*, a dedicated intestinal mucin degrader, and its use in exploring intestinal metagenomes. Plos One *6*, e16876.

Patel, N., Blackmoore, M., Hillman, B., and Kobayashi, D. (2009). Evidence for the role of Type VI secretion during *Lysobacter enzymogenes* pathogenesis of fungal hosts. Phytopath. *99*, S100–S101.

Persson, O.P., Pinhassi, J., Riemann, L., Marklund, B.I., Rhen, M., Normark, S., Gonzalez, J.M., and Hagstrom, A. (2009). High abundance of virulence gene homologues in marine bacteria. Environ. Microbiol. *11*, 1348–1357.

Philippot, L., and Hallin, S. (2005). Finding the missing link between diversity and activity using denitrifying bacteria as a model functional community. Cur. Opin. Microbiol. *8*, 234–239.

Philippot, L., Claysjosserand, A., and Lensi, R. (1995). Use of Tn5 mutants to assess the role of the dissimilatory nitrite reductase in the competitive abilities of 2 *Pseudomonas* strains in soil. Appl. Environ. Microbiol. *61*, 1426–1430.

Porter, S.L., Wadhams, G.H., and Armitage, J.P. (2011). Signal processing in complex chemotaxis pathways. Nat. Rev. Microbiol. *9*, 153–165.

Pothier, J.F., Wisniewski-Dye, F., Weiss-Gayet, M., Moenne-Loccoz, Y., and Prigent-Combaret, C. (2007). Promoter-trap identification of wheat seed extract induced genes in the plant-growth-promoting rhizobacterium *Azospirillum brasilense* Sp245. Microbiology *153*, 3608–3622.

Pothier, J.F., Prigent-Combaret, C., Haurat, J., Moenne-Loccoz, Y., and Wisniewski-Dye, F. (2008). Duplication of plasmid-borne nitrite reductase gene *nirK* in the wheat-associated plant growth-promoting rhizobacterium *Azospirillum brasilense* Sp245. Mol. Plant Microbe In. *21*, 831–842.

Raaijmakers, J.M., and Weller, D.M. (1998). Natural plant protection by 2,4-diacetylphloroglucinol-producing *Pseudomonas* spp. in take-all decline soils. Mol. Plant Microbe In. *11*, 144–152.

Rainey, P.B. (1999). Adaptation of *Pseudomonas fluorescens* to the plant rhizosphere. Environ. Microbiol. *1*, 243–257.

Records, A.R. (2011). The type VI secretion system: A multipurpose delivery system with a phage-like machinery. Mol. Plant Microbe In. *24*, 751–757.

Rediers, H., Vanderleyden, J., and De Mot, R. (2009). Nitrate respiration in *Pseudomonas stutzeri* A15 and its involvement in rice and wheat root colonization. Microbiol. Res. *164*, 461–468.

Rezzonico, F., Defago, G., and Moenne-Loccoz, Y. (2004). Comparison of ATPase-encoding type III secretion system *hrcN* genes in biocontrol fluorescent pseudomonads and in phytopathogenic proteobacteria. Appl. Environ. Microbiol. *70*, 5119–5131.

Rezzonico, F., Binder, C., Defago, G., and Moenne-Loccoz, Y. (2005). The type III secretion system of biocontrol *Pseudomonas fluorescens* KD targets the phytopathogenic chromista *Pythium ultimum* and promotes cucumber protection. Mol. Plant Microbe In. *18*, 991–1001.

Richardson, A.E., Barea, J.M., McNeill, A.M., and Prigent-Combaret, C. (2009). Acquisition of phosphorus and nitrogen in the rhizosphere and plant growth promotion by microorganisms. Plant Soil *321*, 305–339.

Ritz, K. (2007). The plate debate: Cultivable communities have no utility in contemporary environmental microbial ecology. FEMS Microbiol. Ecol. *60*, 358–362.

da Rocha, U.N., van Overbeek, L., and van Elsas, J.D. (2009). Exploration of hitherto-uncultured bacteria from the rhizosphere. FEMS Microbiol. Ecol. *69*, 313–328.

Rodriguez-Navarro, D.N., Dardanelli, M.S., and Ruiz-Sainz, J.E. (2007). Attachment of bacteria to the roots of higher plants. FEMS Microbiol. Lett. *272*, 127–136.

Roesch, L.F., Fulthorpe, R.R., Riva, A., Casella, G., Hadwin, A.K.M., Kent, A.D., Daroub, S.H., Camargo, F.A.O., Farmerie, W.G., and Triplett, E.W. (2007). Pyrosequencing enumerates and contrasts soil microbial diversity. ISME J. *1*, 283–290.

Rojo, F. (2010). Carbon catabolite repression in *Pseudomonas*: optimizing metabolic versatility and interactions with the environment. Fems Microbiol. Rev. *34*, 658–684.

Rudrappa, T., Czymmek, K.J., Pare, P.W., and Bais, H.P. (2008). Root-secreted malic acid recruits beneficial soil bacteria. Plant Physiol. *148*, 1547–1556.

Ryan, R.P., Monchy, S., Cardinale, M., Taghavi, S., Crossman, L., Avison, M.B., Berg, G., van der Lelie, D., and Dow, J.M. (2009). The versatility and adaptation of bacteria from the genus *Stenotrophomonas*. Nat. Rev. Microbiol. *7*, 514–525.

Sait, M., Kamneva, O.K., Fay, D.S., Kirienko, N.V., Polek, J., Shirasu-Hiza, M.M., and Ward, N.L. (2011). Genomic and experimental evidence suggests that *Verrucomicrobium spinosum* interacts with eukaryotes. Front. Microbiol. *2*, 211.

Sakurai, M., Wasaki, J., Tomizawa, Y., Shinano, T., and Osaki, M. (2008). Analysis of bacterial communities on alkaline phosphatase genes in soil supplied with organic matter. Soil Sci. Plant Nut. *54*, 62–71.

Sanguin, H., Remenant, B., Dechesne, A., Thioulouse, J., Vogel, T.M., Nesme, X., Moenne-Loccoz, Y., and Grundmann, G.L. (2006). Potential of a 16S rRNA-based taxonomic microarray for analyzing the rhizosphere effects of maize on *Agrobacterium* spp. and bacterial communities. Appl. Environ. Microbiol. *72*, 4302–4312.

Sanguin, H., Kroneisen, L., Gazengel, K., Kyselkova, M., Remenant, B., Prigent-Combaret, C., Grundmann, G.L., Sarniguet, A., and Moenne-Loccoz, Y. (2008). Development of a 16S rRNA microarray approach for the monitoring of rhizosphere *Pseudomonas* populations associated with the decline of take-all disease of wheat. Soil. Biol. Biochem. *40*, 1028–1039.

Santarella-Mellwig, R., Franke, J., Jaedicke, A., Gorjanacz, M., Bauer, U., Budd, A., Mattaj, I.W., and Devos, D.P. (2010). The compartmentalized bacteria of the *Planctomycetes-Verrucomicrobia-Chlamydiae* superphylum have membrane coat-like proteins. Plos Biol. *8*, e1000281.

Schloss, P.D., and Handelsman, J. (2006). Toward a census of bacteria in soil. Plos Comput. Biol. *2*, 786–793.

Schneemann, I., Wiese, J., Kunz, A.L., and Imhoff, J.F. (2011). Genetic approach for the fast discovery of phenazine producing bacteria. Marine Drugs *9*, 772–789.

Schwarz, S., Hood, R.D., and Mougous, J.D. (2010). What is type VI secretion doing in all those bugs? Trends Microbiol. *18*, 531–537.

Selesi, D., Schmid, M., and Hartmann, A. (2005). Diversity of green-like and red-like ribulose-1,5-bisphosphate carboxylase/oxygenase large-subunit genes (*cbbL*) in differently managed agricultural soils. Appl. Environ. Microbiol. *71*, 175–184.

Silby, M.W., Cerdeno-Tarraga, A.M., Vernikos, G.S., Giddens, S.R., Jackson, R.W., Preston, G.M., Zhang, X.X., Moon, C.D., Gehrig, S.M., *et al.* (2009). Genomic and genetic analyses of diversity and plant interactions of *Pseudomonas fluorescens*. Gen. Biol. *10*, R51.

Sliwinski, M.K., and Goodman, R.M. (2004). Spatial heterogeneity of crenarchaeal assemblages within mesophilic soil ecosystems as revealed by PCR-single-stranded conformation polymorphism profiling. Appl. Environ. Microbiol. *70*, 1811–1820.

Smalla, K., Oros-Sichler, M., Milling, A., Heuer, H., Baumgarte, S., Becker, R., Neuber, G., Kropf, S., Ulrich, A., *et al.* (2007). Bacterial diversity of soils assessed by DGGE, T-RFLP and SSCP fingerprints of PCR-amplified 16S rRNA gene fragments: Do the different methods provide similar results? J. Microbiol. Methods *69*, 470–479.

Sorensen, J., Nicolaisen, M.H., Ron, E., and Simonet, P. (2009). Molecular tools in rhizosphere microbiology-from single-cell to whole-community analysis. Plant Soil *321*, 483–512.

Strous, M., Pelletier, E., Mangenot, S., Rattei, T., Lehner, A., Taylor, M.W., Horn, M., Daims, H., Bartol-Mavel, D., Wincker, P., *et al.* (2006). Deciphering the evolution and metabolism of an anammox bacterium from a community genome. Nature *440*, 790–794.

Svercel, M., Duffy, B., and Defago, G. (2007). PCR amplification of hydrogen cyanide biosynthetic locus *hcnAB* in *Pseudomonas* spp. J. Microbiol. Methods *70*, 209–213.

Throback, I.N., Enwall, K., Jarvis, A., and Hallin, S. (2004). Reassessing PCR primers targeting *nirS*, *nirK* and *nosZ* genes for community surveys of denitrifying bacteria with DGGE. FEMS Microbiol. Ecol. *49*, 401–417.

Torres-Cortes, G., Millan, V., Ramirez-Saad, H.C., Nisa-Martinez, R., Toro, N., and Martinez-Abarca, F. (2011). Characterization of novel antibiotic resistance genes identified by functional metagenomics on soil samples. Environ. Microbiol. *13*, 1101–1114.

Tringe, S.G., von Mering, C., Kobayashi, A., Salamov, A.A., Chen, K., Chang, H.W., Podar, M., Short, J.M., Mathur, E.J., *et al.* (2005). Comparative metagenomics of microbial communities. Science *308*, 554–557.

Troisfontaines, P., and Cornelis, G.R. (2005). Type III secretion: more systems than you think. Physiology *20*, 326–339.

Tseng, T.-T., Tyler, B.M., and Setubal, J.C. (2009). Protein secretion systems in bacterial–host associations, and their description in the Gene Ontology. BMC Microbiol. *9*, S2.

Tyson, G.W., Chapman, J., Hugenholtz, P., Allen, E.E., Ram, R.J., Richardson, P.M., Solovyev, V.V., Rubin, E.M., Rokhsar, D.S., and Banfield, J.F. (2004). Community structure and metabolism through reconstruction of microbial genomes from the environment. Nature *428*, 37–43.

Ulrich, A., and Becker, R. (2006). Soil parent material is a key determinant of the bacterial community structure in arable soils. FEMS Microbiol. Ecol. *56*, 430–443.

Uroz, S., Buee, M., Murat, C., Frey-Klett, P., and Martin, F. (2010). Pyrosequencing reveals a contrasted bacterial diversity between oak rhizosphere and surrounding soil. Environ. Microbiol. Rep. *2*, 281–288.

Verbaendert, I., Boon, N., De Vos, P., and Heylen, K. (2011). Denitrification is a common feature among members of the genus *Bacillus*. Syst. Appl. Microbiol. *34*, 385–391.

Viollet, A., Corberand, T., Mougel, C., Robin, A., Lemanceau, P., and Mazurier, S. (2011). Fluorescent pseudomonads harboring type III secretion genes are enriched in the mycorrhizosphere of *Medicago truncatula*. FEMS Microbiol. Ecol. *75*, 457–467.

Ward, N.L., Challacombe, J.F., Janssen, P.H., Henrissat, B., Coutinho, P.M., Wu, M., Xie, G., Haft, D.H., Sait, M., *et al.* (2009). Three genomes from the phylum *Acidobacteria* provide insight into the lifestyles of these microorganisms in soils. Appl. Environ. Microbiol. *75*, 2046–2056.

Warmink, J.A., and van Elsas, J.D. (2008). Selection of bacterial populations in the mycosphere of *Laccaria proxima*: is type III secretion involved? ISME J. *2*, 887–900.

Weinert, N., Piceno, Y., Ding, G.C., Meincke, R., Heuer, H., Berg, G., Schloter, M., Andersen, G., and Smalla, K. (2011). PhyloChip hybridization uncovered an enormous bacterial diversity in the rhizosphere of different potato cultivars: many common and few cultivar-dependent taxa. FEMS Microbiol. Ecol. *75*, 497–506.

de Weert, S., Vermeiren, H., Mulders, I.H. M., Kuiper, I., Hendrickx, N., Bloemberg, G.V., Vanderleyden, J., De Mot, R., and Lugtenberg, B.J. J. (2002). Flagella-driven chemotaxis towards exudate components is an important trait for tomato root colonization by *Pseudomonas fluorescens*. Mol. Plant Microbe In. *15*, 1173–1180.

Wuichet, K., and Zhulin, I.B. (2010). Origins and diversification of a complex signal transduction system in prokaryotes. Sci. Signal. *3*, ra50.

Xu, Y.X., Wang, G.H., Jin, J., Liu, J.J., Zhang, Q.Y., and Liu, X.B. (2009). Bacterial communities in soybean rhizosphere in response to soil type, soybean genotype, and their growth stage. Soil Biol. Biochem. *41*, 919–925.

Yao, J., and Allen, C. (2006). Chemotaxis is required for virulence and competitive fitness of the bacterial wilt pathogen *Ralstonia solanacearum*. J. Bacteriol. *188*, 3697–3708.

Yousef-Coronado, F., Travieso, M.L., and Espinosa-Urgel, M. (2008). Different, overlapping mechanisms for colonization of abiotic and plant surfaces by *Pseudomonas putida*. FEMS Microbiol. Lett. *288*, 118–124.

Zelenev, V.V., van Bruggen, A.H. C., and Semenov, A.M. (2005). Modeling wave-like dynamics of oligotrophic and copiotrophic bacteria along wheat roots in response to nutrient input from a growing root tip. Ecol. Model. *188*, 404–417.

Soil Metagenomics – Potential Applications and Methodological Problems

Jan Dirk van Elsas, Mariana Silvia Cretoiu, Anna Maria Kielak and Francisco Dini-Andreote

Abstract

Metagenomics has been defined as the study of the collective genomes of the microbiota in given habitat. Soil offers a huge microbial diversity and the use of metagenomics approaches will allow a deeper understanding of soil microbial diversity and function. The two areas, phylogenetically based diversity and functional gene based function, are complementary and may be used side-by-side in order to allow a better understanding of the living soil. Moreover, genes for relevant functions can be cloned into suitable vectors, after which they can be studied and possibly explored for biotechnological purposes. Thus, opportunities for novel product discovery via metagenomics are rapidly rising. However, there are caveats in what metagenomics techniques can tell us about the soil environment and its functioning, and also in the chances of successful exploration of soil.

In this chapter, we review the developments in the metagenomics-based exploitation and exploration of soil and examine how soil metagenomics can enhance our vision about natural functioning and exploration for biotechnological novelty. One major issue, the need for advanced bioinformatics tools, is stressed. We conclude that the rich microbiota of soil offers an astonishing big playground for metagenomics, but that methodological and conceptual problems still hamper its full exploitation.

Introduction

A major hurdle in microbial ecology is the inability to culture most of the microbiota that is present in the system under study. This hurdle is particularly relevant for the living soil. The divergence between the numbers of colony forming units on plates and cell counts from microscopic examination has been denominated 'the great plate count anomaly' (Staley and Konopka, 1985). Thus, only a fraction of the microbial diversity present in most ecosystems, including soil (1–5%), can be accessed through standard cultivation techniques (Staley and Konopka, 1985; Torsvik and Ovreas, 2002; Curtis and Sloan, 2004; Nichols, 2007). Given this lacuna in our knowledge, we can only speculate about the environmental and biotechnological relevance of the majority of organisms that are present in most ecosystems, as these have remained unexplored to date. To access, define and explore this microbiota, genetic material (DNA and/or RNA) needs to be directly extracted from the collective cells in an environmental sample. This nucleic acid pool, which has been named the 'metagenome' of the community in the case of environmental DNA (Handelsman *et al.*, 1998), can be further analysed by modern methods such as PCR- or direct sequencing based assessments of diversity, allowing to analyse the community at the phylogenetic level. Moreover, the DNA can be analysed with respect to function, using direct sequencing, functional microarray hybridizations with the Geochip (He *et al.*, 2012) or genetic or functional screening of metagenomic libraries. All molecular analysis strategies that are used to examine microbial metagenomes have been denoted metagenomics techniques.

Recently, the great potential of metagenomics methods to promote our understanding of the function and diversity of ecosystems has become clear (Riesenfeld *et al.*, 2004; Lefevre *et al.*, 2008; Warnecke and Hess, 2009; Hil and Fenical, 2010; Mocali and Benedetti, 2010; Singh and Macdonald, 2010; Imhoff *et al.*, 2011). It is thought to be very advantageous to be able to 'know everybody' in an ecological system, as knowing the 'players' that are potentially involved in the function of the system may allow us to better predict how the system works and how its components interact. Moreover, the biotechnological exploration of environmental habitats is promoted by the development of metagenomics. For instance, the screening of metagenomic libraries enables the study of genes from hitherto inaccessible microbes, which opens up exiting possibilities for developing novel products (Warnecke *et al.*, 2007; Singh *et al.*, 2008; Uchiyama and Miyazaki, 2009; Fernández-Arrojo *et al.*, 2010).

This chapter aims to discuss the major bottlenecks that pertain to the metagenomics of the microbiota in soil systems for:

1 understanding the functioning and diversity of the living soil; and
2 bioexploration of this environment.

By reviewing the status of the use of metagenomics applied to soil systems, an overview is provided of the most important aspects of the current examination and exploration of natural microbial systems. Then, strategies for future improvements are given. Fig. 3.1 provides a general outline of metagenomics and its application to soil systems.

Metagenomics for fostering our understanding of soil habitats

Most soil habitats are incredibly complex in terms of the diversity of the extant microbiota. Estimates of the numbers of bacterial species per gram (g) of soil range from several thousands (Torsvik and Ovreas, 2002) to up to a million (Gans *et al.*, 2005). In addition, the population density of microorganisms, in particular bacteria, is often very high, with number of 10^8 to 10^{10} cells per g of soil being common. Within this highly diverse and dense microbiota, great redundancy of function often exists. It is likely that, per organismal group, the operons pertaining to a functional class have their own particular way of functioning, in terms of being turned on and off, thus defining the way the function is expressed in the soil (micro)habitat. Given the fact that until today, we have been unable to 'see' all components of that (micro)habitat, great challenges are ahead even when we are equipped with the newest metagenomics techniques. These challenges can be divided into (1) methodological challenges, and (2) computational/bioinformatics/statistical challenges. Hereunder, we will deal with each of these types of challenges in separate. We will then analyse selected key studies that applied metagenomics tools to analyse soil habitats, culminating in a case study performed in our laboratory that addresses the selective effects of the habitat on microbial chitin degraders.

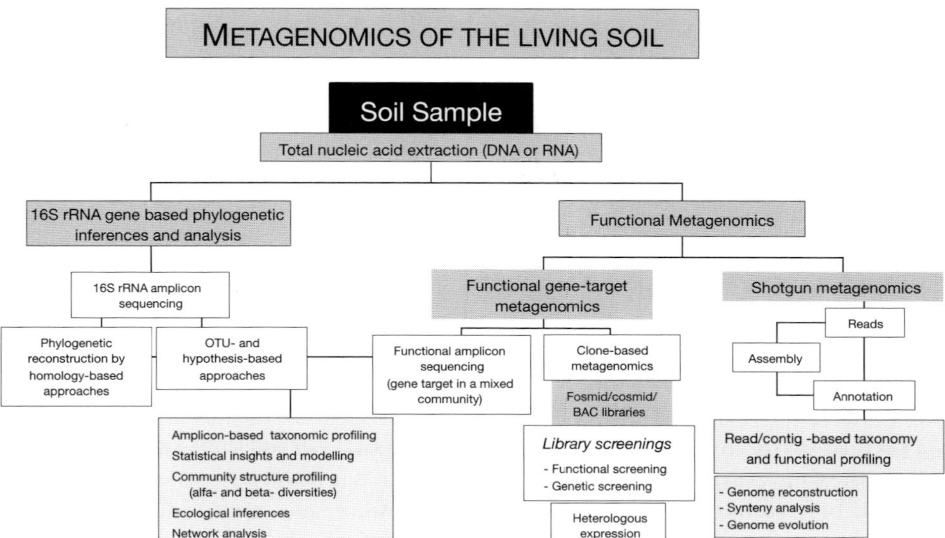

Figure 3.1 General overview of soil metagenomics strategies.

Methodological challenges

A representative metagenomics-based analysis of any environmental habitat depends on the quality of the nucleic acids that are extracted, as well as the quality and reliability (absence of bias) of all subsequent molecular analyses. As we will see in the following, none of the current methodologies fulfil reasonable criteria of representativeness, completeness or lack of bias or distortion. Or, at least, we have no convincing evidence that this is the case. First, soil DNA extraction has been shown to never be complete (Delmont *et al.*, 2011) and thus there will be invariable bias in the ensuing data (Inceoglu *et al.*, 2010). Moreover, extracellular DNA in soil and sediment resulting from lysing or lysed prokaryotic or eukaryotic cells may interfere in studies of the composition of the microbial communities present (Pietramellara *et al.*, 2009). Such released DNA is important, as it can enter bacteria via transformation, However, in cases when microbial communities are considered, removal of this DNA, together with humic acids, may be necessary before any further work, e.g. by cloning, is performed.

Importantly, in spite of about 30 years of development of soil DNA extraction techniques and the current dominance of kit-based approaches, we still ignore whether we are sampling a representative part of the total bacterial community in the soil. Moreover, only small samples are allowed in most extraction protocols and, in the light of the often considerable heterogeneity of soil, it is uncertain how such a sample can represent a whole field, for instance. Also, by sampling a soil microbial community this way, we do away with the spatial structure of and interactions between the soil bacteriota. In addition to the problems right at the forefront of the molecular analyses of soil, any analysis method based on the DNA or RNA extracted from soil has its own caveats and biases. We here will only briefly comment on these, as other reviews extensively describe them (van Elsas and Boersma, 2011; Taberlet *et al.*, 2012; Wang *et al.*, 2012). Very importantly, all methods based on PCR may suffer from preferential amplification and chimera formation. Moreover, hybridization-based methods (for instance, using the Phylochip or Geochip) will suffer from well-known problems of cross-hybridization and variability of signal strength. Direct sequencing methods suffer from selective sequencing bias, sequencing errors, etc. Methods based on cloning will suffer from problems of biased cloning efficiencies. In other words, we are faced with methods that, on the one hand, can elucidate – for the first time in the history of soil biology – the genetic make-up of the soil microbiota, but on the other hand, are fraught with caveats and biases.

Computational challenges. Or how do we deal with megatons of data?

The sheer power of next generation sequencing (NGS), in addition to our current ability to directly apply NGS to environmental habitats, has truly revolutionized our area. The capacity of data generation is overwhelming, but where is our ability to deal with this massive data set? Are there any robust and well-accepted informational and statistical methods that help in the data analysis? We now face a time in which it is possible to sequence megatons of data in only a few hours of machine processing. However, our capability to process these data in the proper way is often limited due to the lack of robust computational 'power' and the proper programs. In this sense, much more investment is required to overcome this stumbling block, which has been denominated the 'computational bubble'.

On the positive side, a series of bioinformatics tools and web-based pipelines have been recently created, which all aim to supply easy tools to handle different formats of NGS data. As an example, a very interesting and user-friendly platform called Galaxy (http://galaxy.psu.edu/ref) (Giardine *et al.*, 2005) has been extensively used for the analysis of large genomic datasets. The most advantageous use of this tool arose in the feasibility to combine and integrate several different scripts (i.e. tools) in one single interface as a framework. The available tools vary from single inter-conversion of file formats, processing different types of raw data, up to data management and analysis. Another, less flexible but broadly used, web-based server for metagenomic analysis is MG-RAST (http://metagenomics.anl.gov/) (Meyer *et al.*, 2008). MG-RAST offers a rapid system tool for data trimming and automatic phylogenetic and functional annotations. As a result, the annotation system produces functional assignments in metagenomes by comparing query sequences to both protein and nucleotide databases, in addition to supplying several tools that allow to compare different metadata. Data visualization in standard view is another key facet of the platform.

For a more directed assessment and to monitor possible shifts in soil microbial community composition, the application of high-throughput amplicon sequencing on the basis of the 16S rRNA gene has been broadly used (Roesch *et al.*, 2007; Caporaso *et al.*, 2012). By such an approach, hundreds to millions of sequences can be generated and computed to cross-compare samples across different sites or treatments. Briefly, the obtained data can be taxonomically assigned and phylogenetically reconstructed by using two main databases (which also encompass web-based tools for sequence data processing): the webpage and server of the Ribosomal Database Project (RDP – available at http://rdp.cme.msu.edu/) and the GreenGenes site (http://greengenes.lbl.gov/cgi-bin/nph-index.cgi). At this point, it is remarkable that, despite the fact that these taxonomical databases have been updated over the past few years, such an effort is far from being obsolete and microbial taxonomy is far from being completely appreciated. For instance, the number of bacterial subdivisions has increased over time, mostly in the past decade, coming from a few tens to up to 84 phyla or candidate phyla recognized at present by the Hugenholtz taxonomic framework (Greengenes database, August 2012).

Last but not least, statistical insights into the microbial communities in soil, e.g. in terms of diversity, community make-up, community distribution (using rank/abundance curves), network analysis and rarefaction to assess coverage, can be obtained by the use of tools that are based on operational taxonomic units (OTUs). Such an approach enables us to depict the community composition by clustering sequences according to a cut-off value of dissimilarity (for the 16S rRNA gene, the value of 3% is broadly accepted). Available tools to perform such types of analysis are mostly command-line software packages such as Mothur (Schloss *et al.*, 2009) and QIIME (Caporaso *et al.*, 2010). The last software package currently encompasses more than one hundred different scripts, varying from sequence trimming, raw data processing and clustering, phylogenetic reconstruction and statistical tests and community analysis.

From single reads to nearly complete genomes

The field of environmental metagenomics has truly exploded in the last 5–10 years, and hence it will be impossible to discuss each and every individual study that has been performed. Rather, we will briefly examine one of the biggest challenges of metagenomics: the endeavour to combine millions of single reads towards the goal of reconstructing nearly complete genomes. As we know, the complexity of the environment to be studied directly affects the sequencing depth and computational effort to be used to properly obtain insight in a microbial community. As an example, a pioneering and famous metagenomic study was performed in acid mine drainage (AMD) (Tyson *et al.*, 2004). Owing to the extremely acid environment, the extant microbial community was constituted of only five major players (three bacterial and two archaeal species), which offered an exciting environment to explore potentially new functions related to ecosystem processes such as nitrogen fixation, sulphur oxidation and iron oxidation. Summarizing, in this study, even using low-depth sequencing, the authors were able to efficiently assemble two nearly complete genomes and partially recovered the three others.

On the other hand, more complex environments such as soil offer a far more challenging community to be genomically reconstructed. For such attempts, ultra-high-throughput sequencing should be used, or instead, the pre-selection and sequencing of clones containing long-fragment inserts. This effort can lead us to elucidate cryptic genome context, like operon assemblages, offering an avenue towards enhancing our understanding of genes synteny and evolution in soil. Indeed, most of the challenges to reconstruct genomic fragments to nearly complete genomes arise from our inability to properly assemble the millions of short reads that are generated. Significant advances have been achieved by the combinatorial use of different assembly tools like those based on the de Bruijn graph, such as Velvet (Zerbino and Birney, 2008), and also tools for *de novo* assembly that are based on sequence overlap and identity (e.g. PHRAP, Newbler). The proper use of such tools varies widely in accordance with ranges of variables (e.g. community complexity, community coverage, read length and sequencing technology). We will not deeply discuss such approaches, as they are far from the scope of this chapter. Instead, we will keep our focus on the functional and applied studies, as detailed below.

A case study – the metagenomics assessment of the chitinolytic process in soil

Recent advances in chitin degradation research indicate that the highest quantity of this biopolymer in soil is turned over by microorganisms (bacteria and fungi), the same being true for marine environments (Massart and Jijakli, 2007; Poulsen *et al.*, 2008; Delpin and Goodman, 2009). However, we understand very little of the putative successions that take place in the chitin degradative process in the soil. Chitinases are very relevant in soil ecology, as they may be involved in the suppression of plant disease caused by fungi and nematodes. In a few studies, the level of disease suppressiveness has been shown to be raised by adding chitin to soils (Mankau and Das, 1969; Spiegel *et al.*, 1989). We also know that the chitinolytic process is quite redundant in soil, with a range of different soil bacteria having chitinolytic capacity. Examples of these are *Enterobacter agglomerans, Serratia marcescens, Pseudomonas fluorescens, Stenotrophomonas maltophilia, Bacillus subtilis* and *Streptomyces* spp. Many of these species contain strains that have been used as biological control agents of phytopathogenic fungi or nematodes and chitinolytic activity has been implicated in this control (Downing and Thomson, 2000; Kobayashi *et al.*, 2002; Kotan *et al.*, 2009). Another ecologically relevant process is the production of chitosan. This compound, derived from chitin via a deacetylation process, is important for the quality of seeds to be planted. In several papers, it has been shown that chitosan enhances the quality of seeds, their tolerance to stress, their germination rate, the induced resistance of plants to particular diseases and the growth of seedlings (Reddy *et al.*, 1999; Ruan and Xue, 2002; Guan *et al.*, 2009; Zeng *et al.*, 2012).

Given the perceived importance of chitin-degrading enzymes, in particular chitinases and chitin deacetylases, in the soil and the lack of knowledge on the successions that take place in the chitin degradation process, we performed a microcosm experiment with moderately acid soil, in which the description of the successional stages of the process was a key objective (Kielak *et al.*, 2013). Hereunder, we describe the salient data that came out of this experiment, in which the level of chitin as well as the pH were altered. Examination of chitinase activities in all systems revealed fast responses of the soil microbiota to the addition of chitin. In fact, the peaks of enzyme activity occurred 7 days after the onset of the experiment. PCR-DGGE based analyses of 16S rRNA and family-18 *chiA* genes revealed clear changes of the bacterial communities as a result of the chitin addition and pH alteration. Moreover, using direct *chiA* gene based pyrosequencing analyses we found that indeed different *chiA* gene types were selected by the addition of chitin at different soil pH. This is further detailed in the section 'Ecological enhancement'. The data obtained in this case study thus provided new insight in the extant diversity of bacterial *chiA* genes in soil and enhance our understanding of the ecology of the chitin degradation process in soil, in particular with respect to successions taking place in the process. It is now important to understand the interplay between the different organisms involved in the process, in order to unravel their relative roles and involvement. In particular, the role played by the different bacteria (see Fig. 3.2) versus the fungi involved is a still open question.

Metagenomics for bioexploration

In the soil metagenomics approaches for bioexploration, clone libraries are most often prepared on the basis of the directly extracted soil DNA. Such libraries, prepared in

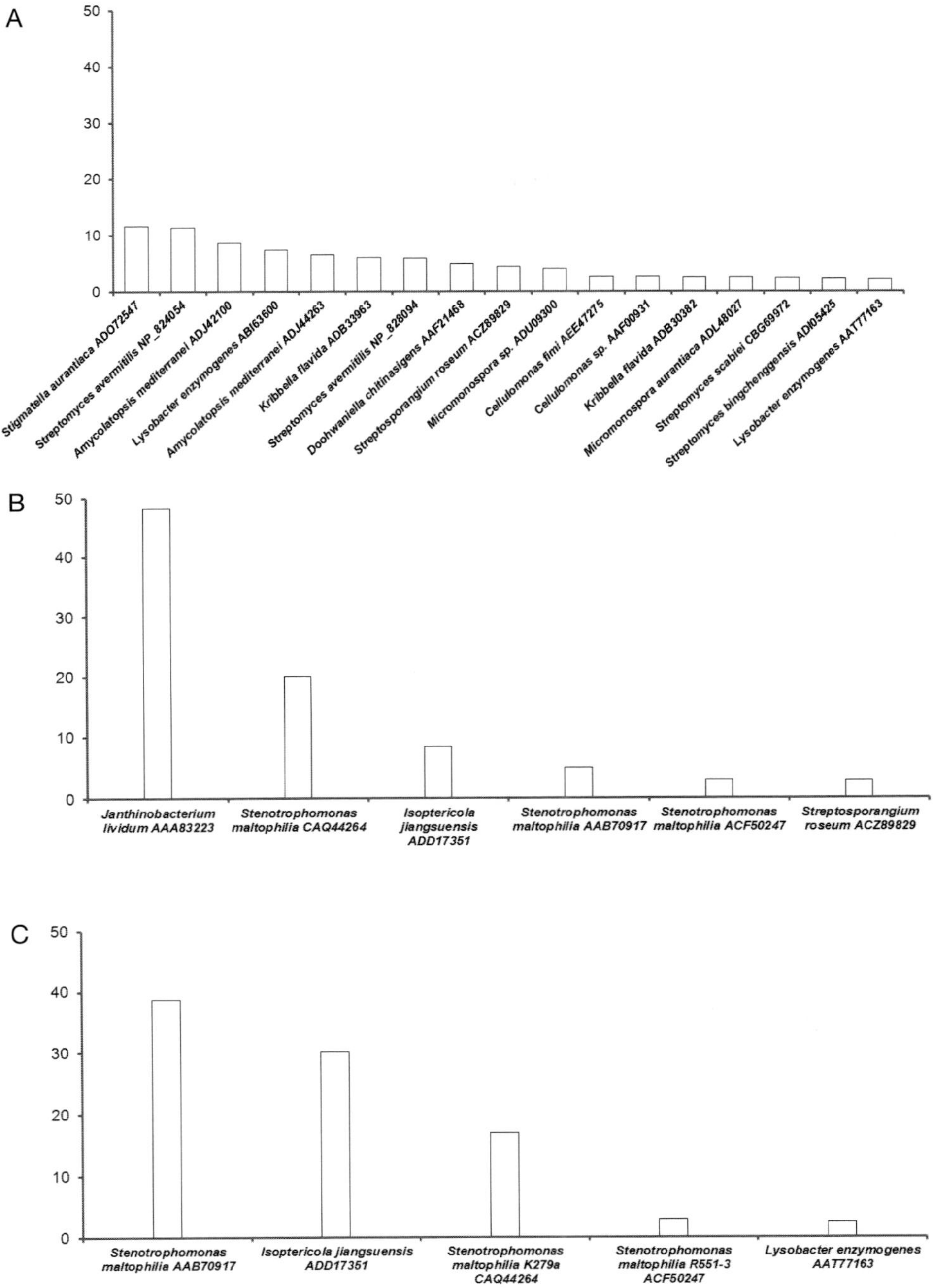

A
50
40
30
20
10
0
Stigmatella aurantiaca ADO72547
Streptomyces avermitilis NP_824054
Amycolatopsis mediterranei ADJ42100
Lysobacter enzymogenes ABI63600
Amycolatopsis mediterranei ADJ44263
Kribbella flavida ADB33963
Streptomyces avermitilis NP_828094
Doohwaniella chitinasigens AAF21468
Streptosporangium roseum ACZ89829
Micromonospora sp. ADU09300
Cellulomonas fimi AEE47275
Cellulomonas sp. AAF00931
Kribbella flavida ADB30382
Micromonospora aurantiaca ADL48027
Streptomyces scabiei CBG69972
Streptomyces bingchenggensis ADI05425
Lysobacter enzymogenes AAT77163
B
50
40
30
20
10
0
Janthinobacterium lividum AAA83223
Stenotrophomonas maltophilia CAQ44264
Isoptericola jiangsuensis ADD17351
Stenotrophomonas maltophilia AAB70917
Stenotrophomonas maltophilia ACF50247
Streptosporangium roseum ACZ89829
C
50
40
30
20
10
0
Stenotrophomonas maltophilia AAB70917
Isoptericola jiangsuensis ADD17351
Stenotrophomonas maltophilia K279a CAQ44264
Stenotrophomonas maltophilia R551-3 ACF50247
Lysobacter enzymogenes AAT77163

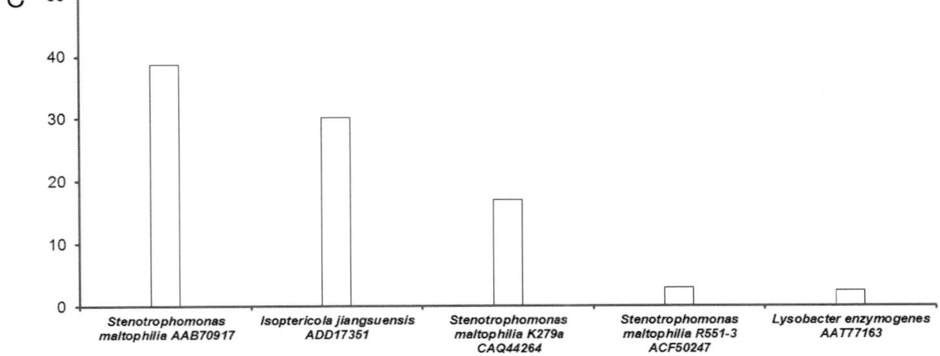

Figure 3.2 Ecological enhancement applied to soil chitin degradation. The relative abundance of main *chiA* gene types in the control (A; T0, pH 5.7) and chitin-treated samples (B; native pH [5.7] or C – raised pH [8.7]) after 7 days of incubation is shown (% of total). The *chiA* gene types were assigned to different bacterial species based on BLAST-P analysis (80% similarity cut-off). Rare sequence types (<2%) are not included in the graphs (T0 – pH 5.7: 12.5%; T7 – pH5.7/chitin: 9.8%; T7 – pH 8.7/chitin: 6.8%).

fosmids, allow the cloning of contiguous stretches of DNA of roughly 35–40 kb, which may encompass the genes or operons that encode useful enzymes. As outlined in van Elsas *et al.* (2008a,b), antibiotic biosynthetic operons often require the cloning of larger pieces of DNA (often up to 200 kb), which is only possible in bacterial artificial chromosomes (BACs). Following library construction, high-throughput screens are required to single out the potentially positive clones. Such screens can be performed by two fundamentally different strategies, i.e. following:

1 a function-based approach; and
2 a sequence-based approach (Schloss and Handelsman, 2003; Kakirde *et al.*, 2010).

In the first strategy, the screening is based on the detection of expression of the required genes in the host. In the second strategy, the focus is on the detection of the relevant genetic sequences, for instance by hybridization or PCR screening. An alternative to this is offered by direct clone sequencing.

In spite of the potential for mining of genetic novelty, the yields from most function-based metagenomics studies have often not yielded sufficiently novel proteins for biotechnology (Beloqui *et al.*, 2008; Hil and Fenical, 2010; Singh and Macdonald, 2010). A key issue here is the often low level of gene expression in the used host (van Elsas *et al.*, 2008a). Alternatively, the screening method used may be rather insensitive, making detection of gene expression difficult (Gabor *et al.*, 2004). A second caveat is the frequent rediscovery of more or less known functions instead of very novel ones, as indicated in a recent paper (Binga *et al.*, 2008). This appears to exacerbate the problem of not getting at the extant genetic diversity through functional metagenomics (Lefevre *et al.*, 2008). Nevertheless, examples of succesful functional screening resulting in successful isolation of enzymes of potential industrial application do exist. Thus, Nacke and co-workers (2012) recently reported on the identification of one cellulase and two xylanase genes derived from a soil metagenome. Expression and characterization of these novel enzymes revealed interesting properties, such as high activity in a broad range of temperatures and pHs. Additionally, the cellulase showed high halotolerance. However, one could question whether this result was good enough, in particular taking into account the fact that more than 350,000 clones were obtained, which contained in total 9.42 Gbp of DNA information (corresponding to about 1,900 prokaryotic genomes).

Considering this last and recent example and the fact that cellulases are likely to be much more numerous than the one positive hit found in a huge amount of cloned DNA, we may here pose the question 'is there such a thing as a 'great screen anomaly'? And, if so, what strategies could be developed to solve this hurdle? In fact, the central question underlying the potential success of metagenomics-based analysis of natural microbial communities is 'How to express genes of unknown origin in high throughput and successfully screen for specific functions?' (Ekkers *et al.*, 2012).

Ecological strategies to overcome limitations of functional metagenomics

To enhance the chances of finding useful target functions, ecological enhancement (also called habitat biasing) has been proposed in order to manipulate the local microbial community prior to the extraction of the metagenomic DNA (Ekkers *et al.*, 2012). Thus, the prevalence of the target functions in the total extracted metagenome is increased *in situ*, and so is the target gene hit rate. In practical terms, an environmental sample is modified and made attractive for specific groups of organisms by adding substrates or modifying its physicochemical conditions (van Elsas *et al.*, 2008b). This then will most likely result in an increase of particular target functions in the resulting metagenome. As an example, such an experimental set-up was used in order to bias soil microbial communities towards organisms that use chitin as the carbon source under conditions of native versus high pH (Kielak *et al.*, 2012; this review, explained in detail below). An advantage of ecological biasing is its low cost and effort, together with the generally low-tech procedures. However, a side effect of ecological enhancement is that organisms that depend on the activities of the target microbes can also proliferate, thus resulting in a potential 'false' enrichment and reduction of the (optimized) target gene hit rate. However, by fine-tuning the selective criteria applied, this problem can be minimized.

In another intelligent approach, specific activities within a microbial community, in particular substrate transformation activities, can be targeted. Thus, stable isotope probing (SIP) has been applied as a method to selectively access particular functions that are involved in an ecological process (Dumont *et al.*, 2006; Cebron *et al.*, 2007). By offering labelled substrate, SIP (which uses the addition of substrate labelled with a stable isotope, i.e. ^{13}C or ^{15}N, to the sample) allows to distinguish the metabolically active members of a microbial community from the inactive ones. If sufficient isotope has been incorporated into the DNA of the active microorganisms, this ('heavy') DNA can be separated from other ('light') DNA by density gradient centrifugation and further analysed. The method thus enables the establishment of a direct link between function and identity (Radajewski *et al.*, 2003; Dumont and Murrell, 2005; Uhlik *et al.*, 2009; Chen and Murrell, 2010; Cupples, 2011). Depending on the type of labelled substrate used, one can additionally bias the sample in much the same way as in ecological enhancement, targeting specific active functional types within a sample. However, a major drawback of SIP remains the fact that quite high concentrations of labelled substrate are often required (next to extended incubation times) to give sufficient yields of labelled DNA in the active organisms. This may result in growth inhibition and an accumulation of the label in the 'wrong' trophic classes. An additional disadvantage is the high cost of labelled substrate. Another problem of SIP is technical, as the differentiation between labelled and unlabelled DNA may be difficult. For instance, unlabelled high G+C% DNA can have a density profile that approaches that of labelled low G+C% DNA (Buckley *et al.*, 2007).

Despite such limitations, SIP is very valuable to reduce sample complexity and increase the hit rates of particular target genes (Chen and Murrell, 2010). It is particularly useful in the search for metabolic genes for biotechnical applications.

The search for novel chitin-degrading enzymes – a case study

Importance of chitin and chitinases

There is great interest of industry in the products of chitin degradation, one of them the production of chitin with a lower level of frayed edges, which can be achieved via partial degradation using the proper enzymes. In addition, the deacetylation of chitin produces chitosan, which is of great use in seed treatment, biocontrol as well as the medical area. Thus, unfrayed chitin and chitin derivatives have broad applications in medicine, one of the latter being an application in wound healing (dressings for burns, surface wounds and skin-graft donor sites). Other medical uses for chitin include anti-bacterial sponges and hospital dressings, artificial blood vessels, contact lenses, tumour inhibition, dental plaque inhibition and blood cholesterol control.

Selecting the proper habitat for exploration by metagenomics

The success of metagenomic approaches in gene mining is dependent on the proper selection of promising sampling environments, in which we have prior knowledge of the prevalence and diversity of the genes or operons of interest. Cretoiu *et al.* (2012) analysed 10 different habitats, i.e. agricultural soil either or not amended with chitin, spent mushroom substrate, wood-based biofilter material and the rhizospheres of the arctic plants *Oxyria digyna* and *Diapensia lapponica,* next to the freshwater sponge *Ephydatia fluviatilis* and the marine sponges *Halichondria panicea, Corticium candelabrum* and *Petrosia ficiformis* for bacterial chitinolytic enzymes. The study was based on the premise that a better understanding of the ecology of chitinolysis across habitats would serve our subsequent bioexploration of selected habitats. Pyrosequencing based analysis of the partial *chiA* gene revealed that limited numbers of OTUs were actually shared between the samples of related and unrelated habitats, pointing at considerable uniqueness of some of the habitats. It also allowed selecting the most promising environment for metagenomic library construction in order to retrieve the whole functional gene with its genomic context.

Based on the information (enzymatic and molecular) obtained for the environments investigated, two soils, namely chitin-treated field soil and rhizosphere soil from *Oxyria digyna,* were selected for the metagenomics-based exploration for chitinases. Such metagenomic libraries, in principle, are based on the random insertion of DNA fragments into suitable vectors. Assuming that the process of insertion is random, such libraries are composed of all genetic material of all genomes that were extracted from the soil microbial community and entered the DNA pool. Long-insert libraries, which in case of fosmid libraries are of approximately 35–40 kb, may help to retrieve not only the genes screened for, but also whole complex operons, which facilitate the analysis of a gene/operon. With larger insert sizes, the chance that entire pathways, in this case involved in chitin degradation, can be cloned becomes larger, as bacterial genes involved in pathways are often grouped in operons. In our study, we produced more than 400,000 clones in fosmids, which are currently being screened for the presence of genes of interest (Cretoiu *et al.*, 2012).

Ecological enhancement of chitin degradation

As indicated in the foregoing, soil DNA based metagenomes often contain one to only a few clones that carry genes/operons of interest. Any analyses of directly prepared soil metagenomes are therefore often tedious, reflecting the predictable low prevalence of the target genes. To specifically modify the hit rate of chitin-degradative genes or operons in soil, a microcosm experiment was set up in our lab (Kielak *et al.*, 2012). We included conditions of native (5.7) versus high (8.7) pH. In this way, the local microbial communities were shifted, over a relatively short time period (i.e. from 7 to up to 30 days), prior to extraction of the metagenomic DNA, thereby changing the prevalence of the dominant types of sequences. An advantage of the ecological enhancement strategy is its low cost and effort, together with the generally low-tech procedures. In the case detailed here, the addition of chitin and the manipulation of the soil pH incited the prevalence of a range of different types of *chiA* genes under the different conditions (Kielak *et al.*, 2012). Fig. 3.2 shows the ecological enhancement that was achieved after only 7 days of incubation with added chitin. The diversity of *chiA* gene types as determined by direct pyrosequencing is shown. Clearly, shifts from a more even *chiA* gene community (Fig. 3.2A) to more uneven ones (Fig. 3.2B and C) were found. In the pool of *chiA* gene types that were found, the enhanced prevalence of typical Gram-negative types was striking (Fig. 3.2B and C), as it contrasts with the widely held belief that actinomycete or streptomycete chitinase types will often prevail in soil in which chitin degradation takes place. Thus, the deliberate manipulation of microbial communities via ecological enhancement, as shown in this example, offers unique possibilities to shift the balance in the functional genes towards rather unexpected types, and can thus enhance the metagenomics hit rates of certain types of genes.

Outlook

Our 'power' to analyse soil systems via metagenomics-based techniques has grown exponentially in recent years (Cretoiu *et al.*, 2012; Delmont *et al.*, 2012; Ekkers *et al.*, 2012; Schmieder and Edwards, 2012; Tamaki *et al.*, 2012). The possibilities offered by metagenomics based analyses of soil to (1) examine the diversity and function of the soil ecosystem and (2) use the genetic information from the soil environment for biotechnology purposes, are numerous. Hence, the statement that metagenomics-based methods are the preferred methods of today and tomorrow that need to be applied to foster our broad understanding of the living soil does not appear as an overstatement. However, as we discussed extensively in this chapter, there are clear caveats to the metagenomics based analysis of soil, and these relate to the technology applied to the environment. A first caveat lies in the incompleteness of sampling a microbiota from the environment on the basis of nucleic acids, mainly due to the incompleteness of extraction. A second very clear hurdle concerns the inadequateness of all molecular analysis methods applied to the soil extracted nucleic acids. A third problem that severely limits our ready understanding of the microbiota of soil lies in the so-called 'computational bubble', which describes our inability to deal with the massive datasets that are generated by current-day NGS. With respect to the biotechnological exploration of environmental habitats, there is a clear hurdle in the 'Great Screen Anomaly' in that positive hit rates are very often extremely low (Ekkers *et al.*, 2012).

So where do we move to from here on? There is a clear need to invest time and effort right at the forefront of soil metagenomics, i.e. in the soil nucleic acid extraction steps. In this phase of the protocol, one faces problems of incomplete sampling, insufficient cell lysis, loss of DNA by adsorptive processes, DNA degradation, etc. Hence, any improvement made will be reflected in the data that are obtained in the following steps. Another clear research need lies at the bioinformatics side, i.e. we need better and more user-friendly tools that allow us to deal with the exploding dataset. Finally, whenever we find trends in our metagenomics-based data set, for instance about a bacterial type that appears to thrive under particular conditions in the soil (and thus could be sampled at high rate), we may need to go back to days of the good old plating technique, which may allow us to better study its dynamics and behaviour in the soil system. It is only with the help of such confirmatory studies that we can place sufficient trust in the metagenomics-derived datasets that are currently obtained from soil.

Acknowledgement

This study was supported by the METAEXPLORE project (FP7 EU) awarded to JDvE.

References

Beloqui, A., de Maria, P.D., Golyshin, P.N., and Ferrer, M. (2008). Recent trends in industrial microbiology. Curr. Opin. Microbiol. *11*, 240–248.

Binga, E.K., Lasken, R.S., and Neufeld, J.D. (2008). Something from (almost) nothing: the impact of multiple displacement amplification on microbial ecology. ISME J. *2*, 233–241.

Buckley, D.H., Huangyutitham, V., Hsu, S.F., and Nelson, T.A. (2007). Stable isotope probing with 15N achieved by disentangling the effects of genome G+C content and isotope enrichment on DNA density. Appl. Environ. Microbiol. *73*, 3189–3195.

Caporaso, J.G., Kuczynski, J., Stombaugh, J., Bittinger, K., Bushman, F.D., Costello, E.K., Fierer, N., Pena, A.G., Goodrich, J.K., Gordon, J.I., *et al.* (2010). QIIME allows analysis of high-throughput community sequencing data. Nat. Methods *7*, 335–336.

Caporaso, J.G., Lauber, L.,Walters, W.A., Berg-Lyons, D., Huntley, J., Fierer, N., Owens, S.M., Betley, J., Fraser, L., Bauer, M., *et al.* (2012). Ultra-high-throughput microbial community analysis on the Illumina HiSeq and MiSeq platforms. ISME J. *6*, 1621–1624.

Cebron, A., Bodrossy, L., Chen, Y., Singer, A.C., Thompson, I.P., Prosser, J.I., and Murrell, J.C. (2007). Identity of active methanotrophs in landfill cover soil as revealed by DNA-stable isotope probing. FEMS Microbiol. Ecol. *62*,12–23.

Chen, Y., and Murrell, J.C. (2010). When metagenomics meets stable-isotope probing: progress and perpectives. Trends Microbiol. *18*, 157–63.

Cretoiu, M.S., Kielak, A.M., Al-Soud, W.A., Sørensen, S.J., and van Elsas, J.D. (2012). Mining of unexplored habitats for novel chitinases – chiA as a helper gene proxy in metagenomics. Appl. Microbiol. Biotechnol. *94*, 1347–1358.

Cupples, A.M. (2011). The use of nucleic acid based stable isotope probing to identify the microorganisms responsible for anaerobic benzene and totuene biodegradation. J. Microbiol. Methods. *85*, 83–91.

Curtis, T.P., and Sloan, W.T. (2004). Prokaryotic diversity and its limits: microbial community structure in nature and implications for microbial ecology. Curr. Opin. Biotechnol. *7*, 221–226.

Delmont, T.O., Robe, P., Clark, I., Simonet. P., and Vogel. T.M. (2011). Metagenomic comparison of direct and indirect soil DNA extraction approaches. J. Microbiol. Methods *86*, 397–400.

Delmont, T.O., Prestat, E., Keegan, K.P., Faubladier. M., Robe, P., Clark, I.M., Pelletier, E., Hirsch, P.R., Meyer, F., Gilbert, J.A., *et al.* (2012). Structure, fluctuation and magnitude of a natural grassland soil metagenome. ISME J. *6*, 1677–1687.

Delpin, M.W., and Goodman, A.E. (2009). Nutrient regime regulates complex transcriptional start site usage within a *Pseudoalteromonas* chitinase gene cluster. ISME J. *3*, 1053–1063.

Downing, K.J., and Thomson, J.A. (2000). Introduction of the *Serratia marcescens* chiA gene into an endophitic *Pseudomonas fluorescens* for the biocontrol of phytopathogenic fungi. Can. J. Microbiol. *46*, 363–369.

Dumont, M.G., and Murrell, J.C. (2005). Stable isotope probing – linking microbial identity to function. Nat. Rev. Microbiol. *3*, 499–504.

Dumont, M.G., Neufeld, J.D., and Murrell, J.C. (2006). Isotopes as tools for microbial ecologists. Curr. Opin. Biotechnol. *17*, 57–58.

Ekkers, D.M., Cretoiu, M.S., Kielak, A.M., and van Elsas, J.D. (2012). The great screen anomaly-a new frontier in product discovery through functional metagenomics. Appl. Microbiol. Biotechnol. *93*, 1005–1020.

Fernández-Arrojo, L., Guazzaroni, M.E., López-Cortés, N., Beloqui, A., and Ferrer, M. (2010). Metagenomic era for biocatalyst identification. Curr. Opin. Biotechnol. *21*, 725–733.

Gabor, E.M., Alkema, W.B., and Janssen, D.B. (2004). Quantifying the accessibility of the metagenome by random expression cloning techniques. Environ. Microbiol. *6*, 879–886.

Gans, J., Wolinsky, M., and Dunbar, J. (2005). Computational improvements reveal great bacterial diversity and high metal toxicity in soil. Science. *309*, 1387–139.

Giardine, B., Riemer, C., Hardison, R.C., Burhans, R., Elnitski, L., Shah, P., Zhang, Y., Blankenberg, D., Albert, I., Taylor, J., *et al.* (2005). Galaxy: a platform for interactive large-scale genome analysis. Genome Res. *15*, 1451–1455.

Guan, Y.J., Hu, J., Wang, X.J., and Shao, C.X. (2009). Seed priming with chitosan improves maize germination and seedling growth in relation to physiological changes under low temperature stress. J. Zhejiang Univ. Sci. B. *10*, 427–433.

Handelsman, J., Rondon, M.R., Brady, S.F., Clardy, J., and Goodman, R.M. (1998). Molecular biological access to the chemisrty of unknown soil microbes, a new frontier for natural products. Chem. Biol. *5*, 245–249.

He, Z., van Nostrand, J.D., and Zhou, J. (2012). Applications of functional gene microarrays for profiling microbial communities. Curr. Opin. Biotechnol. *23*, 460–466.

Hil, R.T., and Fenical, W. (2010). Pharmaceuticals from marine natural products: surge or ebb? Curr. Opin. Biotechnol. *21*, 777–779.

Imhoff, J.F., Labes, A., and Wiese, J. (2011). Bio-mining the microbial treasures of the ocean: new natural products. Biotechnol. Adv. *29*, 468–482.

Inceoglu, O., Hoogwout, E.F., Hill, P., and van Elsas, J.D. (2010). Effect of DNA extraction method on the apparent microbial diversity of soil. Appl. Environ. Microbiol. *76*, 3378–3382.

Kakirde, K.S., Parsley, L.C., and Liles, M.R. (2010). Size does matter: Application-driven approaches for soil metagenomics. Soil. Biol. Biochem. *42*, 1911–1923.

Kielak, A.M., Cretoiu, M.S., Semenov, A.V., Sørensen, S.J., and van Elsas, J.D. (2013). Bacterial chitinolytic communities respond to chitin and pH alteration in soil. Appl. Environ. Microbiol. *79*, 263–272.

Kobayashi, D.Y., Reedy, R.M., Bick, J., and Oudemans, P.V. (2002). Characterization of a chitinase gene from *Stenotrophomonas maltophilia* strain 34S1 and its involvement in biological control. Appl. Environ. Microbiol. *68*, 1047–1054.

Kotan, R., Dikbas, N., and Bostan, H. (2009). Biological control of post harvest disease caused by *Aspergillus flavus* on stored lemon fruits. African J. Biotechnol. *8*, 209–214.

Lefevre, F., Robe, P., Jarrin, C., Ginolhac, A., Zago, C., Auriol, D., Vogel, T.M., Simonet, P., and Nalin, R. (2008). Drug from hidden bugs: their discovery via untapped resources. Res. Microbiol. *159*, 153–161.

Mankau, R., and Das, S. (1969). The influence of chitin amendments on *Meloidogyne incognita*. J. Nematol. *1*, 15–16.

Massart, S., and Jijakli, H.M. (2007). Use of molecular techniques to elucidate the mechanisms of action of fungal biocontrol agents: a review. J. Microbiol. Methods *69*, 229–241.

Meyer, F., D'Souza, P.D., Olson, M.R., Glass, E.M., Kubal, M., Paczian, T., Rodriguez, A., Stevens, R., Wilke, A., Wilkening, J., *et al.* (2008). The Metagenomics RAST server – a public resource for the automatic phylogenetic and functional analysis of metagenomes. BMC Bioinformatics. *9*, 386.

Mocali, S., and Benedetti, A. (2010). Exploring research frontiers in microbiology: the challenge of metagenomics in soil microbiology. Res. Microbiol. *161*, 497–505.

Nacke, H., Engelhaupt, M., Brady, S., Fischer, C., Tautzt, J., and Daniel, R. (2012). Identification and characterization of novel cellulolytic and hemicellulolytic genes and enzymes derived from German grassland soil metagenomes. Biotechnol. Lett. *34*, 663–675.

Nichols, D. (2007). Cultivation gives context to microbial ecologist. FEMS Microbiol. Ecol. *60*, 351–357.

Pietramellara, G., Ascher, J., Borgogni, F., Ceccherini, M.T., Guerri, G., and Nannipieri, P. (2009). Extracellular DNA in soil and sediment: fate and ecological relevance. Biol. Fertility Soils *45*, 219–235.

Poulsen, P.H.B., Moller, J., and Magid, J. (2008). Determination of a relationship between chitinase activity and microbial diversity in chitin amended compost. Bioresource Technol. *99*, 4355–4359.

Radajewski, S., McDonald, I.R., and Murrel, J.C. (2003). Stable-isotope probing of nucleic acids: a window to the function of uncultured microorganisms. Curr. Opin. Biotechnol. *14*, 296–302.

Reddy, B.M.V., Arul, J., Angers, P., and Couture, L. (1999). Chitosan treatment of wheat seeds induces resistance to *Fusarium graminearum* and improves seed quality. J. Agric. Food Chem. *47*, 1208–1216.

Riesenfeld, C.S., Schloss, P.D., and Handelsman, J. (2004). Metagenomics, genomic analysis of microbial communities. Annu. Rev. Genet. *38*, 525–552.

Roesch, L.F.W., Fulthorpe, R.R., Riva, A., Casella, G., Hadwin, A.K.M., Kent, A.D., Daroub, S.H., Camargo, F.A.O., Farmerie, W.G., and Triplett, E.W. (2007). Pyrosequencing enumerates and contrasts soil microbial diversity. ISME J. *1*, 283–290.

Ruan, S.L., and Xue, Q.Z. (2002). Effects of chitosan coating on seed germination and salt tolerance of seedlings in hybrid rice (*Oryza sativa* L.). Acta Agron. Sinica. *28*, 803–808.

Schloss, P.D., and Handelsman, J. (2003). Biotechnological prospects from metagenomics. Curr. Opin. Biotechnol. *14*, 303–310.

Schloss, P.D., Westcott, S.L., Ryabin, T., Hall, J.R., Hartmann, M., Hollister, E.B., Lesniewski, R.A., Oakley. B.B., Parks, D.H., Robinson, C.J., *et al.* (2009). Introducing mothur: open-source, platform-independent, community-supported software for describing and comparing microbial communities. Appl. Environ. Microbiol. *75*, 7537–7541.

Schmieder, R., and Edwards, R. (2012). Insights into antibiotic resistance through metagenomic approaches. Future Microbiol. *7*, 73–89.

Singh, B.K., and Macdonald, C.A. (2010). Drug discovery from uncultivable microorganisms. Drug. Discov. Today *15*, 792–799.

Singh, B., Gautam, S.K., Verma, V., Kumar, M., and Singh, B. (2008). Metagenomics in animal gastrointestinal ecosystem: potential biotechnological prospects. Anaerobe *14*, 138–144.

Spiegel, Y., Cohn, E., and Chet, I. (1989). Use of chitin for controlling *Heterodera avenae* and *Tylenchulus semipenetrans*. J. Nematol. *21*, 419–422.

Staley, J.T., and Konopka, A. (1985). Measurement of *in situ* activities of nonphotosynthetic microorganisms in aquatic and terrestrial habitats. Annu. Rev. Microbiol. *39*, 321–346.

Taberlet, P., Prud'Homme, S.M., Campione, E., Roy, J., Miquel, C., Shehzad, W., Gielly, L., Rioux, D., Choler, P., Clément, J.C., *et al.* (2012). Soil sampling and isolation of extracellular DNA from large amount of starting material suitable for metabarcoding studies. Mol. Ecol. *21*, 1816–1820.

Tamaki, H., Zhang, R., Angly, F.E., Nakamura, S., Hong, P.Y., Yasunaga, T., Kamagata, Y., and Liu, W.T. (2012). Metagenomic analysis of DNA viruses in a wastewater treatment plant in tropical climate. Environ. Microbiol. *14*, 441–452.

Torsvik, V., and Ovreas, L. (2002). Microbial diversity and function in soil: from genes to ecosystems. Curr. Opin. Biotechnol. *5*, 240–245.

Tyson, G.W., Chapman, J., Hugenholtz, P., Allen, E.E., Ram, R.J., Richardson, P.M., Solovyev, V.V., Rubin, E.M., Rokhsar, D.S., and Banfield, J.F. (2004). Community structure and metabolism through reconstruction of microbial genomes from the environment. Nature *428*, 37–43.

Uchiyama, T., and Miyazaki, K. (2009). Functional metagenomics for enzyme discovery: challenges to efficient screening. Curr. Opin. Biotechnol. *20*, 616–622.

Uhlik, O., Jecna, K., Leigh, M.B., Mackova, M., and Macek, T. (2009). DNA-based stable isotope probing: a link between community structure and function. Sci. Total Environ. *407*, 3611–3619.

Van Elsas, J.D., and Boersma, F.G.H. (2011). A review of molecular methods to study the microbiota of soil and the mycosphere. Eur. J. Soil Biol. *47*, 77–87.

Van Elsas, J.D., Speksnijder, A.J., and van Overbeek, L.S. (2008a). A procedure for the metagenomics exploration of disease-suppressive soils. J. Microbiol. Methods. *75*, 515–522.

Van Elsas, D.J., Costa, R., Jansson, J., Sjöling, S., Bailey, M., Nalin, R., Vogel, T.M., and van Overbeek, L. (2008b). The metagenomics of disease-suppressive soils – experiences from the METACONTROL project. Trends Biotechnol. *26*, 591–601.

Wang, Y., and Hayatsu, M., and Fujii, T. (2012). Extraction of bacterial RNA from soil: challenges and solutions. Microbes Environ. *27*, 111–121.

Warnecke, F., and Hess, M. (2009). A perspective: metatranscriptomics as a tool for the discovery of novel biocatalysts. J. Biotechnol. *142*, 91–95.

Warnecke, F., Luginbuhl, P., Ivanova, N., Ghassemian, M., Richardson, T.H., Stege, J.T., Cayouette, M., McHardy, A.C., Djordjevic, G., Aboushadi, N., *et al.* (2007). Metagenomic and functional analysis of hindgut microbiota of a wood-feeding higher termite. Nature *450*, 560–565.

Zeng, D., Luo, X., and Tu, R. (2012). Application of bioactive coatings based on chitosan for soybean seed protection. Int. J. Carbohydrate Chem. ID 104565.

Zerbino, D.R., and Birney, E. (2008). Velvet: algorithms for de novo short read assembly using de Bruijn graphs. Genome Res. *18*, 821–829.

4 Screening Phylogenetic and Functional Marker Genes in Soil Microbial Ecology

Sotirios Vasileiadis, Edoardo Puglisi, PierSandro Cocconcelli and Marco Trevisan

Abstract

Many of the ecosystem services are soil associated with microbes playing a predominant role. Nevertheless, our current knowledge of microbial contribution to ecosystem processes is still limited, partly because in the past centuries research was mostly based on culture-dependent methods, being oblivious of the vast un-cultivable microbial majority as proven during the last decades. Current molecular biology advances provide us with the ability to screen for microbial identities or functions by targeting marker genes in nucleic acid extracts of environmental samples, therefore partly bypassing previous methodological limitations.

Topics addressed here aim at providing an overview of methodologies and concepts related to marker gene screening from environmental samples. Such are the description of marker gene categories, examples of their use in soil environments and the description of marker gene screening state-of-the-art methodologies and specifications. Finally we will exemplify the use of late methodologies for the case of the bacterial small ribosomal subunit screening in soil environments.

Introduction

Soil functionality as identified through ecological processes such as biogeochemical cycles, food webs, disease suppression, degradation of xenobiotics etc., is mainly attributed to the enormous microbial presence and diversity found in soils (Nannipieri *et al.*, 2003). Heterogeneity of resources (Lynch *et al.*, 2004), number and connectivity of trophic micro-sites (Carson *et al.*, 2010), biotic interactions, micro-environmental conditions promoting growing or dormant microbial forms (Jones and Lennon, 2010), are among factors shaping available niches and therefore microbial soil diversity. The ability of microorganisms to occupy available niches and survive in soil is reflected on their genetic content, part of which is expressed through biological activities. Thus far, genetic characterization of certain activities lead to their connection with narrow or wider microbial groups (Fierer *et al.*, 2009) or even to the presence of mobile genetic elements (Frost *et al.*, 2005).

A large body of the existing information about microbial presence and interactions in environmental soil samples is derived from culture-based approaches. In these cases micro-organisms are isolated using nutrient media and analysed in usually axenic assays. Although this methodology has provided important knowledge about microorganisms existing in environmental samples, it has several drawbacks. Two of the most significant are the current lack of ability to cultivate the vast majority (Schloss and Handelsman, 2005) and the lack of information concerning the interaction of microorganisms in their natural environments. A solution to such problems was the adoption of methods related to screening nucleic acids derived from soil environmental samples (directly or after microbial cell enrichment) for presence and activity related coding sequences, the molecular markers. These methods were encompassed the wider field of metagenomics (Handelsman *et al.*, 1998).

Marker genes as biomarkers

The idea of biomarker usage for addressing the identity and activity of living organisms goes way back in centuries. The biomarker concept reflects the usage of virtually anything that may denote the presence of organisms, their environmental interactions, ecological roles and any other biological question of interest. In this manner it may include any substance being product of their activities or structure (e.g. proteins, metabolites, ribonucleic acids, membrane fatty acids), or interacting and introduced substances that are indicative of the identity or allow to assess the activity of the organism (e.g. through up-taken isotopes, or reporter gene activity). Molecular or genomic marker usage involves the screening of genomic elements or their expressed ribonucleic acid sequence (for active molecule coding genes), being informative of organism traits.

Usually, the more the genetic information obtained about an organism, the more detailed are the answers to the questions posed. Characteristic examples reside in screening large nucleic acid stretches of: complete genomes or quality trait loci of animals (Consortium, 2009) and plants (Assunção *et al.*, 2003); complete bacterial genomes (Konstantinidis *et al.*, 2006); environmental microbial consortia through clone library preparation using fosmid, cosmid, bacterial artificial chromosomes (BAC) and yeast artificial chromosomes (YAC) instead of the shorter insert carrying capacity plasmid and bacteriophage vectors (Handelsman *et al.*, 1998; Leveau, 2007). This way not only the information about genes and genomes can be obtained, but also gene expression being the outcome of organization of components of the screened genomic fragments can be elucidated. However, in several cases where a snapshot of environmental samples is necessary and large numbers of microorganisms are analysed, screening of reliable single marker genes provide clearer information and are therefore preferred.

Robust marker genes code for proteins or ribonucleic acid sequences which are central to the activity of interest (e.g. contain the active centre catalysing the a central reaction of the studied pathway). As this activity is conserved throughout members of the microbial group performing it, so are some, key for the activity, protein or ribonucleic acid sequences. For example, in the case of several enzymes these conserved sites correspond to the enzyme active centre where catalysis of a certain reaction takes place. On the other hand, sites not essential to the catalytic reaction are usually less conserved among organisms and more prone to mutations. The less conserved sites of a protein or nucleic acid chain are considered to have a structural role, they are not subjected to strong selective pressure and therefore alteration in their composition can rarely confer loss of fitness (neutral) (Kimura, 1968, 1991). In several cases, evidence was found that the mutation rates of less conserved sites in time, can be calculated and this information can be used for inferring relations among organisms or genetic elements in building genealogies. This way they may serve as 'molecular clocks' with characteristic example the case of (hyper-)variable sequence regions (V-regions) among microorganisms in the ribosomal RNA coding gene (Woese, 1987; Olsen and Woese, 1993;

De Rijk *et al.*, 1995; Tourasse and Gouy, 1997). Owing to the ability to distinguish different genotypic groups by analysing the referred less conserved amino acid or nucleic acid sequence sites, they are also called signature sequences. Apart from this over-simplified introduction to sequence conservation, that can be useful for comprehending the following sections, we are not going to review evolutionary concepts in depth. Interested readers are encouraged to look into the large existing body of bibliography like for example 'The phylogenetic handbook' (2009).

Marker genes may provide phylogenetic information when used for inferring evolutionary relations and also address qualitative or quantitative functional relations between environmental samples and/or gene owners. They have been also previously divided among (i) the ones naturally occurring within genetic elements of an organism (or intrinsic markers) and (ii) the ones that are inserted to an organism by genetic engineering (or recombinant markers) (Tebbe, 2005). Recombinant marker genes are out of the scope of this chapter and therefore are not going to be further discussed. Moreover, we will refer to the intergenic spacer DNA fragment analysis as a phylogenetic marker, which although does not reside in the gene category, it will be encompassed in the chapter due to its importance and for avoiding to devote a separate paragraph.

Phylogenetic and functional marker genes

Two of the most prominent questions posed while screening microbial assemblages of environmental samples, are 'who is there?' and 'what are they doing?' (Handelsman, 2004).

For answering the first question, marker genes coding for conserved activities throughout the studied microbial groups are used. Suitable genes as phylogenetic markers provide representative information about the rest microbial genome they are derived from. Comparison of these markers between microorganisms, approximate the results obtained after comparing the 'essential genome' which includes the collection of genes mostly responsible for the observed microbial phenotypes. Such marker gene is the small ribosomal subunit (SSU) coding gene (Woese *et al.*, 1990; Jordan *et al.*, 2002; Konstantinidis *et al.*, 2006). Although the preciseness of the usage of a single marker gene for inferring evolutionary relations particularly for eukaryotic taxa is generally reduced (Forney *et al.*, 2004; Kuramae *et al.*, 2006), consistency of the SSU phylogeny has established its use in contemporary systematics. Moreover, the significant methodological advantages provided by a single marker gene usage particularly when addressing microbial ecology concepts in soil, will be discussed further on in this chapter. Other marker genes studied for their potential use in phylogenetic relations in the past, encompass genes coding for the large ribosomal subunit (LSU), the large collection of ribosomal proteins, aminoacyl t-RNA synthases, t-RNAs, elongation factors, ATPases, gyrases (Tourasse and Gouy, 1997; Ludwig *et al.*, 1998; Roger *et al.*, 1999; Woese *et al.*, 2000; Roberts *et al.*, 2008; Widmann *et al.*, 2010). Another quite often used DNA marker, not coding for a function in its entire length, is the intergenic spacer (ITS) between the small the large ribosomal subunit coding genes locus (Borneman and Triplett, 1997; Anderson and Cairney, 2004; Fechner *et al.*, 2010). Some of the referred molecular markers have provided a higher resolution compared to the SSU due to increased variability of the signature sequences among different organisms (as is the case of the ITS particularly for eukaryotic taxa). However, the numerous studies in which the SSU marker gene is used assisted in creating far more extensive databases for this gene compared to other marker genes. Examples include the Ribosomal Database Project database (Cole *et al.*, 2005), the Greengenes database (DeSantis *et al.*, 2006) and the SILVA database (Pruesse *et al.*, 2007). The extensive database serves as benchmark for microbial ecology studies and therefore makes the SSU a so far preferred choice particularly for the highly diverse soil environments.

The second question posed in the beginning of this section ('what are they doing?') requires deeper knowledge of protein structures and functions. Robust functional marker genes usually code for protein subunits encompassing sequences under strong selective pressure like the active centres of the enzymes of interest. In the case of microbial groups with deep existing current knowledge the functional identity may be well correlated with the phylogenetic identity. Such example is derived from microbial groups mostly responsible for prokaryotic ammonia oxidation, the rate-limiting step of nitrification, which is a central pathway of the nitrogen cycle. The bacterial group encompasses populations of the β-proteobacterial ammonia oxidizers and was successfully monitored (to current knowledge) in several environments using as biomarker the gene coding for the A subunit of the ammonia monooxygenase enzyme (AMO) (Rotthauwe *et al.*, 1997; Stephen *et al.*, 1999; Avrahami *et al.*, 2003; Aoi *et al.*, 2004). The archaeal group has been recently proposed to comprise the distinct phylum of *Thaumarchaeota* according to functional and phylogenetic information derived from the SSU and other phylogenetic marker genes (Brochier-Armanet *et al.*, 2008; Brochier-Armanet *et al.*, 2012). However, besides the growing evidence of consistency between the presence of the referred archaeal marker genes and the activity, the level of participation of *Taumarchaeota* in ammonia oxidation and the conditions under which this group performs the activity in natural environments are currently under investigation (Levičnik-Höfferle *et al.*, 2011; Pester *et al.*, 2011).

Examples of functional groups which can be identified by phylogenetic markers are not very common (Fierer *et al.*, 2007). In many cases phylogenetics are not able to define guilds, because for example an activity may be widely spread among taxa during evolution, or due to horizontal gene transfer. Characteristic is the case of nitrate reduction as its identified to be performed from members of all cellular life domains (Moreno-Vivián *et al.*, 1999; Philippot, 2002). Therefore, the use of phylogenetic markers on their own is of little use for characterizing the activity potentials in environmental samples. In such cases metagenomic strategies involving library generation with large DNA inserts have provided a solution for functionally characterizing taxa using both phylogenetic and functional marker gene screening. When a phylogenetic marker gene coincides in the same DNA insert with a functional marker gene, this provides the ability to connect phylogenetic relations with functions and was previously termed phylogenetic anchoring of functional genes (Riesenfeld *et al.*, 2004). As mentioned above, lateral gene transfer is responsible for acquired activities in several microorganisms. Lateral gene transfer is often occurring in cases of strong selective environmental pressures posed by e.g. stressors that force microorganisms to uptake energy costly related resistance genes. Multiple examples are derived from contaminated environments with the most characteristic one, that of mercuric reductase marker genes (Smets and Barkay, 2005; Puglisi *et al.*, 2011).

Methodologies for marker gene screening in soil samples

One of the greatest advents for microbial ecology of marker gene usage was the ability to bypass the restricting culture based methods. Microbial cells would be screened for markers *in situ* or they would be subjected to DNA extraction, or environmental DNA and/or reverse transcribed RNA extracts (cDNA) would be either used for generating metagenomic libraries or would be screened directly for markers (Handelsman *et al.*, 1998; Leveau, 2007).

Particularly in the case of direct nucleic acid extract screening for signature sequences from cells or environmental samples, two major approaches improved the screening throughput: (i) the hybridization-based methods, including phylogenetic marker screening with whole-cell fluorescence *in situ* hybridization (FISH) (Amann *et al.*, 1995; Amann and Fuchs, 2008) or functional marker gene screening using catalysed reporter deposition FISH (CARD-FISH) (Pernthaler and Amann, 2004) and also microarrays for screening phylogenetic and functional markers (Sessitsch *et al.*, 2006; He *et al.*, 2007, 2010; Yergeau *et al.*, 2007, 2009); and (ii) the polymerase chain reaction (PCR)-based methods, distinguishing marker gene polymorphisms according to genomic composition as performed by denaturant or temperature gradient gel electrophoresis (DGGE or TGGE), or distinguishing marker gene polymorphisms according to PCR product length with or without concomitant performance of restriction digestion steps [terminal restriction fragment length polymorphism (T-RFLP), length heterogeneity PCR (LH-PCR), ribosomal intergenic spacer analysis (RISA), etc.] (Muyzer *et al.*, 1993; Schütte *et al.*, 2008), or quantifying a marker gene via real-time quantitative PCR (qPCR). Rough comparisons between related methodologies are shown in Fig. 4.1 and described in the text below.

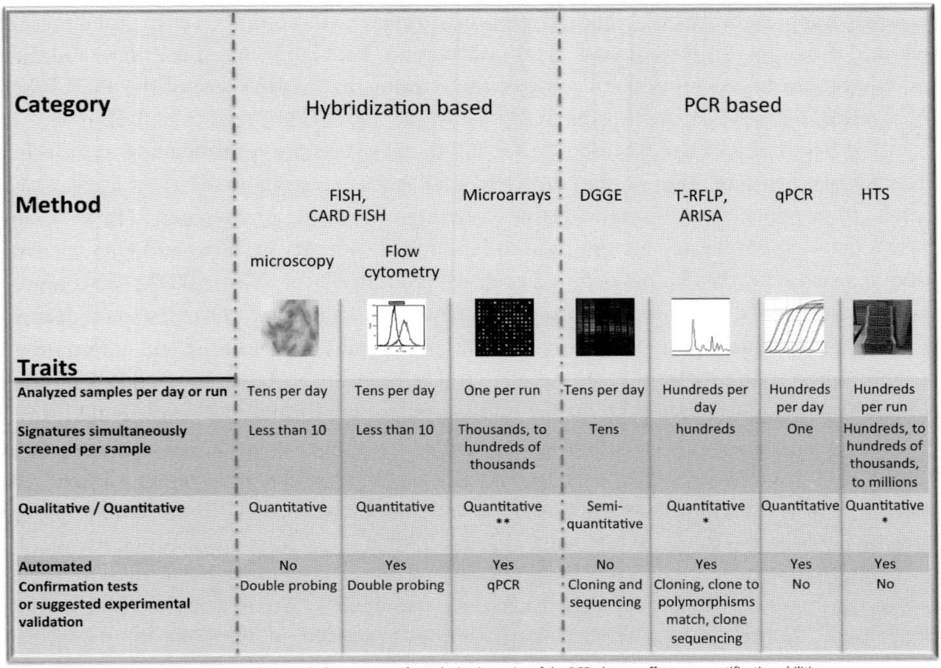

Category	Hybridization based			PCR based			
Method	FISH, CARD FISH		Microarrays	DGGE	T-RFLP, ARISA	qPCR	HTS
	microscopy	Flow cytometry					
Traits							
Analyzed samples per day or run	Tens per day	Tens per day	One per run	Tens per day	Hundreds per day	Hundreds per day	Hundreds per run
Signatures simultaneously screened per sample	Less than 10	Less than 10	Thousands, to hundreds of thousands	Tens	hundreds	One	Hundreds, to hundreds of thousands, to millions
Qualitative / Quantitative	Quantitative	Quantitative	Quantitative **	Semi-quantitative	Quantitative *	Quantitative	Quantitative *
Automated	No	Yes	Yes	No	Yes	Yes	Yes
Confirmation tests or suggested experimental validation	Double probing	Double probing	qPCR	Cloning and sequencing	Cloning, clone to polymorphisms match, clone sequencing	No	No

*: careful preparation (e.g. low number of PCR cycles) is necessary for reducing intensity of the PCR plateau effect on quantification abilities
**: the PCR plateau effect introduced bias is applicable in case a single marker gene like the SSU is screened through multiple taxa after PCR

Figure 4.1 Comparison of examples of hybridization and PCR-based marker screening methods. A colour version of this figure is available in the plate section at the back of the book.

Both described approaches have pros and cons. For example, PCR-based approaches suffer from preferential amplification in multi-template environmental samples and also the plateau effect (thus, apart from qPCR approaches, having reduced quantitative value), while hybridization methods suffer from result variability and probe specificity issues (Polz and Cavanaugh, 1998; Draghici *et al.*, 2006; McIlroy *et al.*, 2010). Both approaches employ oligonucleotides (probes, primers) complementary to target marker gene positions, which allow the identification of signature sequences. Specificity (e.g. ratio of positive to total events) and the sensitivity (e.g. ratio of identified positive to total positive events) of the oligonucleotides are two important features for assessing the success of the methodologies. Although the sensitivity is somewhat difficult to estimate in environmental samples, specificity in many cases is not. PCR based approaches when coupled to sequence reading post cloning of PCR products can provide an assessment means. In hybridization approaches several methods have been proposed with some of which being laborious. In the case of FISH for example confirmation tests include double probing application (Amann and Fuchs, 2008) or coupling with fluorescently activated cell sorting (FISH-FACS) (Amann and Fuchs, 2008) and concomitant PCR based screening, or in the case of microarrays via qPCR verification, assessment becomes quite laborious for multiple sample screening. Next to that, the costs of necessary equipment are also sometimes discouraging concerning use of these applications (flow cytometer, fluorescence screening microscopes, microarray chips and reading equipment). However, the greatest advantage of FISH based approaches against the PCR based approaches described further on, is the lack of any necessity to increase the amounts of partial sequence reads to enable their screening. Therefore, amplification related biases are not present and each studied microbial group can be compared to the total present cells and providing important ecological information. Less laborious specificity validation and equipment accessibility were some of the reasons that made some of the PCR-based approaches (e.g. DGGE) quite popular.

In the beginning of the first decade for 2000, next generation sequencing approaches allowing massive screening of short nucleic acid sequence reads made their appearance. Results of previous years, along with the screening of the SSU phylogenetic marker using these technologies revealed an immense soil microbial diversity (Schloss and Handelsman, 2006; Roesch *et al.*, 2007). According to estimates a number of above 500,000 SSU amplicon sequence reads would be necessary for analysing the diversity of the bacterial domain of a soil gram (Schloss and Handelsman, 2006). Taking into account the soil environment variability, rendering the replication of analysed samples important (Prosser, 2010), such burden became quite heavy even for the quite popular high throughput sequencing (HTS) technology of pyrosequencing (maximum of 150,000 sequence reads analysed per sample). Another HTS technology which has recently been preferred for its screening abilities in massive parallel screening of amplicons, is that one of Illumina with more than 25 million reads analysed per sample (Wu *et al.*, 2010; Bartram *et al.*, 2011; Vasileiadis *et al.*, 2013). However, pyrosequencing has still a significant advantage over other HTS methodologies due to the increased length of individual reads screened (currently about 700 bp much higher than Illumina sequence read length abilities being about 230 using the paired end read module, v4 chemistry and the Genome Analyser apparatus). Another technology currently approximating the specifications of Illumina (about one order of magnitude less reads of the same size) and is expected to enter dynamically the field of amplicon screening with increased sequenced fragment lengths and further cost reduction, is Ion-torrent (Rothberg *et al.*, 2011).

The characteristics of the HTS technologies described above render them suitable for different applications. For example in cases where marker gene database support is high enough (like the prokaryotic SSU), phylogenetically informative marker gene sites are included within 230 bp and the urge for high read numbers exists (e.g. high-complexity soil environments), the multimillion read producing technologies as described suffice (Liu *et al.*, 2007; Claesson *et al.*, 2010). In the case of low complexity environments and low database support (like in the case of most functional gene markers), the phylogenetic information included in larger sequence stretches is important. An overview of the technological status of high throughput sequencing technologies for interested readers is made available by Glenn (2011).

Primer and probe designing (non-protein-coding sequences and protein-coding sequences) strategies

Signature sequences of gene markers comprise the basis of the screening process. In the hybridization based approaches the signature sequences of interest are targeted and positive hits by hybridization of designed complementary oligonucleotide probes are desired. In the PCR-based approaches the targets are highly conserved sequences (signature sequences for the whole group) for which complementary oligonucleotides are designed which flank the signature sequences for group individuals. For example, in the case of the prokaryotic SSU marker gene the V region sequence parts comprise the individual specific signature sequences, having increased variability among several *Bacteria*. On the other hand, the highly conserved for all bacteria SSU sequences between V-regions may serve as total bacteria signature sequences (Fig. 4.2).

For designing probes and primers there are certain strategies devised, taking into account the fact that some DNA or RNA targets are best aimed when considering conserved nucleic acid sequences, while others when considering conserved amino-acid sequences (e.g. when nucleic acid substitutions among targeted sequences are synonymous and therefore do not result in protein function change). Quite often targeted sequences in environmental samples do not have conserved stretches long enough (e.g. oligonucleotides longer than 15–18 bp) that would allow designing of sequence group specific oligonucleotides. In such cases and for exploiting sequence conservation of vicinal sequence regions, degeneracies are introduced in the position of variable nucleic acid bases among targeted groups (Linhart and Shamir, 2002).

Alignment-based primer and probe designing

The usual strategy followed for suitable primer and probe sequence sites identification is aligning multiple target sequences and using the consensus sequence as a guide. A very crucial step in this approach is the performance of appropriate alignments. The consensus sequence usually reflects the proportion of occurrences of residues per position. Therefore, a careful selection of the targeted template sequences for the multiple sequence alignment (MSA) should be performed, in order to verify the representativeness of the states of each position in the consensus sequence. Several alignment strategies and algorithms are

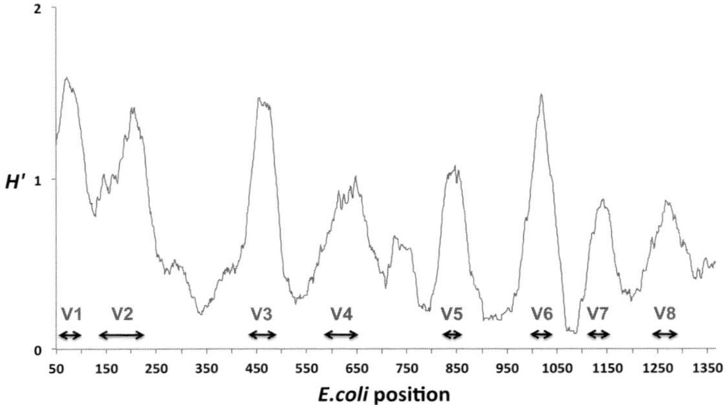

Figure 4.2 Shannon entropy (*H'*) plot of bacterial SSU multiple sequence alignment created using 42,109 soil-derived sequences existing in the RDP database, previously published by Vasileiadis *et al.* (2012) (figure use approved by all co-authors). For the entropy calculation, alignment columns were removed according to reference sequence gaps (sequence derived from *Escherichia coli*, GeneBank accession number 1VS5_A), while the hypervariable regions indicated as designated by Baker *et al.* (2003), according to *E. coli* nucleotide position numbering. Poorly supported areas of the beginning and the end of the sequences were not considered (due to usage of nearly full sequences), excluding this way one more existing V-region the V9. A colour version of this figure is available in the plate section at the back of the book.

available, the selection of which depends on the experimental requirements and usually such selection involves in trade-offs between speed and accuracy.

The very popular in the past method used for building MASs was the progressive pairwise comparison method (of either sequences or alignments) performed by Clustal W (Thompson *et al.*, 1994). Due to reduced accuracy and number of sequences possible to process Clustal W use for building alignments is not much preferred nowadays. More accurate existing methods take into account also secondary structure information apart from progressively performing pairwise alignments. Secondary structure information is used by the ARB software (Ludwig *et al.*, 2004), which is originally designed for building SSU marker gene MSAs and currently used in the curation of SILVA database (Pruesse *et al.*, 2007). Another software using several approaches like progressive alignment method, automated database search and secondary structure information for proteins is T-Coffee (Notredame *et al.*, 2000). Although quite accurate, T-Coffee cannot facilitate enormous alignments that are sometimes required for microbial ecology-related primer designing. Quite faster (being able to handle thousands of sequences) but less accurate than T-Coffee are the iterative refinement methods performed by MAFFT (Katoh *et al.*, 2002) and MUSCLE (Edgar, 2004). Another relatively fast method suitable for protein sequences whose accuracy relies on provided alignment profiles of closely related sequences uses the probabilistic hidden Markov models algorithms. Such is the HMMER software (Finn *et al.*, 2011) used to curate the Pfam protein families database (Finn *et al.*, 2008). In the case of the well supported by databases SSU marker gene, quite fast algorithms like the one applied by the NAST aligner can be used (DeSantis *et al.*, 2006). The NAST aligner requires an aligned reference database whose alignment quality is quite definitive for the quality of the final alignment. The NAST aligner has the ability of aligning hundreds of thousands of sequences in a few hours.

Multiple software options exist for primer and probe designing which requires alignments as input. The Primer3 (Rozen and Skaletsky, 2000) and the ARB (Ludwig *et al.*, 2004) software are some of the most popular with the later specialized for single probe designing. However they lack the ability to design primers or probes with degeneracies. More recent software applications in primer designing take into account also false positives by providing a negative background of sequences like the PrimerProspector (Walters *et al.*, 2011). In the case that protein alignments are provided as template alignments, software like the iCODEHOP (Boyce *et al.*, 2009) searches for conserved MSA blocks and designs degenerate primers on these regions.

K-mer search based primer and probe designing

Another method for primer designing suitable for markers where there is strong database support is the one using *k-mer* search. The basic principle of the method is to screen by *k*-mers (short words) the targeted sequences for conserved sites per individual sequence. Concomitantly the conserved fragments existing in a satisfactory percentage of targeted sequences are compared against non-intended targets and are maintained or rejected accordingly. This method bypasses the necessity for sequence alignment performance (and therefore the respective potential uncertainties), but is totally dependent on the provided sequences and is suitable for nucleic acid sequences and not amino-acid sequences. Therefore, in the case of low database support the primers and probes generated may have increased specificity, while potential ribonucleic acid or protein higher than primary-level structures are disregarded. Software using this method has been previously designated for generating primers for virology applications (Gardner *et al.*, 2009); however, designing primers for cellular life-derived sequences is also possible.

Both the above general approaches in primer and probe design, require further processing of the designed primers. This processing involves besides specificity and sensitivity tests, also screening of the generated oligonucleotides for compatibility with melting and annealing temperatures, hairpin formations, self- or hetero- (for primers) dimerization phenomena, guanine or cytosine (GC) primer content and strength of hydrogen bonds of the 3′ ends of primers (important for specificity during the PCR annealing step). Some of the primer designing software mentioned above along with a large amount of free software and web-based applications, perform such tests using proposed algorithms for calculating the related energetics.

Experimental design for screening of the bacterial 16S rDNA marker gene with short read producing high-throughput sequencing technologies

In this section we are going to look in brief in the way that existing database knowledge for bacterial SSU marker gene may assist in designing high-throughput sequencing approaches for screening soil bacterial diversity. As explained earlier in this chapter, the most prominent sequencing technologies concerning produced read numbers are suffering from available reduced read lengths (less than 230 bp). Therefore, screened sequence parts of the marker gene total stretch (approximately 1500 bp long) should be compatible with the technology used and encompass information representative as possible of the complete marker. Moreover, the necessity for reduction of experimental biases and replication of biological treatments, involves in several cases multiplexing of PCR products. Several actions for achieving these goals include computational approaches addressed in the following paragraphs.

Initial step is to evaluate the database bias on the experimental design. Existing bacterial SSU databases contain sequences derived by microorganisms from several environments. Therefore, sequence conservation found among sequences of the same gene among organisms provides the necessary information for designing of appropriate primers. Such examples exist in extensive past studies in bacterial SSU sequence conservation like that of Wang and Qian (2009) which was based on the ribosomal database project (RDP) database. However, their investigation depth (based on the large sequence collection of the RDP database) might also be a pitfall concerning screening of specific environments like soil. That is because as indicated by a simple keyword search of the RDP database, about 56% of the ~1,000,000 SSU sequences longer than 1200 bp deposited in are derived from human body related environments, while less than 5% of the sequences are derived from soil. The identified richness in these two environments in several studies is totally different compared to the corresponding richness found in the RPD database, with soil being by far more rich in estimated species numbers (about 15,000 different species were identified for the complete human microbiome and more than 50,000 species are estimated to exist per soil gram (Torsvik and Ovreas, 2002; Gans *et al.*, 2005; Tringe *et al.*, 2005; Schloss and Handelsman, 2006; Huse *et al.*, 2008; Tamames *et al.*, 2010). Therefore, when using the complete database information one may falsely conclude to higher conservation than the one expected in soil, due to actual lower sequence representation of taxa not existing in the human microbiome but abundant in soil. One solution to this problem is to reject sequences not derived from soil environments. In case that environment-wise more generic primers are desired, a solution could be to add weights to the sequences according to their diversity in the environment they are derived from.

Concomitant steps involve identifying informative and representative sequence regions according to sequencing specifications flanked by conserved sequences, with primer sites of high sensitivity. As mentioned above, the required information should be included within 230 bp according to current high throughput sequencing specifications (with this length being further reduced considering the multiplexing option described below). Therefore, primers with desired specificities and sensitivities should be searched throughout the selected database part, which amplify these lengths.

Screening of multiple PCR products in a single sequencing run (multiplexing) is another step for exploiting the possibilities provided by high throughput sequencing technologies. This involves the indexing of samples by addition of short oligonucleotides that are used during analysis as sample identifiers for sorting PCR product derived sequences. Previously proposed multiplexing methods involve: (a) primer indexing by addition of a few unique bases in the 5′ end of one (or both) of the amplification primers plus a 2 bp linker sequence for reducing bias effects caused by the index sequence during environmental sample PCR amplification (Wu *et al.*, 2010; Degnan and Ochman, 2011); (b) usage of primers during environmental PCR amplification with 5′ extensions having the complete Illumina sequencing adapters plus an index sequence (Bartram *et al.*, 2011), which allows a third sequence read (in paired-end reads usage) for identification of barcodes (similar philosophy to that of Illumina multiplexing kits; Meyer and Kircher, 2010). The second approach has the advantage that barcode index reading does not interfere with the operational read length (like in the first approach), but has the restriction of the number of samples that can be multiplexed (currently up to 96 error correcting barcodes – no such restriction exists for the first approach). In the case that the costs of screening are to be reduced dramatically, the first option allowing screening of more than 96 samples is the apparent choice. However, operational amplicon screening length is also reduced according to the number of barcode bases plus the linker sequence length. One pitfall of the multiplexing concept is the potential amplification bias caused when environmental samples are directly amplified with barcoded primers. Berry *et al.* (2011) proposed a two-step amplification of environmental nucleic acid targets for reducing this bias. According to the followed protocol, non-indexed primers were used for the initial PCR amplification performed with 20 cycles, while five PCR cycles were performed on the first cycle titrated PCR product using the barcoded primers. Another advantage of this approach compered to the (a) indexing method described above is the lack of necessity of using the 2 bp linker sequence. Popular tool for designing multiplexing index sequences is the Barcrawl software (Frank, 2009). Besides designing the index sequences *per se*, Barcrawl also performs selection taking into account desired dissimilarities among indexes and discards indexes according to homopolymers, hairpin and heteroduplex formations. Going one step further, the PrimerProspector software (Walters *et al.*, 2011) supports optimization against biases during amplification caused by index sequences.

Finally, potential amplicons found in a certain environment can be tested for representation of the complete length sequences concerning sequence distance estimation or taxonomical classification. The first one can be achieved by comparing sequence distances generated by alignments of sequence variants according to sequence positions for corresponding sequences. Such example is provided in a previous study for soil derived sequences (Vasileiadis *et al.*, 2012) obtained from the RDP database for the V3, V4, V5 and V6 hypervariable regions (Fig. 4.3). Comparison of taxonomic annotations of partial reads and full-length sequences can also serve in identifying the amount of remaining information and/or loss of confidence by using partial 16S rDNA reads in combination with several annotation pipelines (e.g. local alignment based or pattern search based like that of the naive Bayesian classifier (Wang *et al.*, 2007)). These two steps can be further improved concerning the selection of primer sites, when considering the relative abundance of sequences found in related environments (provided that such information is available) according to previous studies of microbial diversity.

Concluding remarks and potentials

Soil microbial ecology has moved forwards with marker gene screening in the late years in an attempt to elucidate microbial presence and activity in natural environments. This effort has greatly improved current knowledge about marker variability in natural environments and iterative processes have provided feedback about the amount of information provided by gene markers and other biomarkers. Current available technological advances like microarrays and high throughput sequencing technologies have scaled down some previously considered as immense tasks. Database enhancement due to mainly the application of high throughput sequencing technologies of random environmental DNA and RNA, has sometimes been the object of scepticism in the late years (Baveye, 2009). However great

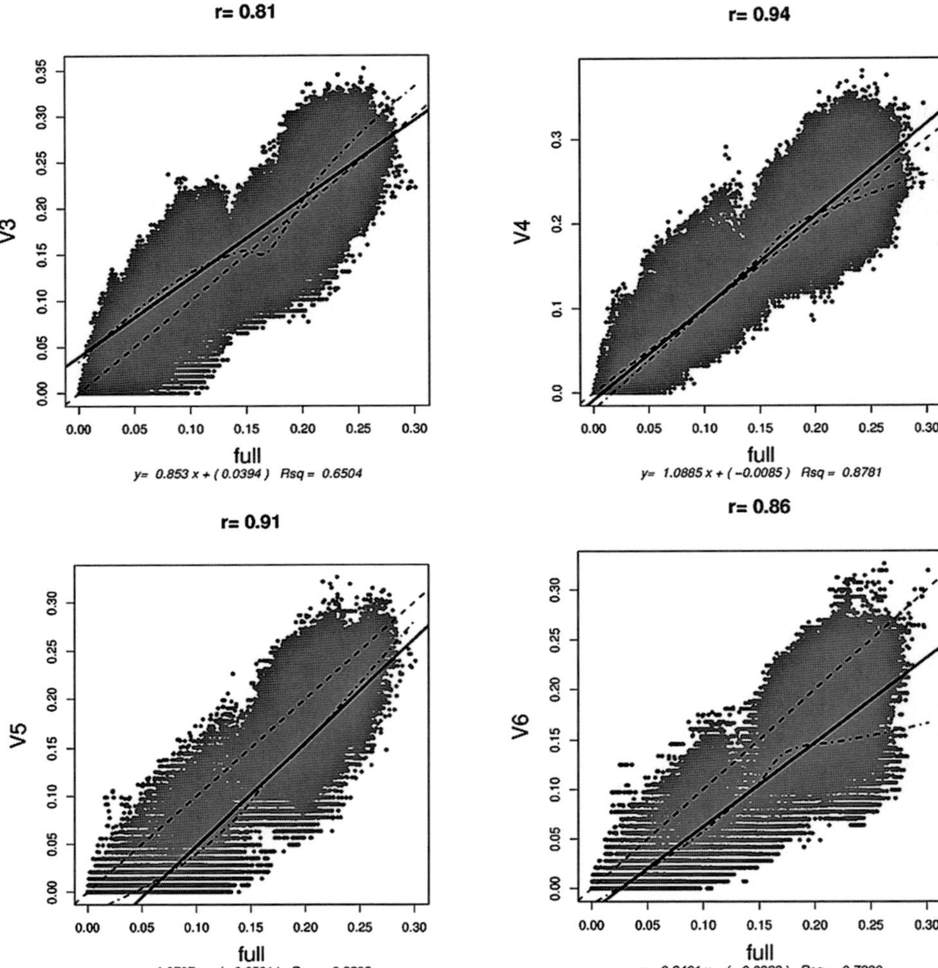

Figure 4.3 Pearson correlation tests between corresponding sequence distances of the examined V regions and the full-length variants, as published by Vasileiadis *et al.* (2012) (figure use approved by all co-authors). Test correlation index (*r*) values and linear models (presented with solid lines) used to describe overall trends are provided above and below each plot. Local relationships between corresponding sequence distances of the FL and the rest datasets are expressed with the non-parametric LOWESS (locally weighted regression and smoothing scatterplots) regression analysis plotting (dot-dashed lines), while the ideal y = x correlation is also plotted (dashed lines).

benefits were obtained by these efforts in multidisciplinary approaches of 'soil -omics' for marker gene screening with the undeniable support of computational biology in gene and protein prediction. The improvement of existing along with the development of several other genotyping tools are promising for addressing current pitfalls and rendering marker gene screening applications into 'daily use' diagnostic tools. Improvements like the referred include the ability to enrich mRNA in environmental samples for *Eukaryotes* but till more recently also for the dominating *Prokaryotes* (Rio *et al.*, 2010; Stewart *et al.*, 2010). This biotechnological advancement can increase RNA screening resolution (which was hampered by rRNA molecule domination of RNA extracts) and is very valuable for unlocking functional gene marker screening in environmental samples. Another example includes multidisciplinary approaches where phylogenetic marker gene screening maybe used as an initial step in obtaining whole genome information about non-yet cultivable microorganisms dominating e.g. heavily contaminated environments with remediation prospects (Puglisi *et al.*, 2011). In this case, massive sample screening may pinpoint microorganisms with remediation abilities and be used for whole-cell and/or functional gene FISH probes designing, which in turn can be used for fluorescently activated cell sorting and concomitant whole genome amplification and sequencing (Hutchison and Venter, 2006; Kalyuzhnaya *et al.*, 2006; Amann and Fuchs, 2008; Rajendhran and Gunasekaran, 2008).

Such technologies and approaches, taking their infant steps, owe their existence to the enormous amount of recent research efforts. Moreover, these examples do point out the potential contribution of marker gene studies in the near future to both soil microbial ecology fundamental knowledge and daily applications.

References

Amann, R., and Fuchs, B.M. (2008). Single-cell identification in microbial communities by improved fluorescence *in situ* hybridization techniques. Nat. Rev. Micro. *6*, 339–348.

Amann, R.I., Ludwig, W., and Schleifer, K.H. (1995). Phylogenetic identification and *in situ* detection of individual microbial cells without cultivation. Microbiol. Rev. *59*, 143–169.

Anderson, I.C., and Cairney, J.W.G. (2004). Diversity and ecology of soil fungal communities: increased understanding through the application of molecular techniques. Environ. Microbiol. *6*, 769–779.

Aoi, Y., Masaki, Y., Tsuneda, S., and Hirata, A. (2004). Quantitative analysis of amoA mRNA expression as a new biomarker of ammonia oxidation activities in a complex microbial community. Lett. Appl. Microbiol. *39*, 477–482.

Assunção, A.G.L., Schat, H., and Aarts, M.G.M. (2003). Thlaspi caerulescens, an attractive model species to study heavy metal hyperaccumulation in plants. New. Phytol. *159*, 351–360.

Avrahami, S., Liesack, W., and Conrad, R. (2003). Effects of temperature and fertilizer on activity and community structure of soil ammonia oxidizers. Environ. Microbiol. *5*, 691–705.

Baker, G.C., Smith, J.J., and Cowan, D.A. (2003). Review and re-analysis of domain-specific 16S primers. J. Microbiol. Methods *55*, 541–555.

Bartram, A.K., Lynch, M.D.J., Stearns, J.C., Moreno-Hagelsieb, G., and Neufeld, J.D. (2011). Generation of multi-million 16S rRNA gene libraries from complex microbial communities by assembling paired-end Illumina reads. Appl. Environ. Microbiol. *77*, 3846–3852.

Baveye, P.C. (2009). To sequence or not to sequence the whole-soil metagenome? Nat. Rev. Microbiol. *7*, 756.

Berry, D., Mahfoudh, K.B., Wagner, M., and Loy, A. (2011). Barcoded Primers Used in Multiplex Amplicon Pyrosequencing Bias Amplification. Appl. Environ. Microbiol. *77*, 7846–7849.

Borneman, J., and Triplett, E.W. (1997). Molecular microbial diversity in soils from eastern Amazonia: evidence for unusual microorganisms and microbial population shifts associated with deforestation. Appl. Environ. Microbiol. *63*, 2647–2653.

Boyce, R., Chilana, P., and Rose, T.M. (2009). iCODEHOP: a new interactive program for designing COnsensus-DEgenerate Hybrid Oligonucleotide Primers from multiply aligned protein sequences. Nucleic Acids Res. *37*, W222-W228.

Brochier-Armanet, C., Boussau, B., Gribaldo, S., and Forterre, P. (2008). Mesophilic crenarchaeota: proposal for a third archaeal phylum, the Thaumarchaeota. Nat. Rev. Microbiol. *6*, 245–252.

Brochier-Armanet, C., Gribaldo, S., and Forterre, P. (2012). Spotlight on the Thaumarchaeota. ISME J. *6*, 227–230.

Carson, J.K., Gonzalez-Quinones, V., Murphy, D.V., Hinz, C., Shaw, J.A., and Gleeson, D.B. (2010). Low pore connectivity increases bacterial diversity in soil. Appl. Environ. Microbiol. *76*, 3936–3942.

Claesson, M.J., Wang, Q., O'Sullivan, O., Greene-Diniz, R., Cole, J.R., Ross, R.P., and O'Toole, P.W. (2010). Comparison of two next-generation sequencing technologies for resolving highly complex microbiota composition using tandem variable 16S rRNA gene regions. Nucleic Acids Res. *38*, e200.

Cole, J.R., Chai, B., Farris, R.J., Wang, Q., Kulam, S.A., McGarrell, D.M., Garrity, G.M., and Tiedje, J.M. (2005). The Ribosomal Database Project (RDP-II): sequences and tools for high-throughput rRNA analysis. Nucleic Acids Res. 33, D294-D296.

Consortium, T.B.H. (2009). Genome-Wide Survey of SNP Variation Uncovers the Genetic Structure of Cattle Breeds. Science *324*, 528–532.

De Rijk, P., Van de Peer, Y., Van den Broeck, I., and De Wachter, R. (1995). Evolution according to large ribosomal subunit RNA. J. Mol. Evol. *41*, 366–375.

Degnan, P.H., and Ochman, H. (2011). Illumina-based analysis of microbial community diversity. ISME J. *6*, 183–194.

DeSantis, T.Z., Hugenholtz, P., Keller, K., Brodie, E.L., Larsen, N., Piceno, Y.M., Phan, R., and Andersen, G.L. (2006). NAST: a multiple sequence alignment server for comparative analysis of 16S rRNA genes. Nucleic Acids Res. *34*, W394–399.

DeSantis, T.Z., Hugenholtz, P., Larsen, N., Rojas, M., Brodie, E.L., Keller, K., Huber, T., Dalevi, D., Hu, P., and Andersen, G.L. (2006). Greengenes, a chimera-checked 16S rRNA gene database and workbench compatible with ARB. Appl. Environ. Microbiol. *72*, 5069–5072.

Draghici, S., Khatri, P., Eklund, A.C., and Szallasi, Z. (2006). Reliability and reproducibility issues in DNA microarray measurements. Trends Genet. *22*, 101–109.

Edgar, R.C. (2004). MUSCLE: multiple sequence alignment with high accuracy and high throughput. Nucleic Acids Res. *32*, 1792–1797.

Fechner, L.C., Vincent-Hubert, F., Gaubert, P., Bouchez, T., Gourlay-Francé, C., and Tusseau-Vuillemin, M.-H. (2010). Combined eukaryotic and bacterial community fingerprinting of natural freshwater biofilms using automated ribosomal intergenic spacer analysis. FEMS Microbiol. Ecol. *74*, 542–553.

Fierer, N., Bradford, M.A., and Jackson, R.B. (2007). Toward an Ecological Classification of Soil Bacteria. Ecology *88*, 1354–1364.

Fierer, N., Grandy, A.S., Six, J., and Paul, E.A. (2009). Searching for unifying principles in soil ecology. Soil Biol. Biochem. *41*, 2249–2256.

Finn, R.D., Clements, J., and Eddy, S.R. (2011). HMMER web server: interactive sequence similarity searching. Nucleic Acids Res. *39*, W29-W37.

Finn, R.D., Tate, J., Mistry, J., Coggill, P.C., Sammut, S.J., Hotz, H.-R., Ceric, G., Forslund, K., Eddy, S.R., Sonnhammer, E.L.L., and Bateman, A. (2008). The Pfam protein families database. Nucleic Acids Res. *36*, D281-D288.

Forney, L.J., Zhou, X., and Brown, C.J. (2004). Molecular microbial ecology: land of the one-eyed king. Curr. Opin. Microbiol. *7*, 210–220.

Frank, D.N. (2009). BARCRAWL and BARTAB: software tools for the design and implementation of barcoded primers for highly multiplexed DNA sequencing. BMC Bioinformatics *10*, 362.

Frost, L.S., Leplae, R., Summers, A.O., and Toussaint, A. (2005). Mobile genetic elements: The agents of open source evolution. Nat. Rev. Microbiol. *3*, 722–732.

Gans, J., Wolinsky, M., and Dunbar, J. (2005). Computational improvements reveal great bacterial diversity and high metal toxicity in soil. Science *309*, 1387–1390.

Gardner, S.N., Hiddessen, A.L., Williams, P.L., Hara, C., Wagner, M.C., and Colston, B.W. (2009). Multiplex primer prediction software for divergent targets. Nucleic Acids Res. *37*, 6291–6304.

Glenn, T.C. (2011). Field guide to next-generation DNA sequencers. Mol. Ecol. Resour. *11*, 759–769.

Handelsman, J. (2004). Metagenomics: application of genomics to uncultured microorganisms. Microbiol. Mol. Biol. Rev. *68*, 669–685.

Handelsman, J., Rondon, M.R., Brady, S.F., Clardy, J., and Goodman, R.M. (1998). Molecular biological access to the chemistry of unknown soil microbes: a new frontier for natural products. Chem. Biol. *5*, R245–249.

He, Z., Deng, Y., Van Nostrand, J.D., Tu, Q., Xu, M., Hemme, C.L., Li, X., Wu, L., Gentry, T.J., Yin, Y., Liebich, J., Hazen, T.C., and Zhou, J. (2010). GeoChip 3.0 as a high-throughput tool for analyzing microbial community composition, structure and functional activity. ISME J. *4*, 1167–1179.

He, Z.L., Gentry, T.J., Schadt, C.W., Wu, L.Y., Liebich, J., Chong, S.C., Huang, Z.J., Wu, W.M., Gu, B.H., Jardine, P., Criddle, C., and Zhou, J. (2007). GeoChip: a comprehensive microarray for investigating biogeochemical, ecological and environmental processes. ISME J. *1*, 67–77.

Huse, S.M., Dethlefsen, L., Huber, J.A., Welch, D.M., Relman, D.A., and Sogin, M.L. (2008). Exploring microbial diversity and taxonomy using SSU rRNA hypervariable tag sequencing. PLoS Genet. *4*, e1000255.

Hutchison, C.A., and Venter, J.C. (2006). Single-cell genomics. Nat. Biotechnol. *24*, 657–658.

Jones, S.E., and Lennon, J.T. (2010). Dormancy contributes to the maintenance of microbial diversity. Proc. Natl. Acad. Sci. U.S.A. *107*, 5881–5886.

Jordan, I.K., Rogozin, I.B., Wolf, Y.I., and Koonin, E.V. (2002). Essential genes are more evolutionarily conserved than are nonessential genes in bacteria. Genome Res. *12*, 962–968.

Kalyuzhnaya, M.G., Zabinsky, R., Bowerman, S., Baker, D.R., Lidstrom, M.E., and Chistoserdova, L. (2006). Fluorescence *in situ* hybridization-flow cytometry-cell sorting-based method for separation and enrichment of type I and type II methanotroph populations. Appl. Environ. Microbiol. *72*, 4293–4301.

Katoh, K., Misawa, K., Kuma, K.Ä., and Miyata, T. (2002). MAFFT: a novel method for rapid multiple sequence alignment based on fast Fourier transform. Nucleic Acids Res. *30*, 3059–3066.

Kimura, M. (1968). Evolutionary Rate at the Molecular Level. Nature *217*, 624–626.

Kimura, M. (1991). The neutral theory of molecular evolution: A review of recent evidence. Jpn. J. Genet *66*, 367–386.

Konstantinidis, K.T., Ramette, A., and Tiedje, J.M. (2006). The bacterial species definition in the genomic era. Philos. Trans. R. Soc. Lond. B. Biol. Sci. *361*, 1929–1940.

Kuramae, E.E., Robert, V., Snel, B., Weiss, M., and Boekhout, T. (2006). Phylogenomics reveal a robust fungal tree of life. FEMS Yeast Res. *6*, 1213–1220.

Lemey, P., Salemi, M., and Vandamme, A.M. (2009). The phylogenetic handbook: a practical approach to phylogenetic analysis and hypothesis testing. 2nd Edn. (Cambridge, UK: Cambridge University Press).

Leveau, J.H.J. (2007). The magic and menace of metagenomics: prospects for the study of plant growth-promoting rhizobacteria. Eur. J. Plant Pathol. *119*, 279–300.

Levičnik-Höfferle, Š., Nicol, G.W., Ausec, L., Mandić-Mulec, I., and Prosser, J.I. (2011). Stimulation of thaumarchaeal ammonia oxidation by ammonia derived from organic nitrogen but not added inorganic nitrogen. FEMS Microbiol. Ecol. *80*, 114–123.

Linhart, C., and Shamir, R. (2002). The degenerate primer design problem. Bioinformatics *18*, S172-S181.

Liu, Z., Lozupone, C., Hamady, M., Bushman, F.D., and Knight, R. (2007). Short pyrosequencing reads suffice for accurate microbial community analysis. Nucleic Acids Res. *35*, e120.

Ludwig, W., Strunk, O., Klugbauer, S., Klugbauer, N., Weizenegger, M., Neumaier, J., Bachleitner, M., and Schleifer, K.H. (1998). Bacterial phylogeny based on comparative sequence analysis (review). Electrophoresis *19*, 554–568.

Ludwig, W., Strunk, O., Westram, R., Richter, L., Meier, H., Yadhukumar, Buchner, A., Lai, T., Steppi, S., Jobb, G., *et al.* (2004). ARB: a software environment for sequence data. Nucleic Acids Res. *32*, 1363–1371.

Lynch, J.M., Benedetti, A., Insam, H., Nuti, M.P., Smalla, K., Torsvik, V., and Nannipieri, P. (2004). Microbial diversity in soil: ecological theories, the contribution of molecular techniques and the impact of transgenic plants and transgenic microorganisms. Biol. Fertil. Soils *40*, 363–385.

McIlroy, S.J., Tillett, D., Petrovski, S., and Seviour, R.J. (2010). Non-target sites with single nucleotide insertions or deletions are frequently found in 16S rRNA sequences and can lead to false positives in fluorescence *in situ* hybridization (FISH). Environ. Microbiol. *13*, 33–47.

Meyer, M., and Kircher, M. (2010). Illumina Sequencing Library Preparation for Highly Multiplexed Target Capture and Sequencing. Cold Spring Harb. Protoc. 2010, pdb.prot5448.

Moreno-Vivián, C., Cabello, P., Martínez-Luque, M., Blasco, R., and Castillo, F. (1999). Prokaryotic Nitrate Reduction: Molecular Properties and Functional Distinction among Bacterial Nitrate Reductases. J. Bacteriol. *181*, 6573–6584.

Muyzer, G., Dewaal, E.C., and Uitterlinden, A.G. (1993). Profiling of complex microbial-populations by denaturing gradient gel-electrophoresis analysis of polymerase chain reaction-amplified genes-coding for 16S ribosomal-RNA. Appl. Environ. Microbiol. *59*, 695–700.

Nannipieri, P., Ascher, J., Ceccherini, M.T., Landi, L., Pietramellara, G., and Renella, G. (2003). Microbial diversity and soil functions. Eur. J. Soil Sci. *54*, 655–670.

Notredame, C.d., Higgins, D.G., and Heringa, J. (2000). T-coffee: a novel method for fast and accurate multiple sequence alignment. J. Mol. Biol. *302*, 205–217.

Olsen, G., and Woese, C. (1993). Ribosomal RNA: a key to phylogeny. FASEB J. *7*, 113–123.

Pernthaler, A., and Amann, R. (2004). Simultaneous Fluorescence *In situ* Hybridization of mRNA and rRNA in Environmental Bacteria. Appl. Environ. Microbiol. *70*, 5426–5433.

Pester, M., Schleper, C., and Wagner, M. (2011). The Thaumarchaeota: an emerging view of their phylogeny and ecophysiology. Curr. Opin. Microbiol. *14*, 300–306.

Philippot, L. (2002). Denitrifying genes in bacterial and archaeal genomes. BBA-Gene Struct. Expr. *1577*, 355–376.

Polz, M.F., and Cavanaugh, C.M. (1998). Bias in Template-to-Product Ratios in Multitemplate PCR. Appl. Environ. Microbiol. *64*, 3724–3730.

Prosser, J.I. (2010). Replicate or lie. Environ. Microbiol. *12*, 1806–1810.

Pruesse, E., Quast, C., Knittel, K., Fuchs, B.M., Ludwig, W., Peplies, J., and Glockner, F.O. (2007). SILVA: a comprehensive online resource for quality checked and aligned ribosomal RNA sequence data compatible with ARB. Nucleic Acids Res. *35*, 7188–7196.

Puglisi, E., Hamon, R., Vasileiadis, S., Coppolecchia, D., and Trevisan, M. (2011). Adaptation of soil microorganisms to trace element contamination: a review of mechanisms, methodologies, and consequences for risk assessment and remediation. Crit. Rev. Environ. Sci. Technol. *42*, 2435–2470.

Rajendhran, J., and Gunasekaran, P. (2008). Strategies for accessing soil metagenome for desired applications. Biotechnol. Adv. *26*, 576–590.

Riesenfeld, C.S., Schloss, P.D., and Handelsman, J. (2004). Metagenomics: genomic analysis of microbial communities. Annu. Rev. Genet. *38*, 525–552.

Rio, D.C., Ares, M., Jr, Hannon, G.J., and Nilsen, T.W. (2010). Enrichment of Poly(A)+ mRNA using immobilized oligo(dT). Cold Spring Harb. Protoc. 2010, pdb.prot5454-.

Roberts, E., Sethi, A., Montoya, J., Woese, C.R., and Luthey-Schulten, Z. (2008). Molecular signatures of ribosomal evolution. Proc. Natl. Acad. Sci. U.S.A. *105*, 13953–13958.

Roesch, L.F., Fulthorpe, R.R., Riva, A., Casella, G., Hadwin, A.K.M., Kent, A.D., Daroub, S.H., Camargo, F.A.O., Farmerie, W.G., and Triplett, E.W. (2007). Pyrosequencing enumerates and contrasts soil microbial diversity. ISME J. *1*, 283–290.

Roger, A.J., Sandblom, O., Doolittle, W.F., and Philippe, H. (1999). An evaluation of elongation factor 1 alpha as a phylogenetic marker for eukaryotes. Mol. Biol. Evol. *16*, 218–233.

Rothberg, J.M., Hinz, W., Rearick, T.M., Schultz, J., Mileski, W., Davey, M., Leamon, J.H., Johnson, K., Milgrew, M.J., Edwards, M., *et al.* (2011). An integrated semiconductor device enabling non-optical genome sequencing. Nature *475*, 348–352.

Rotthauwe, J.H., Witzel, K.P., and Liesack, W. (1997). The ammonia monooxygenase structural gene amoA as a functional marker: Molecular fine-scale analysis of natural ammonia-oxidizing populations. Appl. Environ. Microbiol. *63*, 4704–4712.

Rozen, S., and Skaletsky, J.H. (2000). Primer3 on the WWW for general users and for biologist programmers In Bioinformatics Methods and Protocols: Methods in Molecular Biology. S. Krawetz, and S. Misener, eds. (Totowa, NJ: Humana Press), pp. 365–386.

Schloss, P.D., and Handelsman, J. (2005). Metagenomics for studying unculturable microorganisms: cutting the Gordian knot. Genome Biol. *6*, 229.

Schloss, P.D., and Handelsman, J. (2006). Toward a census of bacteria in soil. PLoS Comput. Biol. *2*, 786–793.

Schütte, U., Abdo, Z., Bent, S., Shyu, C., Williams, C., Pierson, J., and Forney, L. (2008). Advances in the use of terminal restriction fragment length polymorphism (T-RFLP) analysis of 16S rRNA genes to characterize microbial communities. Appl. Microbiol. Biotechnol. *80*, 365–380.

Sessitsch, A., Hackl, E., Wenzl, P., Kilian, A., Kostic, T., Stralis-Pavese, N., Sandjong, B.T., and Bodrossy, L. (2006). Diagnostic microbial microarrays in soil ecology. New Phytol. *171*, 719–736.

Smets, B.F., and Barkay, T. (2005). Horizontal gene transfer: perspectives at a crossroads of scientific disciplines. Nat. Rev. Micro. *3*, 675–678.

Stephen, J.R., Chang, Y.J., Macnaughton, S.J., Kowalchuk, G.A., Leung, K.T., Flemming, C.A., and White, D.C. (1999). Effect of toxic metals on indigenous soil p-subgroup proteobacterium ammonia oxidizer community structure and protection against toxicity by inoculated metal-resistant bacteria. Appl. Environ. Microbiol. *65*, 95–101.

Stewart, F.J., Ottesen, E.A., and DeLong, E.F. (2010). Development and quantitative analyses of a universal rRNA-subtraction protocol for microbial metatranscriptomics. ISME J. *4*, 896–907.

Tamames, J., Abellan, J.J., Pignatelli, M., Camacho, A., and Moya, A. (2010). Environmental distribution of prokaryotic taxa. BMC Microbiol. *10*, 85.

Tebbe, C.C. (2005). Marker Genes in Soil Microbiology. In Microorganisms in Soils: Roles in Genesis and Functions. A. Varma, and F. Buscot, eds.(Heidelberg, Berlin: Springer), Vol. 3, pp. 359–382.

Thompson, J.D., Higgins, D.G., and Gibson, T.J. (1994). CLUSTAL W: improving the sensitivity of progressive multiple sequence alignment through sequence weighting, position-specific gap penalties and weight matrix choice. Nucleic Acids Res. *22*, 4673–4680.

Torsvik, V., and Ovreas, L. (2002). Microbial diversity and function in soil: from genes to ecosystems. Curr. Opin. Microbiol. *5*, 240–245.

Tourasse, N.J., and Gouy, M. (1997). Evolutionary distances between nucleotide sequences based on the distribution of substitution rates among sites as estimated by parsimony. Mol. Biol. Evol. *14*, 287–298.

Tringe, S.G., von Mering, C., Kobayashi, A., Salamov, A.A., Chen, K., Chang, H.W., Podar, M., Short, J.M., Mathur, E.J., Detter, J.C.,*et al.* (2005). Comparative metagenomics of microbial communities. Science *308*, 554–557.

Vasileiadis, S., Puglisi, E., Arena, M., Cappa, F., Cocconcelli, P.S., and Trevisan, M. (2012). Soil bacterial diversity screening using single 16S rRNA gene V regions coupled with multi-million read generating sequencing technologies. PLoS One *7*, e42671.

Vasileiadis, S., Puglisi, E., Arena, M., Cappa, F., van Veen, J.A., Cocconcelli, P.S., and Trevisan, M. (2013). Soil microbial diversity patterns of a lowland spring environment. FEMS Microbiol. Ecol. In press, DOI: 10.1111/1574-6941.12150.

Walters, W.A., Caporaso, J.G., Lauber, C.L., Berg-Lyons, D., Fierer, N., and Knight, R. (2011). PrimerProspector: de novo design and taxonomic analysis of barcoded PCR primers. Bioinformatics *27*, 1159–1161.

Wang, Q., Garrity, G.M., Tiedje, J.M., and Cole, J.R. (2007). Naive Bayesian classifier for rapid assignment of rRNA sequences into the new bacterial taxonomy. Appl. Environ. Microbiol. *73*, 5261–5267.

Wang, Y., and Qian, P.Y. (2009). Conservative fragments in bacterial 16S rRNA genes and primer design for 16S ribosomal DNA amplicons in metagenomic studies. PLoS One *4*, e7401.

Widmann, J., Harris, J.K., Lozupone, C., Wolfson, A., and Knight, R. (2010). Stable tRNA-based phylogenies using only 76 nucleotides. RNA *16*, 1469–1477.

Woese, C.R. (1987). Bacterial evolution. Microbiol. Rev. *51*, 221–271.

Woese, C.R., Kandler, O., and Wheelis, M.L. (1990). Towards a natural system of organisms: proposal for the domains Archaea, Bacteria, and Eucarya. Proc. Natl. Acad. Sci. U.S.A. *87*, 4576–4579.

Woese, C.R., Olsen, G.J., Ibba, M., and Söll, D. (2000). Aminoacyl-tRNA Synthetases, the Genetic Code, and the Evolutionary Process. Microbiol. Mol. Biol. Rev. *64*, 202–236.

Wu, J.Y., Jiang, X.T., Jiang, Y.X., Lu, S.Y., Zou, F., and Zhou, H.W. (2010). Effects of polymerase, template dilution and cycle number on PCR based 16 S rRNA diversity analysis using the deep sequencing method. BMC Microbiol. *10*, 255.

Yergeau, E., Kang, S., He, Z., Zhou, J., and Kowalchuk, G.A. (2007). Functional microarray analysis of nitrogen and carbon cycling genes across an Antarctic latitudinal transect. ISME J. *1*, 163–179.

Yergeau, E., Schoondermark-Stolk, S.A., Brodie, E.L., Dejean, S., DeSantis, T.Z., Goncalves, O., Piceno, Y.M., Andersen, G.L., and Kowalchuk, G.A. (2009). Environmental microarray analyses of Antarctic soil microbial communities. ISME J. *3*, 340–351.

Soil Metatranscriptomics

Yongkyu Kim, Carl-Eric Wegner and Werner Liesack

Abstract

Metatranscriptomics is defined as the analysis of microbial community gene expression in a particular environment, as opposed to metagenomics which is the study of the genomic content of entire microbial communities. Massively parallel sequencing of RNA is the key component of metatranscriptomics. The analysis of mRNA has the potential to discover novel genes and to uncover functional adaptations of microbial communities to local environmental conditions. Alternatively, total RNA may be used for analysis. This approach provides insight into the taxonomic composition of microbial communities. The subject of this review is soil metatranscriptomics. We discuss the experimental and bioinformatic workflow that can be applied to the metatranscriptomic analysis of soil microbial communities. Since research in the field of soil metatranscriptomics is still in its infancy, we also review the recent advances in marine metatranscriptomics.

Introduction

While metagenomics is a tool to inventory the gene content of microbial communities, metatranscriptomics provides an avenue to access their activities. The key feature of metatranscriptomic analysis is random sequencing of microbial community RNA. Metatranscriptomics thus overcomes limitations inherent to PCR- and microarray-based studies that rely on known sequence information for adequate experimental design. As such, studies that are based on PCR or microarrays are closed approaches for which the detectable range of sequence diversity is predefined by primers and probes. By contrast, metatranscriptomics, an open approach, has a great potential to discover novel genes. While conventional Sanger sequencing of cloned cDNA is still being applied, the use of massively parallel 454 pyrosequencing is now common practice of most metatranscriptomic studies. A single sequencing run generates up to 100,000 or 1,000,000 cDNA reads, depending on whether the GS Junior system or the GS FLX platform (Roche 454 Life Sciences, Branford, USA) is used for analysis. Over recent years, the average read length achieved by 454 pyrosequencing has increased from 100 to 750 nucleotides (Fig. 5.1).

Metatranscriptomic analysis can be performed using either total RNA or enriched mRNA as template for cDNA synthesis and 454 pyrosequencing. The use of total RNA results in a large data set of 16S and 23S rRNA-tags. These provide insight into the taxonomic composition of microbial communities and allow us to identify the active community members, given that ribosome abundance is correlated with metabolic activity (Rehman *et al.*, 2010).

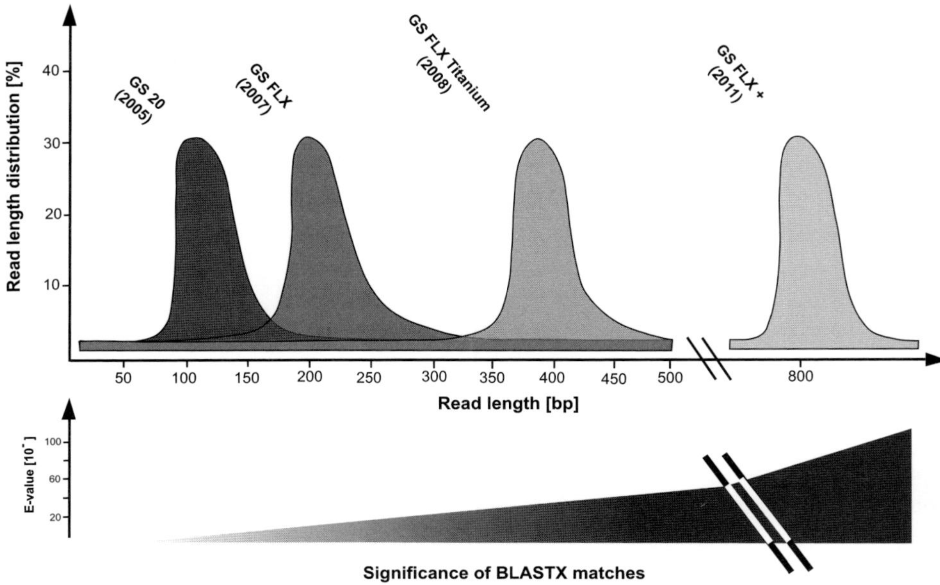

Figure 5.1 Read length distribution achieved by the chemistries available for 454 pyrosequencing. The year in which Roche's 454 Life Sciences launched the different chemistries is given in parenthesis. Statistical significance of BLASTX matches increases with read length (Shrestha *et al.*, 2009).

Taxonomic profiling of complex microbial communities is done using 16S rRNA-tags mainly due to the more comprehensive databases available for 16S rRNA than for 23S rRNA (Radax *et al.*, 2012). Analysis of enriched mRNA is defined by taxonomic assignment and functional annotation of putative mRNA-tags, thereby providing information on the functional genes and metabolic pathways expressed at a certain time and place.

Here, we discuss the experimental and bioinformatic workflow of soil metatranscriptomics. Among other criteria, preservation of the RNA expression profiles at the time of sampling, sufficient yield and high integrity of total RNA, and effective enrichment of mRNA are crucial steps in soil metatranscriptomics.

To date, there are only a handful of studies that explored the soil metatranscriptome. While research in the field of soil metatranscriptomics has just started to emerge, microbial community gene expression in marine waters has already been investigated extensively. A short overview of recent advances in marine metatranscriptomics is therefore part of this review.

The experimental and bioinformatic workflow

Accessing the soil metatranscriptome requires the sequential application of several processing steps. These can be summarized into (i) generation of metatranscriptomic data sets and (ii) bioinformatic analysis. The discussion of the individual processing steps follows the scheme in Fig. 5.2.

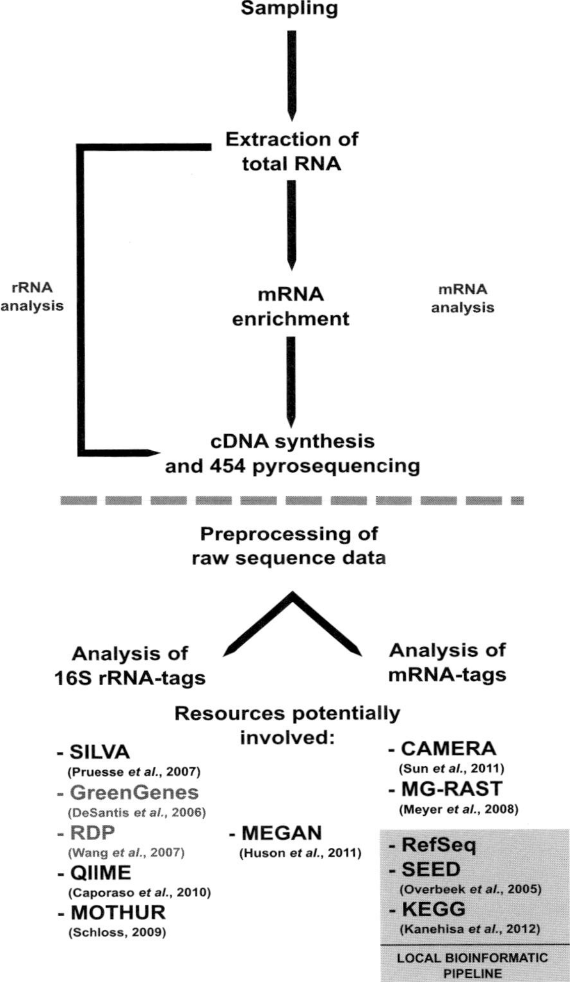

Figure 5.2 Schematic workflow of processing steps involved in soil metatranscriptomics.

Sampling

Preservation of the RNA expression profiles at the time of sampling is critical in metatranscriptomic analysis. Shock-freezing in liquid nitrogen immediately after sampling is one method to preserve metatranscriptomic profiles and RNA integrity. The samples will be stored at –70°C until RNA extraction (Hurt *et al.*, 2001; Sessitsch *et al.*, 2002). However, liquid nitrogen may not be available if samples are taken in the field. Therefore, an alternative method is the use of RNA-stabilizing reagents in which the soil is suspended immediately after sampling. Several reagents are commercially available, such as RNA*later*® (Ambion, Carlsbad, USA), RNAprotect® Bacteria Reagent (Qiagen, Hilden, Germany), and LifeGuard™ Soil Preservation solutions (MO-BIO, Carlsbad, USA). Their basic principle is the use of high-salt buffers. For example, RNA*later*® contains 25 mM sodium citrate, 10 mM EDTA, and 70 g ammonium sulphate per 100 ml solution (pH 5.2). This reagent rapidly infiltrates microbial cells. The high concentration of ammonium sulphate causes a salt-induced denaturation of protein, resulting in the mass precipitation of cellular proteins and associated RNA. However, the cellular structure remains intact. Presumably, RNase will also be precipitated and rendered inaccessible to cellular RNA. Stabilization and preservation reagents are widely used to extract RNA from tissues, bacterial cultures, marine samples, and soil (Frias-Lopez *et al.*, 2004; Mutter *et al.*, 2004; Ricke *et al.*, 2004; Meyer *et al.*, 2006).

Extraction of total RNA

Over the last two decades, various methods have been published on the extraction of total RNA from soil, using different strategies. Among these were: (i) recovery of ribosomes by ultracentrifugation using a lysis buffer that contains polyvinylpolypyrrolidone and $MgCl_2$ (Felske *et al.*, 1996); (ii) pretreatment with aluminium sulphate (Persoh *et al.*, 2008); (iii) use of cetyltrimethylammonium bromide in the extraction buffer (Griffiths *et al.*, 2000; Bürgmann *et al.*, 2003); (iv) sodium dodecyl sulphate-based lysis combined with the use of high-salt extraction buffers (Hurt *et al.*, 2001); and (v) the use of size-exclusion spin columns (Wang *et al.*, 2009). Total RNA obtained by these methods was mostly used in reverse transcription (RT)-PCR to detect and quantify the expression of particular genes such as, for example, *nifD* or *mcrA* (Holmes *et al.*, 2004; Yuan *et al.*, 2011; Deangelis and Firestone, 2012). RT-PCR requires only a low amount of template RNA, thus making it possible to use diluted extracts of total RNA for analysis. Dilution reduces the amount of remaining contaminants and inhibitory substances (e.g. Bürgmann *et al.*, 2003). None of these extraction methods was explicitly developed and optimized for metatranscriptomic studies.

More recently, Mettel *et al.* (2010) assessed various procedural steps to optimize the extraction and purification of total RNA from soil. Optimization was done using four soil types differing in humic acid content: grassland, rice paddy, forest, and agricultural soils. The criteria for assessment were purity, integrity, and yield. In agreement with previous reports (Noonberg *et al.*, 1995), low-pH extraction was highly effective in stabilizing RNA. This involved the use of a phenol-based lysis buffer (pH 5.0) during bead beating and phenol-chloroform (pH 4.5–5.0) in subsequent extraction steps. The raw extract of total RNA was obtained after DNase digestion. While the RNA yield was comparable, a greater RNA integrity and lower humic acid content (reduced by approximately 40%) were observed in low-pH (4.5–5.0) than in high-pH (7.0–8.0) extracts.

Among the purification methods tested, Q-Sepharose column chromatography showed the best performance in humic acid removal from raw extracts of total RNA. The removal efficiency was assessed spectrophotometrically at 400 nm and determined to be 95% regardless of soil type. Sodium chloride is used to elute the RNA from the sepharose column. A concentration of 1.5 M NaCl was found to be optimal to achieve high RNA purity and sufficient yield. Depending on the humic acid content, the balance between RNA purity and yield can be refined by appropriate changes in the counter ion concentration. After purification, small RNAs and partially degraded RNA (< 200 nt) were removed using either RNeasy® MinElute® Cleanup kit (Qiagen) or RNA Clean and Concentrator™ (Zymo Research, Irvine, USA). RNA purity was determined spectrophotometrically using the 260/280 nm absorbance ratio, in addition to the spectrophotometric measurement of humic acids at 400 nm. RNA integrity and yield were assessed simultaneously using microfluidic chip-based electrophoresis devices, such as Bioanalyzer 2100 (Agilent, Santa Clara, CA, USA) and Experion (Bio-Rad, Hercules, USA). The extracts of total RNA were of sufficient quality for downstream applications, including enrichment of mRNA and cDNA synthesis. Enriched mRNA ranged from 200 bp to 4 kb in size.

Meanwhile, various commercial kits are available for the extraction of total RNA from soils such as, for example, RNA PowerSoil® total RNA isolation kit (MO-BIO), FastRNA® Pro Soil-Direct kit (qbiogene, Santa Ana, USA), and E.Z.N.A. Soil RNA kit (OMEGA Bio-tek, Norcross, USA). All three kits are based on phenol extraction followed by kit-specific RNA purification methods involving the use of either column or bead applications to capture the RNA. Isolation of total RNA by the ZR Soil/Fecal RNA MicroPrep™ kit (Zymo Research) is based on a series of column treatments, thereby avoiding phenol extraction. Optionally, DNase treatment may be included. While convenient handling is a major advantage of the commercial extraction kits, their use may not yield sufficient amount of RNA for metatranscriptomic studies. When using, for example, the ZR Soil/Fecal RNA MicroPrep™ kit, the maximum amount of soil that can be processed in each reaction is 0.25 g of soil (fresh weight). However, a total of 6–8 g of soil need to be processed to isolate sufficient amounts of enriched mRNA for 454 pyrosequencing, given that the rapid library preparation method is employed for cDNA production (see below). While approximately half of the reagents supplied in a single ZR Soil/Fecal RNA MicroPrep™ kit is needed to process 6 to 8 g of soil, the method by Mettel *et al.* (2010) can be easily scaled up to extract a sufficient amount of enriched mRNA from a single sample.

mRNA enrichment

Effective enrichment of mRNA is a crucial step in the functional analysis of microbial community gene expression, given that mRNA comprises only 1–3% of total cellular RNA. The procedure chosen for mRNA enrichment depends on the exact objectives of the study. Polyadenylation at the 3′-end is a specific feature of eukaryotic mRNA. If the metatranscriptome of eukaryotic microorganisms (fungi, metazoa) shall be analysed, 3′-polyadenlyated mRNA will be selectively captured using poly-dT-coated paramagnetic beads, following the instructions of the manufacturer (e.g. Dynabeads® mRNA kits [Invitrogen, Carlsbad, USA], and PolyATract® mRNA Isolation system [Promega, Madison, USA]). The use of an oligo-dT primer in first-strand cDNA synthesis even increases the selectivity for eukaryotic mRNA (Bailly *et al.*, 2007; Damon *et al.*, 2012).

Enzymatic digestion and subtractive hybridization of rRNA are the methods of choice if the total soil metatranscriptome, but in particular that of bacteria, shall be analysed. The mRNA-ONLY™ mRNA isolation kit (Epicentre, Madison, USA) makes use of a 5′-monophosphate-dependent exonuclease to enzymatically degrade rRNA (Fig. 5.3). While mature rRNA is 5′-monophosphorylated, eukaryotic mRNA is protected by cap structure and bacterial mRNA carries a triphosphate group. The feature of rRNA to be 5`-monophosphorylated was therefore thought to be a means to specifically degrade rRNA by 5′ to 3′ processive exonuclease activity, thereby enriching mRNA relative to rRNA. However, when applied to soil RNA extracts, the 5′-monophosphate-dependent exonuclease was found to degrade not only rRNA but also a considerable amount of soil mRNA (Mettel *et al.*, 2010). The most likely explanation for this observation is that the triphosphate group at the 5′-end is converted to a monophosphate form during mRNA decay in bacteria (Celesnik *et al.*, 2007). Hence, the exonuclease treatment may direct metatranscriptomic studies towards the analysis of fresh mRNA that is 5′-triphosphorylated (He *et al.*, 2010; Mettel *et al.*, 2010). Moreover, the activity of 5′-monophosphate-dependent exonuclease

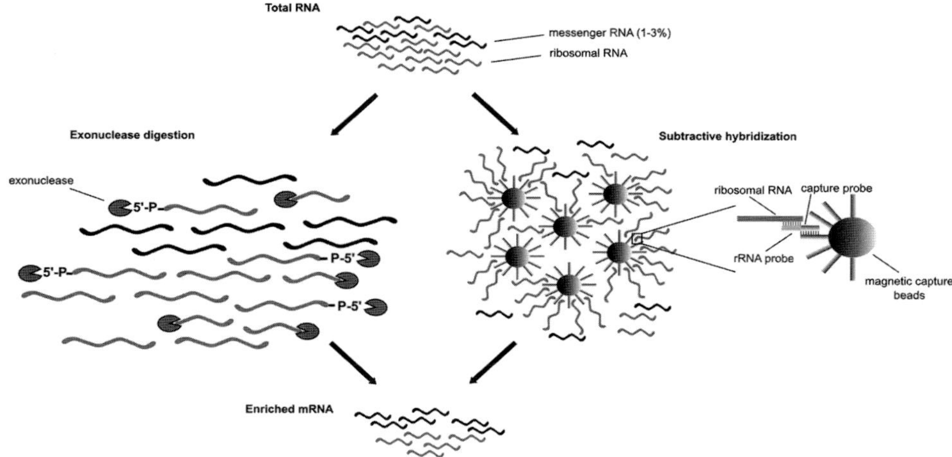

Figure 5.3 Selective enrichment of mRNA by exonuclease digestion and subtractive hybridization of rRNA. See text for details.

is affected by even trace amounts of residual humic acids. Humic substances are powerful enzymatic inhibitors that have to be quantitatively removed during nucleic acid extraction from soil (Mettel *et al.*, 2010).

In contrast to the exonuclease treatment, subtractive hybridization of rRNA preserves the full diversity of mRNA transcripts, resulting in the analysis of processed and unprocessed bacterial mRNA. The key component of subtractive hybridization is a set of capture probes that are complementary to highly conserved sequence regions within rRNA (Fig. 5.3). Several commercial kits are available that specifically remove bacterial rRNA by subtractive hybridization: MICROB*Express*™ (Ambion), RiboMinus™ (Invitrogen), and Ribo-Zero™ (Epicentre). Although these kits perform well in pure culture studies (Giannoukos *et al.*, 2012), they have certain shortcomings when applied to soil RNA. The vast microbial diversity in soil makes the phylogenetic range of the rRNA-targeted capture probes a critical factor in mRNA enrichment. For example, the oligonucleotide probes used by MICROB*Express*™ do not capture rRNA of various bacterial taxa such as *Thermotoga* and *Deinococcus*. A complete list is available on the manufacturer's website (Invitrogen). Another limitation is that the rRNA removal efficiency of subtractive hybridization declines with increasing fragmentation of the RNA used for mRNA enrichment (He *et al.*, 2010). Physical fragmentation during the extraction of total RNA from soil cannot be completely avoided. Moreover, depending on growth stage and environment, fragmentation of rRNA, particularly of 23S rRNA, occurs in many bacteria naturally (Selenska-Pobell and Evguenieva-Hackenberg, 1995; Klein *et al.*, 2002). The Ribo-Zero™ rRNA removal kit (version Meta-Bacteria) that has recently been introduced by Epicentre is claimed to capture both intact and partially degraded rRNA. The proportion of putative mRNA-tags in a data set derived from total RNA was found to be 1% of total reads. The use of MICROB*Express*™ and Ribo-Zero™ to eliminate rRNA increased the proportion of putative mRNA-tags to respectively 10–23% and 50–70% (Fig. 5.4). This result is consistent with the findings of a recent study (Giannoukos *et al.*, 2012), which revealed that Ribo-Zero™ is more effective in enriching bacterial mRNA than other commercially available kits, including MICROB*Express*™ and mRNA-ONLY™.

RiboMinus™ (Invitrogen) and Ribo-Zero™ (Epicentre) offer a series of subtractive hybridization kits that allow specific removal of human, mammalian, plant, or bacterial rRNA. However, none of the commercial kits available to date for the enrichment of mRNA offer the removal of archaeal rRNA. Therefore, the development of habitat-specific capture probes and their use in subtractive hybridization of rRNA may be an important supplementary approach when metatranscriptomics shall be applied to environments in which archaea are present in high abundance. Habitat-specific archaea probes may be developed by PCR amplification of archaeal 16S rRNA genes, resulting in PCR products that contain a T7 promoter sequence at the 3′-end (Stewart *et al.*, 2010). Biotinylated antisense rRNA is generated by *in vitro* transcription using the archaeal rRNA gene amplicons as templates. The *in vitro* produced antisense rRNA will then be used as capture probe in subtractive hybridization of the archaeal rRNA.

Another option to separate mRNA from rRNA may be the use of agarose gel electrophoresis. The non-rRNA will be recovered by precise excision of the agarose between the major rRNA bands. This method may allow separation of putative mRNA simultaneously from both rRNA and humic acids, because the latter migrate faster in the electric field than RNA molecules (McGrath *et al.*, 2008). However, as yet, this extraction method was not assessed for its potential use in metatranscriptomics, while both exonuclease digestion and subtractive hybridization of rRNA were repeatedly the methods of choice in studies on the bacterial metatranscriptome (Frias-Lopez *et al.*, 2008; Poretsky *et al.*, 2009; Vila-Costa *et al.*, 2010).

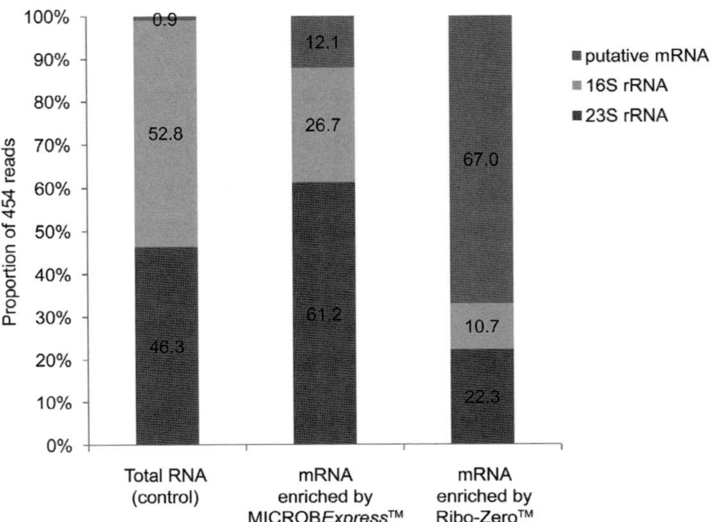

Figure 5.4 Effect of applying MICROB*Express*™ and Ribo-Zero™ on the enrichment of mRNA, using the same batch of total RNA as starting material. Total RNA was obtained from flooded paddy soil (Mettel *et al.*, 2010). The increase in putative mRNA-tags relative to the control (total RNA) was determined by 454 pyrosequencing. Archaeal 16S and 23S rRNA sequences made a strong contribution to the remaining rRNA-tags in the mRNA pools enriched by either MICROB*Express*™ or Ribo-Zero™.

cDNA synthesis and 454 pyrosequencing

The procedural steps involved in cDNA synthesis and library preparation depend on the Roche/454 platform used for metatranscriptomic analysis. To date, nearly all the marine studies used either GS20 or GS FLX pyrosequencing. These chemistries require a starting amount of as much as 4 μg cDNA. To produce this amount of cDNA, total RNA and, in particular, enriched mRNA need to be amplified by *in vitro* transcription. In contrast to the logarithmic amplification by PCR, *in vitro* transcription is a linear amplification method using cDNA as a template. The first step, however, is polyadenylation of template RNA. First-strand cDNA synthesis is achieved by priming with a specific oligo-dT primer containing a T7 promoter at the 5′-end, followed by RNase H-mediated second-strand cDNA synthesis. Homopolymers significantly lower the quality of 454 pyrosequencing. Therefore, the oligo-dT primer may be modified, so that a BpmI restriction site is inserted next to the 16-dT stretch (Frias-Lopez *et al.*, 2008). The cDNA flanked by T7 promoter is used to produce antisense RNA by RNA polymerase-mediated amplification. Commercial kits, such as the MessageAmp™ II aRNA Amplification kit (Ambion), are designed to produce up to 100 μg of antisense RNA from as low as a few nanograms of input RNA. Antisense RNA is converted into double-stranded cDNA by random priming and, if present, the poly A/T stretch is removed by BpmI digestion (Frias-Lopez *et al.*, 2008).

These days, the GS FLX Titanium and GS FLX+ chemistries are commonly used in 454 pyrosequencing. cDNA subjected to GS FLX Titanium or GS FLX+ pyrosequencing may be produced via *in vitro* transcription as described above (de Menezes *et al.*, 2012). However, Roche launched a new method for cDNA synthesis in 2011. This cDNA rapid library preparation method is compatible with GS FLX Titanium and GS FLX+ pyrosequencing. The method requires only a minimum amount of 200 ng of input RNA and uses random priming to directly produce double-stranded cDNA without the need of *in vitro* transcription. Preparation of the cDNA involves 454 RL adaptor ligation. The excess of non-ligated adaptors and cDNA fragments < 400–500 bp are eliminated by a sizing step using AMPure® XP beads (Beckman Coulter, Brea, USA). The vast majority of cDNA fragments should be in the size range from 500 to 1200 bp. 454 pyrosequencing is based on light emission that is produced by a cascade of enzymatic reactions during each nucleotide incorporation (Ronaghi *et al.*, 1996). The light is captured by a CCD (charge-coupled device) camera and converted into digital images that are collected during the sequencing run.

Taken together, the use of *in vitro* transcription allows to produce high amounts of cDNA, but pyrosequencing will be biased towards the 3`-end of the input RNA. By contrast, cDNA produced by rapid library preparation will collectively span the complete size range of the input RNA. Therefore, sequence data obtained via the cDNA rapid library preparation method may reflect the composition of the metatranscriptome more accurately than those generated via the use of *in vitro* transcription.

Preprocessing of raw sequence data

In order to obtain reliable sequences, 454 raw data have to be preprocessed by two procedural steps prior to rRNA-tag or mRNA-tag analysis. First step is the automated preprocessing by the 454 operating software. The second step is quality filtration using third-party applications.

454 raw data are composed of a collection of 800 (GS FLX Titanium) or 1600 (GS FLX+) digital images representing the consecutive incorporation of nucleotides during sequencing.

The raw data undergo image and signal processing to generate a Standard Flowgram Format (SFF) file. It contains information on the quality and flow light intensity of each base call. The set of parameters used for image and signal preprocessing depends on whether amplicon or shotgun sequencing was performed. Preprocessing includes filtering of reads that (i) are too short, (ii) have too many poor incorporations or nucleotide interruptions or (iii) have too many nucleotide incorporations. The latter type of reads is possibly produced on beads carrying more than two DNA templates.

Using the produced SFF files, the sequence data and their quality information will be extracted to FASTA+QUAL or FASTQ format using SFF tools provided either by Roche 454 Life Sciences or by third parties, such as Flower (Malde, 2011), SFF Workbench (Heracle BioSoft SRL), sff2fastq, and sff_extract. FASTA+QUAL and FASTQ are standard formats that are accepted by most bioinformatic tools used for downstream analysis. If multiple samples are analysed simultaneously in a single 454 sequencing run, these tools split the original SFF file into sample-specific SFF files based on multiple identifiers. These are barcodes added during cDNA library preparation. The next steps apply for each SFF file individually and involve (i) the automated removal of the multiple identifiers, and (ii) the export of sequence data and quality information to the FASTA+QUAL or FASTQ format.

At this preprocessing stage, 454 sequence data sets still contain artifacts, sequence contaminations, artificial replicates, and reads with ambiguous sequence motifs. These poor-quality reads have to be removed prior to rRNA-tag or mRNA-tag data analysis. Otherwise, they could lead to false annotation or incorrect mapping to reference genomes. This preprocessing step involves both filtering of poor-quality reads and sequence trimming, using online available stand-alone tools. Quality filtration of 454 reads takes into account sequence length, average quality score, contamination by ambiguous base calls, and complexity scores to detect simple repeat sequences, such as homopolymers and dinucleotide repeats. End-trimming improves the quality of sequence data, given that sequencing errors occur more frequently at the 3′-end (Balzer *et al.*, 2011).

Various tools are available to perform quality filtration. FastQC summarizes the sequence statistics of the 454 raw data for quality control. The NGS QC Toolkit (Patel and Jain, 2012) and PRINSEQ (Schmieder and Edwards, 2011) provide a graphical summary of raw and preprocessed sequence data. After final preprocessing, the sequence data sets will be subjected to rRNA-tag or mRNA-tag analysis, depending on whether they are derived from total RNA or enriched mRNA.

Analysis of rRNA-tags

Preprocessed sequence data sets that are derived from total RNA are compared against SILVA small and large subunit ribosomal RNA reference databases (SSURef and LSURef, respectively) (Urich *et al.*, 2008; Radax *et al.*, 2012). The SILVA database is a comprehensive online resource for quality checked and aligned ribosomal RNA sequences (Pruesse *et al.*, 2007). In BLASTN, an *E*-value cutoff of 1e-10 is used to confidently identify rRNA sequences. Alternatively, a bit score value of 50 can be applied, in particular when short reads of 100 to 200 nucleotides are analysed (Frias-Lopez *et al.*, 2008; Urich *et al.*, 2008; Stewart *et al.*, 2011). Some 454 reads may be assigned to putative homologues in both SSURef and LSURef databases at a given cutoff level. In that case, these 454 reads can be classified as either 16S or 23S rRNA-tags based on the highest alignment score in either database. 16S rRNA-tags will be taxonomically assigned by the lowest common ancestor algorithm based on the BLAST outputs (Urich *et al.*, 2008). They may be unambiguously assigned to a single species or genus. However, if a given 16S rRNA-tag is assigned with similar probability to different species or genera, this rRNA-tag will be placed by the lowest common ancestor algorithm to the next higher taxon that includes all the species or genera to which the rRNA-tag was assigned. The taxonomic assignments will be collectively displayed in the format of a hierarchical dendrogram using MEGAN (Huson *et al.*, 2011). MEGAN is a bioinformatic tool for the analysis and visualization of the taxonomic content of metagenomic and metatranscriptomic data sets.

Alternative tools for the analysis of 16S rRNA-tags are QIIME and MOTHUR (Schloss, 2009; Caporaso *et al.*, 2010). QIIME provides a collection of Python script libraries that were originally developed for the analysis of 16S rRNA gene amplicons. Once the 16S rRNA-tags are clustered into operational taxonomic units, modules implemented in QIIME can be used to quantify microbial diversity within and between samples based on diversity indices (e.g. Chao1, Shannon) and distance matrices (e.g. weighted and unweighted UniFrac). A comparable functionality is provided by MOTHUR, a command-line based tool solely written in C++. Alternatively to the databases provided by SILVA, those of the Ribosomal Database Project (RDP) (Cole *et al.*, 2009) and Greengenes (McDonald *et al.*, 2012) may be used for taxonomic assignment of bacterial and archaeal rRNA-tags. However, the use of RDP and Greengenes may be problematic because both database projects do not offer reference data sets of eukaryotic rRNA.

Given that a sufficient number of 454 reads is assigned to the same taxonomic group and overlap each other, these rRNA-tags can be assembled to an almost full-length 16S rRNA sequence. Assembly tools, such as CAP3 (Huang, 1999), Geneious (Biomatters), and CLC Genomics Workbench (CLC Bio), create an rRNA consensus sequence that can be used in phylogenetic treeing analysis. In a study by Radax *et al.* (2012), such assembled consensus sequences showed high sequence identity (95–99%) to 16S ribosomal cDNA clones obtained by RT-PCR from the same environment.

Analysis of mRNA-tags

Even after efficient depletion of rRNA by subtractive hybridization or other techniques, metatranscriptomic data sets still contain a large proportion of rRNA sequences, ranging from 30% to 80% of total reads (He *et al.*, 2010). These rRNA reads have to be identified and removed. Otherwise, they could be misclassified as protein-coding transcripts and then be added to the protein database as putative or hypothetical proteins. For example, it is reported that one conserved region of 23S rRNA was consistently misclassified to create a spurious Pfam protein family (Pfam ID PF10695) with the function of cell wall hydrolase (Tripp *et al.*, 2011). Therefore, all the preprocessed 454 reads derived from enriched mRNA have to be checked against the rRNA databases. Those 454 reads that match entries in either the SSURef database or the LSURef database will be eliminated. Subsequently, non-rRNA reads will be compared against the Rfam database that contains a collection of non-coding RNA families (Gardner *et al.*, 2011). Identification of non-coding RNA reads is done using the INFERNAL program, which takes sequence and structural similarity into account (Nawrocki *et al.*, 2009; Shi *et al.*, 2009). Most of the cDNAs derived from non-coding RNA are eliminated during 454 library preparation for GS-FLX Titanium or GS FLX+ pyrosequencing. However, some groups of bacterial non-coding RNA, such as OLE (Rfam ID RF01071) and snR30/84 (Rfam IDs RF01271 and RF01270), have sizes > 500 nucleotides and 454 reads of these non-coding RNAs can therefore be present in metatranscriptomic libraries. In order to avoid false annotation of non-coding RNA-derived cDNAs to putative or hypothetical proteins, such 454 reads also have to be eliminated. All the 454 reads that survive preprocessing and removal steps are regarded to be putative mRNA-tags and subjected to further bioinformatic analysis.

Taxonomic assignment and functional annotation of putative mRNA-tags are generally done by searching translated cDNA sequences against protein databases, because protein sequences are evolutionarily more conserved than nucleotide sequences (Mount, 2001). Searching translated cDNA sequences takes more time and needs more computational power than searches at the nucleotide level. This is due to the fact that a single protein-coding read contains six potential open reading frames. In addition, the scoring algorithm used to generate amino acid sequence alignments is far more complicated than the algorithm applied to align nucleic acid sequences (Altschul *et al.*, 1997). Metatranscriptomic data sets can be analysed using either online available tools or, if sufficient computational power is available, a local bioinformatic pipeline. The great advantage of online tools, such as CAMERA and MG-RAST, is that they provide the computational power for handling large sequence data sets and the complete pipeline for downstream analysis. These tools can be operated even by someone who has limited knowledge of applied bioinformatics. However, they only make use of selected databases, which affects access to the full range of homologues that may be deposited in public databases for a given query sequence. By contrast, local pipelines make it possible to select any database and extract any sequence information of interest for in-depth analysis, but require skill and experience in applied bioinformatics.

CAMERA and MG-RAST

Over the last couple of years, various online portals and tools for the analysis of complex omics data have been released. Two of those are CAMERA and MG-RAST. Although CAMERA (Sun *et al.*, 2011) and MG-RAST (Meyer *et al.*, 2008) were originally developed for the analysis and comparison of metagenomic data, their ORF annotation pipelines can also be used for the analysis of metatranscriptomic data. Both CAMERA and MG-RAST provide complete toolkits, ranging from quality control of raw sequence data to the taxonomic assignment and functional annotation of putative mRNA-tags. This includes advanced graphical representation of the annotation results. They also offer access to publicly available metagenomic data sets for comparative analyses and allow for the analysis of data sets shared among groups of collaborators. CAMERA was initially developed for marine microbial ecology research but, like MG-RAST, can be used to analyse metagenomic and metatranscriptomic data sets of any environmental study. Both CAMERA and MG-RAST make use of the same public resources, but structure and algorithms of their data analysis pipelines differ.

CAMERA (Cyber infrastructure for Advanced Marine Microbial Ecology Research and Analysis) was developed to maintain a comprehensive set of associated bioinformatics tools and a collaborative online environment (Seshadri *et al.*, 2007; Sun *et al.*, 2011). The first step in the analysis workflow is the submission of sequence and metadata to the CAMERA online environment through a standardized interface. Metadata will have to be submitted according to standards called MIGS (Minimum Information about a Genome Sequence), MIENS (Minimum Information about an Environmental Sequence), and MIMARKS/MixS (Minimum Information about a Marker Gene Sequence, Minimum Information about any X Sequence). Their implementation into comparative analysis of sequence data is promulgated by the Genomics Standards Consortium (Field *et al.*, 2008; Yilmaz *et al.*, 2011). This international consortium aims to establish a catalogue of information required to precisely describe sequence data. Depending on the source of the sequence data, this may include sampling details, sequencing parameters and sequence-specific features such as taxonomic affiliations. Optionally, literature information may also be uploaded. Combining all the information submitted to CAMERA creates a semantically aware database that can be subjected to downstream analysis using bioinformatic tools implemented in CAMERA. The analysis workflow may involve quality control of raw sequence data, elimination of rRNA and tRNA sequences, and functional annotation of protein-coding sequences by BLAST against a single user-selected database (Fig. 5.5).

MG-RAST (Metagenomics Rapid Annotation Server using Subsystem Technology) uses the SEED annotation environment for comparative genomics. The SEED aims to produce high-quality annotation of sequence data, similar to what is done by KEGG (Kyoto

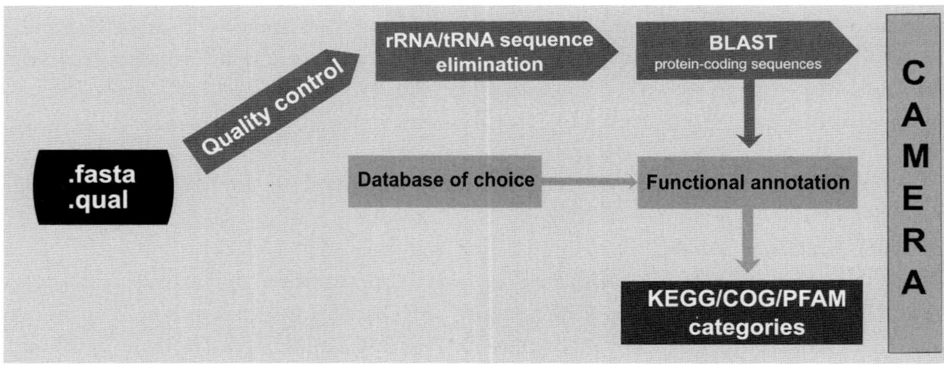

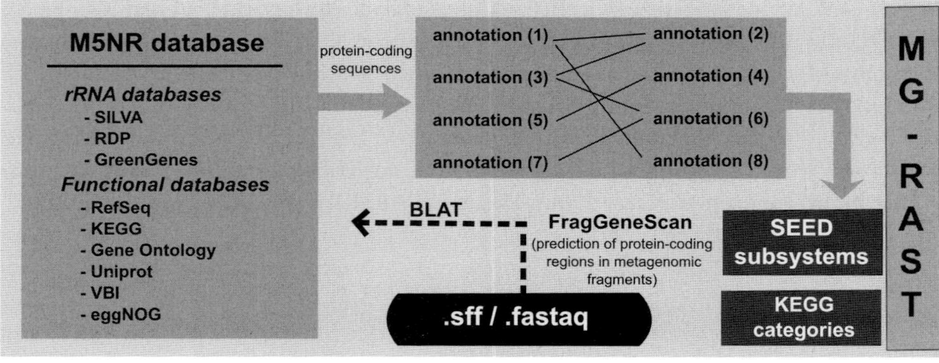

Figure 5.5 Data flow and analysis in CAMERA and MG-RAST. CAMERA makes use of individual user-selected bioinformatic pipelines, which may include quality control followed by the elimination of rRNA/tRNA sequences, and functional and taxonomic classification of protein-coding sequences. By contrast, MG-RAST involves a more comprehensive data analysis. Metatranscriptomic sequences are automatically searched by BLAT against multiple databases implemented in M5NR. The multiple search results for protein-coding sequences are cross-compared and the most likely one is assumed to be valid. See text for further details.

Encyclopedia of Genes and Genomes) (Ogata *et al.*, 1999) and MetaCyc (Caspi *et al.*, 2008). It is based on hierarchically organized subsystems which are used for automated annotation and curation of genomic data. Subsystems represent a collection of functional roles that make up a metabolic pathway, a complex (e.g. the ribosome), or a class of proteins (e.g. two component signal-transduction proteins within *Staphylococcus aureus*). Each of these subsystems is supervised by a group of annotators that are experts in their field of research. Genomic information related to a particular subsystem is only annotated and curated by this specific group of annotators. In the first release of SEED, 173 subsystems have been defined with 2133 distinct functional roles, taking the genomic information of 383 organisms into account (Overbeek *et al.*, 2005).

A key component of MG-RAST is the M5NR database representing a cross-referenced, indexed database that links all the functional databases implemented in MG-RAST intimately (Fig. 5.5). After uploading sequence and metadata to the MG-RAST server, the first analysis step is automated data normalization, meaning the removal of artificial replicate sequences from raw data sets. In metagenomic analysis, potential protein-coding regions are identified by FragGeneScan that uses a gene prediction algorithm based on the Hidden Markov Model and takes into account error models of next-generation sequencing (Rho *et al.*, 2010). In metatranscriptomic analysis, sequences may be directly searched by BLAT (Kent, 2002) against all the M5NR databases (Fig. 5.5). This may result in multiple annotations for protein-coding sequences. These multiple annotations are compared to each other. The final annotation results are classified into the SEED subsystems. Alternatively to this default approach, the KEGG functional categories may be used for data representation, particularly if the focus is on the analysis of metabolic pathways.

When the same public metagenome data set is analysed through CAMERA and MG-RAST, different patterns of functional annotation are observed (Table 5.1). This can be explained by the fact that their analysis pipelines are differently organized. CAMERA allows the access to and use of individual tools and databases implemented in this analysis package. For example, CAMERA enables the export of quality-trimmed sequence data for further analysis using third-party applications, while MG-RAST does not. MG-RAST makes use of a single but comprehensive pipeline in which sequence data will be automatically processed from quality trimming to taxonomic assignment and functional annotation. As a consequence, taxonomic and functional classification via CAMERA made use of a single preselected database pipeline, while assignments obtained through MG-RAST represent consensus findings based on all the functional databases implemented in M5NR.

In summary, both platforms, CAMERA and MG-RAST, offer comprehensive toolkits

Table 5.1 Functional categorization of the Waseca farm soil metagenome (Tringe *et al.*, 2005) via MG-RAST and CAMERA

	KEGG			COG		
Categories	Methane metabolism	Nitrogen metabolism	Sulphur metabolism	Energy production	RNA processing	Cell cycle control, cell division
MG-RAST	678	294	84	5929	7	471
CAMERA	390	49	/	7341	48	956

for metatranscriptomic data analysis. However, the number of parameters that can be specified by the user is limited compared to a local bioinformatic pipeline. Moreover, the global patterns of taxonomic and functional assignments offered by CAMERA and MG-RAST as results output do not allow in-depth analysis of specific microbial groups.

Local bioinformatic pipeline

Local bioinformatic pipeline means data analysis using standalone tools on local computer resources. It is basically divided into BLAST search of metatranscriptomic data against a public protein database and parsing the BLAST output for taxonomic assignment and functional annotation. The bottleneck and most time-consuming step of the local pipeline is the sum of all the individual BLAST searches to find homologues of putative mRNA-tags. For example, on a dual-core desktop, it took almost a week to run BLASTX of about 7,500 putative mRNA-tags against the NCBI Reference Sequence (RefSeq) database containing 14,090,554 protein sequences originating from 16,609 different microorganisms (released in January 2012). In contrast, a cluster of 600 quad-core nodes finished the same BLASTX analysis in a half an hour. If sufficient computational power is available to perform BLAST searches massively in parallel, data analysis via local pipelines may require even less time than through CAMERA or MG-RAST. One of the great advantages of a local bioinformatic pipeline is that depending on the research objectives, any database of interest can be explored such as, for example, environmental sequence databases, protein databases of particular organisms or custom-made databases. In contrast to CAMERA and MG-RAST, the local pipeline makes it possible to access the complete list of BLAST sequence homologues of each putative mRNA-tag.

Thus, local pipelines offer flexibility and reliability in the analysis and data interpretation of BLAST outputs. Taxonomic and functional information can be extracted from the BLAST output using bioinformatic packages that are available in various programming languages, such as Biopython, Bioperl, BioJava, BioRuby, and Bioconductor [R]. Alternatively, the BLAST output may be further analysed using MEGAN (Huson *et al.*, 2011). For example, correct taxonomic assignment of mRNA-tags is difficult if they are derived from genes that are highly conserved (e.g. housekeeping genes) or distributed among many phylogenetically diverse organisms by lateral gene transfer. This is true, in particular, if the attempt is made to assign these mRNA-tags at the genus or even species level (Burke *et al.*, 2011). When CAMERA and MG-RAST is used for data analysis, taxonomic assignment of putative mRNA-tags is based simply on the best-matching homologue. By contrast, local bioinformatic pipelines allow to collectively analyse all the BLASTX hits of each mRNA-tag using the lowest common ancestor algorithm. In the data output, widely conserved sequences are placed at the node of a given phylum or class and thus close to the root, while species-specific sequences are positioned at the tip of the taxonomic scheme produced by MEGAN. This type of data analysis helps to avoid best match-derived bias as shown in Fig. 5.6.

Differential gene expression can be statistically detected based on the number of mRNA-tags which are mapped to reference genes. Various statistical tools have been developed using R packages, such as DESeq (Anders and Huber, 2010), EdgeR (Robinson and Oshlack, 2010), and baySeq (Hardcastle and Kelly, 2010). In metatranscriptomic analysis, the number of mRNA-tags, however, that are assigned to a certain reference gene is low, owing to the enormous complexity of microbial communities. Most functional assignments

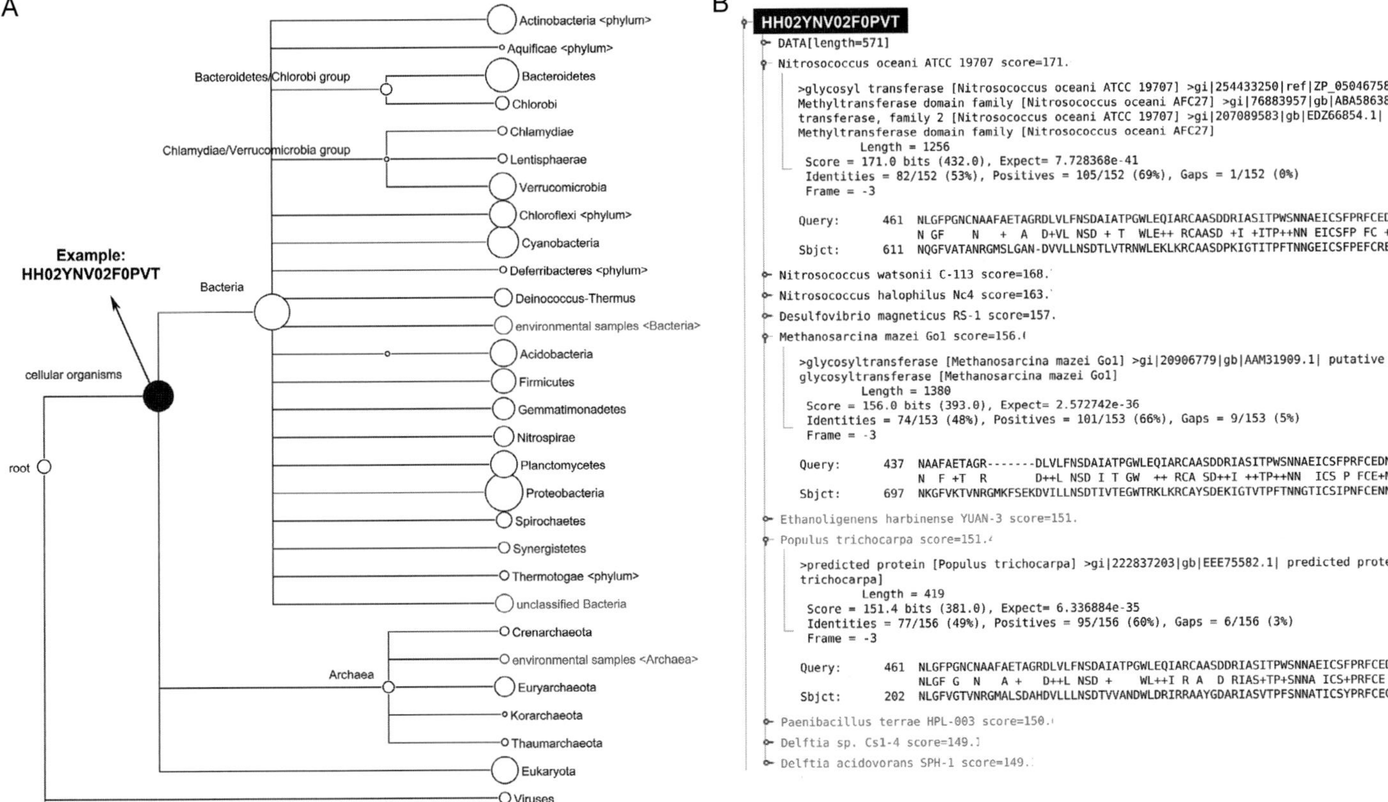

Figure 5.6 Taxonomic scheme produced by MEGAN4, exemplarily shown for a single mRNA-tag (HH02YNVO2FOPVT). This mRNA-tag was assigned at the level of 'cellular organisms' (A). BLASTX search against NCBI RefSeq database suggests a wide taxonomic distribution, including bacteria (e.g. *Nitrosococcus oceani*), archaea (*Methanosarcina mazei*), and eukarya (*Populus trichocarpa*). The sequence similarity scores among the top 10 hits (out of 100 matches) are nearly identical (B).

may even be represented by singletons, despite the fact that mRNA-tag data sets may comprise several ten or hundred thousand 454 reads. mRNA-tags that were assigned to different reference genes may, however, be grouped into the same functional category. Therefore, gene category overrepresentation analysis, which is widely used in systems biology, can be applied to detect biological processes that are expressed at different levels. SEED subsystems, KEGG pathways, and gene ontology are commonly used to categorize biological sequence information (Shi *et al.*, 2011; Stewart *et al.*, 2012).

Each functional category is examined whether the number of mRNA-tags assigned to it is overrepresented within a given sample or between samples. Metastat (White *et al.*, 2009) and STAMP (Parks and Beiko, 2010) are statistical software packages that allow to detect the functional categories which are related to different levels of expression among metatranscriptomic profiles. Among other, EasyGO (Zhou and Su, 2007), GOminer (Zeeberg *et al.*, 2003), GOstat (Beissbarth and Speed, 2004), GSEA (Subramanian *et al.*, 2005), and DAVID (Huang *et al.*, 2009) are statistical tools to perform gene ontology analyses.

Recent achievements in metatranscriptomics

To date, there are only a handful of studies in the field of soil metatranscriptomics (Table 5.2). These applied different methods to access the soil metatranscriptome. Two studies analysed global gene expression patterns of eukaryotic microorganisms, using conventional Sanger sequencing (Bailly *et al.*, 2007; Damon *et al.*, 2012). Polyadenylated eukaryotic mRNAs were enriched for cDNA synthesis by affinity capture on poly-dT-coated paramagnetic beads. The libraries were dominated by fungal cDNA with more than 50% of total sequences. The use of random or semi-random priming for synthesis of environmental cDNA primarily resulted in surveys of bacterial and archaeal soil metatranscriptomes (Urich *et al.*, 2008; Shrestha *et al.*, 2009; de Menezes *et al.*, 2012). In another two studies, soil metatranscriptome libraries were used in functional expression assays, resulting in the identification of previously unknown genes encoding an acid phosphatase and a novel family of fungal dipeptide transporters (Damon *et al.*, 2011; Kellner *et al.*, 2011).

Research to date in marine metatranscriptomics is focused on microbial assemblages in ocean waters. Throughout, GS 20 or GS FLX pyrosequencing was used to analyse cDNA libraries, resulting in average read lengths between 98 and 223 nucleotides (Table 5.3). Microbial assemblages in marine waters are less diverse than microbial communities in soil. Based on the Preston`s canonical hypothesis, mathematical models were developed that relate the total number of bacteria in a given sample to the number of bacteria in the least abundant species. Using these models, predictions were made that the ocean harbours 160 different bacterial species per millilitre, while 6,400 to 38,000 bacterial species are present per gram of soil. It was extrapolated that the ocean contains 2×10^6 different bacterial species, which can be found in half a ton of soil (Curtis *et al.*, 2002, 2006). Marine microbial diversity is dominated by some groups of bacteria, including *Pelagibacter, Roseobacter* and *Prochlorococcus* (Rappé *et al.*, 2000). *Pelagibacter ubique,* which is a member of the SAR11 clade in the *Alphaproteobacteria,* may even be the most abundant bacterial species on earth (Rappé *et al.*, 2002). During recent years, various international research initiatives were focused on comparative genomics of marine microbes, among those 'The Global Ocean Sampling Campaign' (Shaw *et al.*, 2008), 'The Tara Oceans Project' (Karsenti *et al.*, 2011), 'The Hawaiian Ocean Time Series Project' (DeLong *et al.*, 2006), and 'The Bermudan Ocean

Table 5.2 List of soil metatranscriptomic studies[a]

Study	Soil	Template molecules	mRNA enrichment	% of non-rRNA	Template amplification	cDNA synthesis	Sequencing platform	Average read length	Total number of reads
Bailly *et al.* (2007)	Forest	Eukaryotic mRNA	Poly-dT beads	100	RT-PCR	Oligo-dT primer	Sanger	Not indicated	119
Urich *et al.* (2008)	Sandy lawn	Total RNA	Not applied	8.2	Not applied	Random priming	GS 20	107	258,411
Shrestha *et al.* (2009)	Rice paddy	Enriched mRNA	SH	95	RT-PCR	SD14[b]	Sanger	534	805
Damon *et al.* (2012)	Forest	Eukaryotic mRNA	Poly-dT beads	92.8	RT-PCR	Oligo-dT primer	Sanger	Not indicated	20,000
de Mendenez *et al.* (2012)	Timber treatment facility	Enriched mRNA	SH	83.5	IVT	Random priming	GS FLX Titanium	361	534,668

IVT, *in vitro* transcription; RT-PCR, reverse-transcription PCR; SH, subtractive hybridization of rRNA.
[a]Two additional studies used soil metatranscriptomic libraries in functional assays (Damon *et al.*, 2011; Kellner *et al.*, 2011).
[b]SD14: oligonucleotide primer (14-mer) designed to target the Shine–Dalgarno sequence of bacterial mRNA transcripts (Fleming *et al.*, 1998).

Table 5.3 List of marine metatranscriptomic studies

References	Poretsky *et al.* (2005)	Frias-Lopez *et al.* (2008)	Gilbert *et al.* (2008)	Poretsky *et al.* (2009)	Shi *et al.* (2009)	Gilbert *et al.* (2010)	Poretsky *et al.* (2010)	Hewson *et al.* (2010)	Vila-Costa *et al.* (2010)	McCarren *et al.* (2010)	Stewart *et al.* (2011)	Mou *et al.* (2011)	Shi *et al.* (2011)
Study site	Ocean and lake	Ocean surface water	Coastal water	Ocean surface water	Open ocean	Ocean surface water	Coastal water	Open ocean	Ocean surface water	Ocean water microcosms	Marine oxygen minimum zone	Surface water microcosms	Ocean water microcosms
Template molecules	Enriched mRNA	Enriched mRNA	Enriched mRNA	Enriched mRNA	Enriched mRNA	Enriched mRNA	Enriched mRNA	Enriched mRNA	Enriched mRNA	Total RNA	Enriched mRNA	Enriched mRNA	Enriched mRNA
mRNA enrichment	SH	SH	SH	Exo[a] + SH	SH	SH	Exo[a] + SH	Exo[a] + SH	Exo[a] + SH	Not applied	SH	Exo[a] + SH	SH
% of non-rRNA	80	47.1	99.9	47.1	53.2	38.1	47.1	36.5	51	3.6	38.7	31.2	18-40
Template amplification	RT-PCR	IVT	MDA	IVT	IVT	IVT	IVT	IVT	IVT	IVT	IVT	IVT	IVT
cDNA synthesis	SD14[b]	Random priming	Random priming	Random priming	Random priming	Random priming	Random priming	Random priming	Random priming	Random priming	Random priming	Random priming	Random priming
Sequencing platform	Sanger	GS 20	GS 20	GS 20	GS 20	GS FLX	GS FLX	GS 20 GS FLX	GS FLX	GS FLX	GS FLX	GS FLX	GS FLX
Average read length (bp)	Not indicated	114	98	99	103	~300	184	168	209	~200	172	153	186
Total number of reads	282	128,324	506,353	240,422	388,738	1,030,617	1,576,184	1,174,917	606,286	2,945,424	1,562,754	597,735	3,149,768

IVT, *in vitro* transcription; MDA, multiple displacement amplification; RT-PCR: reverse-transcription PCR; SH, subtractive hybridization of rRNA.
[a]Exo: 5′-monophosphate dependent exonuclease treatment.
[b]SD14: oligonucleotide primer (14-mer) designed to target the Shine–Dalgarno sequence of bacterial mRNA transcripts (Fleming *et al.*, 1998).

Time Series Project' (Venter *et al.*, 2004). Their collective research efforts created a public database of representative genomes and metagenomes from marine environments. This may explain why, despite the short read lengths, a relatively large proportion of putative mRNA-tags could be functionally annotated with high significance in marine metatranscriptomic studies.

Soil metatranscriptomics

Fungi and bacteria are the two main decomposer groups in soil. They share the function of decomposing organic matter and are responsible for most nutrient transformations in soil, regenerating minerals that limit plant productivity (Six *et al.*, 2006; Thiet *et al.*, 2006; Rousk *et al.*, 2009). In oxic soils, the competitive (negative) interaction between fungal and bacterial growth is believed to be attributable to a combination of exploitation competition and direct interference of one group by the other (Rousk *et al.*, 2008; Rousk *et al.*, 2010a). One of the most influential factors that determines bacterial growth is soil pH. While both relative abundance and diversity of bacteria are positively related to pH, the relative abundance of fungi is not affected by pH and fungal diversity is only weakly related with pH (Rousk *et al.*, 2010a, 2010b). The direct effect of pH on bacterial community composition is presumably due to the narrow pH ranges for optimal growth of bacteria. In contrast, fungi generally have wider pH ranges for optimal growth than bacteria. This may explain why fungal growth, relative to bacterial growth, increases with decreasing pH (Rousk *et al.*, 2010b). In a pH gradient from 8.3 to 4.5, activity-based measurements showed an approximately 30-fold increase in fungal importance, as indicated by the fungal growth/bacterial growth ratio. The corresponding effects on biomass marker for fungi and bacteria revealed a two- to threefold difference in fungal importance in the same pH interval (Rousk *et al.*, 2009). Other environmental factors that affect fungal and bacterial growth differently include substrate quality, salinity, temperature, and metal toxicity. For example, barley straw was found to primarily promote fungal growth, while alfalfa especially stimulated bacterial growth. The differential growth response of fungi and bacteria was, at least to some extent, related to the different C/N ratios of straw (C/N = 15) and alfalfa (C/N = 75) (Rousk and Bååth, 2007).

The first soil metatranscriptome study focused on the development of a method to elucidate the functional diversity of eukaryotic microorganisms in forest soil (Bailly *et al.*, 2007). This method relied on the ability to specifically enrich polyadenylated eukaryotic mRNAs by affinity capture. Study site was a nutrient-poor, non-calcareous sand of pH 5.5. The phylogenetic diversity was analysed using 520-bp PCR fragments amplified from 18S rRNA genes and reverse-transcribed 18S rRNA. The majority of Sanger sequences (> 70%) were from fungi and unicellular eukaryotes (protists). Functional analysis of 119 mRNA-derived cDNA clones identified genes with no homologues in public databases (32%) and genes encoding hypothetical proteins of unknown function (35%). Among the cDNAs that could be assigned to functional categories, most were associated with housekeeping functions such as protein synthesis and post-translational modification. However, a few cDNA sequences were predicted to encode genes related to biogeochemical processes, such as cytochrome P450, phosphate transporter, and glutamine synthetase. Cytochrome P450 plays an important role in the breakdown of phenolics and xenobiotics, while phosphate transporter and glutamine synthetase are involved in the utilization of soil nutrients. These findings led to the conclusion that metatranscriptomics has the potential to uncover functional activities of complex soil microbial communities.

A more comprehensive metatranscriptomic analysis of soil eukaryotes was performed by Damon *et al.* (2012). The study was based on the assumption that the initial degradation of lignocellulosic materials in acidic forest soils is carried out primarily by fungi. The aim was to get a global view of their functional activities in soils sampled from beech (*Fagus sylvatica*) and spruce (*Picea abies*) forests. Both study sites were a sandy clay with a pH of 3.9. A total of 10,000 cDNAs were randomly analysed for each soil. Taxonomic affiliation of cDNAs and 18S rRNA revealed a dominance of sequences from fungi and metazoans, while protists represented less than 12% of the 18S rRNA sequences. Forty percent of cDNA sequences from beech forest soil and 50% from spruce forest soil had homologues in the NCBI non-redundant protein database. Most of the annotated cDNA sequences were predicted to encode proteins involved in cell maintenance metabolism, such as glycolysis, tricarboxylic acid cycle, and amino acid biosynthesis and degradation. Homologue searches in KEGG pathways and in specialized protein databases revealed the expression of enzymes implicated in soil ecosystem functioning and, more specifically, in the turnover of plant biomass and in soil nutrient cycling and utilization. Enzymes putatively involved in the degradation of plant cell wall components (cellulose, hemicelluloses, pectin, and lignin) represented 0.5% (beech soil) and 0.8% (spruce soil) of cDNA sequences. In addition, enzymes responsible for the breakdown of non-cell wall organic molecules were identified, such as different classes of proteases, a phytase, a cutinase and a putative carotenoid ester lipase. These were all highly similar to enzymes known to be excreted by fungi.

In another study, structure and function of a grassland soil microbial community were simultaneously characterized by GS 20 pyrosequencing of random-primed cDNA. Study site was a nutrient-poor sandy soil with a pH of 7.1. Starting material for cDNA synthesis was total RNA (Urich *et al.*, 2008). Of 258,411 RNA-tags, 193,219 rRNA-tags contained valid taxonomic information, while 21,133 tags were derived from mRNA (Table 5.1). The frequency with which bacteria and fungi were detected agrees well with the finding that bacteria have a competitive advantage over fungi at neutral pH (Rousk *et al.*, 2010b). A total of 85.5% of the rRNA-tags were derived from members of bacterial phyla, in particular from *Actinobacteria* (47,513 rRNA-tags) and *Proteobacteria* (38,477 rRNA-tags). Their members are typically present in high abundance in soil microbial communities. Eukaryotes accounted for 22,740 rRNA-tags (~ 11% of total reads), with more than 80% of the tags being taxonomically assigned at least at the kingdom level. The tags were affiliated to the kingdoms Fungi (~50%), plants (Viridiplantae, ~20%) and Metazoa (~10%). Among fungi, rRNA-tags related to members of the phylum *Ascomycota* were most numerous (~66%), followed by those of the phyla *Glomeromycota* and *Basidiomycota*. The crenarchaeal candidate division Group I.1b was identified as the predominant archaeal taxon. A total of 80 mRNA-tags were taxonomically affiliated with this group, thereby providing a glimpse into their *in situ* activity. While most of these mRNA-tags were associated with housekeeping functions, several cDNAs were identified to encode crenarchaeal *amoA* and *amoC*. High expression level of crenarchaeal ammonia monooxygenase was seen as further evidence that ammonia oxidation represents the main energy metabolism of crenarchaeal Group I.1b (Karner *et al.*, 2001; Leininger *et al.*, 2006). Various other crenarchaeal mRNA-tags were predicted to encode methyl-malonyl-CoA mutase and 4-hydroxybutyryl-CoA dehydratase. Both enzymes are indicative of a CO_2 fixation pathway characterized in hyperthermophilic crenarchaeota (Berg *et al.*, 2007). Members of the crenarchaeal Group I.1b were therefore assumed to have a similar CO_2 fixation pathway, suggesting that they are chemolithoautotrophs. When the total set of mRNA-tags was compared with metagenomic data obtained from the same soil, several functional categories in SEED subsystems displayed differential representation. Particularly, categories related to RNA and protein metabolism were significantly overrepresented in the metatranscriptomic data set, a finding characteristic of metabolically active populations. Subcategories in carbohydrate metabolism related to the aerobic degradation of mono-, di-, and oligosaccharides and of amino sugars were found to be underrepresented among the mRNA-tags. Meanwhile, the mesophilic ammonia-oxidizing crenarchaeota were reclassified as *Thaumarchaeota*, including Group I.1b (Brochier-Armanet *et al.*, 2011; Pester *et al.*, 2011). Their major role in soil ammonia oxidation was further evidenced (Abell *et al.*, 2010; Gubry-Rangin *et al.*, 2010). Although emerging genomic data are available for members of the *Thaumarchaeota*, it remains unclear whether these microorganisms are autotrophs or mixotrophs (Hallam *et al.*, 2006; Walker *et al.*, 2010).

Shrestha *et al.* (2009) compared the transcriptional activity of aerobic and anaerobic bacterial communities in a paddy soil oxygen gradient, using Sanger sequencing. Comparative analysis of 805 random cDNAs resulted in 179 (oxic zone) and 155 (anoxic zone) different cDNA clusters and singletons. These were defined as environmentally expressed sequence tags. A total of 116 cDNA sequences were predicted novel, while 218 sequence tags could be assigned by BLASTX. Their taxonomic affiliations reflected the dominance of bacteria over archaea and eukaryotes in flooded paddy soil. At the class level, mRNA sequences affiliated with *Alpha-* and *Betaproteobacteria* were more frequently detected in the oxic surface layer, while those assigned to *Deltaproteobacteria* were overrepresented in the anoxic soil. The taxonomic pattern of the mRNA transcripts was largely consistent with those of 16S rRNA and 16S rRNA gene sequences, for both the oxic and anoxic zones. This finding indicated that, at least at the phylum and subphylum levels, the taxonomic affiliation of mRNA transcripts collectively provides valuable information on community composition. The mRNA sequences were BLAST searched against the Conserved Domain database and classified into COG functional categories. A few deduced protein sequences were characteristic of community functions specifically expressed in the oxic zone, such as β-oxidation of long-chain fatty acids. However, significant differences in gene expression between the oxic and anoxic zones were not observed, most likely due to the fact that only a relatively low number of random mRNA sequences was analysed. Most of these were affiliated with housekeeping functions.

Microbial community response to phenanthrene was evaluated by metatranscriptomics (de Menezes *et al.*, 2012). Phenanthrene was degraded most rapidly between 5 and 10 days after amendment to soil. Samples for the extraction of total RNA were taken on day 7, at which 65% of the initially amended phenanthrene had been degraded. Total RNA was subjected to mRNA enrichment using MICROB*Express*™, followed by *in vitro* transcription and conversion of amplified RNA into random-primed cDNA. GS FLX Titanium pyrosequencing was used to obtain putative mRNA-tags from phenanthrene-amended soils and unamended control soils (35,728 and 52,728 reads, respectively). Phenanthrene led to a marked increase in transcripts associated with aromatic compound metabolism, respiration, and stress response (5.7-, 2-, and 1.5-fold, respectively, relative to the control). Concurrently the abundance of transcripts involved in virulence, carbohydrate metabolism, DNA metabolism and phosphorus metabolism decreased by 1.7-, 2.0-,

2.1-, and 3.1-fold, respectively. In particular, a 1.8- to 33-fold increase in the abundance of transcripts related to dioxygenases, stress response and detoxification was observed, among those transcripts for thioredoxin and P-type ATPases. Taxonomic assignment of the transcripts showed that actinobacteria were most responsible for mRNA de novo expression. The increase in transcript abundance of various dioxygenases reflected their central role in the initial stages of phenanthrene metabolism. The phenanthrene uptake mechanisms of bacterial cells are not well understood, but the high expression level of heavy metal P-type ATPases in phenanthrene-treated soil suggested a potential role of these proteins in active phenanthrene uptake. Heavy metal P-type ATPases are known to catalyse the translocation of heavy metal cations across membranes. They play a role in heavy metal detoxification (Rosen, 2002), as well as in acid stress response in *Lactobacillus* spp. (Penaud *et al.*, 2006). However, they have never been associated with phenanthrene metabolism in previous studies.

In addition to sequencing-based surveys, metatranscriptomic libraries were used for functional screening. A metatranscriptomic library obtained from eukaryotic microorganisms in a sugar maple forest soil was exploited to identify unknown fungal genes encoding an acid phosphatase (PHO5) and an imidazoleglycerol-phosphate dehydratase (HIS3) (Kellner *et al.*, 2011). The screening involved the construction of a yeast secretion vector carrying metatranscriptomic cDNA, followed by transformation into two knockout strains of *Saccharomyces cerevisiae* (*his3*$^-$ and *pho5*$^-$ mutations). The transformants were screened in two ways: (i) incubation on histidine-deficient minimal medium and (ii) by plating on low-P_i medium using an overlaid staining agar. Among 13,000 clones containing soil metatranscriptome inserts, one cDNA from each screening was found to complement the histidine auxotrophy of *S. cerevisiae* DSY-5 and the *pho5*$^-$ mutation in *S. cerevisiae* ML20C. By combining soil metatranscriptomics with functional screening, new enzymes were discovered by complementation of auxotrophic mutants. This approach has the potential to provide novel information on fungal metabolic activities in soil.

A novel family of fungal dipeptide transporters was identified using soil-extracted polyadenylated eukaryotic mRNA for the construction of an environmental cDNA expression library (Damon *et al.*, 2011). The library was transferred into W303 *S. cerevisiae* mutant strain defective in two known dipeptide transporter genes, *PTR2* and *DAL5*. The mutant was thus unable to utilize most dipeptides as a nitrogen source. The *S. cerevisiae* mutant was screened for functional complementation by plating transformed cells on a yeast nitrogen base minimal medium lacking uracil and containing the three dipeptides Tyr-Ala, Ala-Leu and Ala-Tyr as sole nitrogen sources. A total of 7.7×10^5 transformants were screened. Sequence analysis identified six different putative polypeptides among the 25 transformants that grew on the selective medium. The six environmental oligopeptide transporter proteins were homologous to each other and related to sequences in *Ascomycota* and *Basidiomycota* fungal species, but were not homologous to either Ptr2p or Dal5p. High-throughput phenotyping of yeast mutants expressing two different environmental transporters revealed that they both had broad substrate specificity and could transport more than 60–80 different dipeptides. Expression studies in *Xenopus* oocytes showed that one of the environmental transporters induced currents upon dipeptide addition, suggesting proton-coupled cotransport of dipeptides. The findings of this study led to the conclusion that the newly identified transporter family allows fungi to efficiently scavenge oligopeptides generated in soils by proteolysis.

Marine metatranscriptomics

Conventional Sanger sequencing of 282 cDNA clones was employed in the first marine metatranscriptomic study, resulting in the detection of a few transcripts associated with environmentally important processes such as sulphur oxidation, C1 carbon assimilation, and nitrogen fixation (Poretsky et al., 2005). All subsequent metatranscriptomic studies were based on 454 pyrosequencing of random-primed cDNA and involved a steady increase in the total number of 454 reads analysed (Table 5.3). A large number of novel highly expressed genes was frequently found in metatranscriptomic libraries derived from phytoplankton induced to bloom by the addition of nitrate and phosphate. Analysis of corresponding metagenomes confirmed much higher levels of assembly and overrepresentation of some gene families in the metatranscriptomic libraries (Gilbert et al., 2008). Metatranscriptomic data of a natural microbial community in ocean surface waters showed an ecotype distribution of *Prochlorococcus* consistent with the distribution pattern inferred from the analysis of the internal transcribed spacer region of 18S–26S nuclear ribosomal DNA. The relative gene expression levels determined by the frequency of pyrosequencing reads assigned to particular transcripts have proven to be comparable to abundance data obtained by qPCR (Frias-Lopez et al., 2008). In the metatranscriptomic studies by Gilbert *et al.* (2008) and Frias-Lopez *et al.* (2008), more than a half of the non-rRNA-tags were found to have no significant matches in current protein databases and were therefore classified as novel. Shi *et al.* (2009), however, discovered that a large fraction of these novel cDNA sequences were comprised of well-known small RNAs, as well as new groups of previously unrecognized putative small RNAs. These findings were achieved primarily by genome-specific mapping of the novel cDNAs. The mapping approach also revealed that depth-dependent variations in small RNAs were consistent with depth distributions of broad taxonomic groups, as previously reported by DeLong *et al.* (2006).

Marine metatranscriptomics has been widely used to investigate spatial and temporal patterns of microbial gene expression. Metatranscriptomes collected day and night from geographically distinct areas showed a great similarity in their metabolic functional profiles (74% on average), while the remaining portion of putative mRNA-tags was functionally and taxonomically unique to each sampling site (Hewson *et al.*, 2010). Cyanobacterial transcripts involved in photosynthesis, C1 metabolism and oxidative phosphorylation were overrepresented during daytime, while transcripts related to housekeeping activities, such as amino acid and membrane biosynthesis, were found to be overrepresented during night-time (Poretsky *et al.*, 2009). The first 'multi-omics' approach combined 16S rRNA amplicon sequencing with metagenomic and metatranscriptomic profiling. Eight samples were taken from a temperate coastal site (Western English Channel) in 2008, representing three seasons (winter, spring, summer) and covering day and night. Several conclusions were drawn from the bioinformatic analysis of the 'omics' data. Among these were that (i) higher 16S rRNA diversity also reflects a higher diversity of mRNA transcripts, (ii) community-level changes in both 16S rRNA-based diversity and metagenomic profiles are better explained by seasonal patterns, while those in metatranscriptomic profiles are better explained by diel patterns and shifts in particular functional categories of genes, (iii) changes in the expression of key genes related to photosynthesis occurred between seasons and between day and night, and (iv) the samples contained a substantial proportion of orphan sequences with unknown function and these appeared to contribute most to differences in the microbial gene expression patterns between samples (Gilbert *et al.*, 2010). Metatranscriptomic analyses along the depth profile of a permanent marine oxygen minimum zone revealed that transcripts affiliated with the ammonia-oxidizing archaeon *Nitrosopumilus maritimus* (aerobic nitrification) were most abundant at a depth of 85 m. In contrast, transcripts matching the anammox bacterium *Kuenenia stuttgartiensis* (anaerobic nitrification) dominated at the core of the oxygen minimum zone in a depth of 200 m (Stewart *et al.*, 2012).

Another set of studies examined the transcriptional response of marine microbial assemblages to nutrient amendments. In response to dissolved organic matter, transcripts encoding transporters and those associated with two-component sensor system and the assimilation of phosphate and nitrogen were significantly enriched relative to the control. The transcripts related to phosphate and nitrogen assimilation matched various taxonomic groups, but in particular *Idiomarina* and *Alteromonas* spp. (McCarren *et al.*, 2010; Poretsky *et al.*, 2010). The addition of dimethylsulphoniopropionate (DMSP) to ocean surface waters triggered an increase in mRNA transcripts derived from *Gammaproteobacteria* and *Bacteroidetes*, while there was little contribution of bacterioplankton groups, *Roseobacter* and SAR11, which are known to harbour DMSP degradation genes (Vila-Costa *et al.*, 2010). The treatment with DMSP led to an increase in transcripts involved in heterotrophic activity and the degradation of C_3 compounds, reflecting the metabolism of DMSP. The polyamines putrescine and spermidine are important nitrogen sources that are ubiquitously present in seawater. Their addition to surface waters stimulated the expression of genes encoding the cellular translation machinery and the metabolism of organic nitrogen and carbon (Mou *et al.*, 2011). Of the three known pathways for polyamine degradation in bacteria, the transamination pathway was the only one to be up-regulated in response to the amendment of putrescine and spermidine. Taxonomic assignment of significantly enriched transcripts suggested that *Roseobacter*- and SAR11-affiliated bacteria were the predominant taxa driving the transformation of polyamines in coastal ocean water. The study by Shi *et al.* (2012) simulated deep-water nutrient injection in the North Pacific Subtropical Gyre. These injections represent an aperiodic yet significant source of inorganic nutrients to surface water picoplankton assemblages, which has implications for phytoplankton bloom formation. The mixing of deep seawater collected from 700 m depth and nutrient-limiting surface waters stimulated the transcriptional activity of an *Alteromonas*-like population within 12 h, relative to the control. In particular, genes related to chemotaxis, cell motility and carbon metabolism were most highly expressed. The expression patterns of the *Alteromonas* populations responsive to deep sea water were different from those reported for dissolved organic matter-responding alteromonads (McCarren *et al.*, 2010), suggesting a perturbation-specific metabolic response. In addition to *Alteromonas*, *Prochlorococcus* responded to deep sea water. Significantly higher levels of transcripts associated with carbon fixation and photosynthesis were observed, relative to the control.

Conclusions and outlook

Studies on global gene expression of soil microbial communities are still in their infancy, particularly in comparison to marine metatranscriptomics. This may have various reasons.

One of these is that the extraction of RNA of sufficient purity and integrity for use in soil metatranscriptomics remains challenging. There is no single method that can be efficiently applied to all different types of soil. In particular, the adsorption of RNA to soil high in clay content is a problem not yet solved (Wang *et al.*, 2012). Obviously, the methods developed to extract total RNA and enriched mRNA need to be further improved. This is true, in particular, for the isolation of enriched mRNA from soil bacterial and archaeal communities. Eukaryotic mRNA can be selectively isolated by affinity capture.

Another reason is that soil microbial communities are among the most diverse and species-rich assemblages known (Curtis *et al.*, 2002, 2006; Torsvik *et al.*, 2002; Gans *et al.*, 2005). In addition, soil microbial diversity may greatly vary at a small spatial scale in response to changes in the physico-chemical characteristics of soil. This makes it difficult to create an adequate database of representative genomes and metagenomes for transcript mapping, similar to that achieved for microbial assemblages in marine waters.

The lack of a representative database of well-annotated genomes and metagenomes implicates that only 20–40% of total reads in soil metatranscriptome data sets may be assigned to known functions, even after effective enrichment of mRNA prior to cDNA synthesis. The majority of random reads are putative mRNA-tags that have no homologues of known function or remaining rRNA-tags. In particular, the proportion of mRNA-tags that can be assigned a putative function is low in the metatranscriptomic analysis of anaerobic soil microbial communities (Y. Kim and W. Liesack, unpublished results).

Among those mRNA-tags that are assigned a putative function, only a minor fraction (< 15%) provides information on genes that are expressed in response to particular environmental conditions or cues. The vast majority of mRNA-tags (> 85%) are associated with housekeeping functions such as transcription and translation (Stewart *et al.*, 2010). This explains the need for deep sequencing of each cDNA library. Otherwise, it is not possible to detect differential gene expression in response to environmental change with sufficient significance. For example, transcripts associated with heavy metal P-type ATPases and thioredoxin were reported to show the greatest changes in abundance between unamended control and phenanthren-amended soils (de Menezes *et al.*, 2012). Among approximately 80,000 non-rRNA-tags analysed, the absolute transcript numbers were as follows (control vs. phenanthren-amended soil): 46 versus 122 (heavy metal P-type ATPases) and 22 versus 77 (thioredoxin). This situation calls for new experimental strategies, making it possible to increase sequencing depth of each sample and to allow for proper replication. One strategy may be to combine 454 pyrosequencing with metatranscriptomic analysis by Illumina deep sequencing. Mapping hundred of thousands or even millions of short Illumina reads against the metatranscriptomic dataset obtained by 454 pyrosequencing may greatly increase the statistical significance of changes observed in transcript abundance between samples. Further progress in sequencing technologies may combine sufficient read length with increased sequencing depth at affordable costs. Candidates are the next-generation sequencing platforms offered by Illumina and Ion Torrent. An alternative strategy may be to combine mRNA-SIP with metatranscriptomic analysis. The advent of mRNA-SIP provides a direct mechanism to enrich for transcripts that are expressed during the incubation with particular labelled carbon sources (Dumont *et al.*, 2011; Jansson *et al.*, 2012).

Acknowledgements

This work was supported by the Max Planck Society and the Deutsche Forschungsgemeinschaft (SFB 987). CEW is a member of the International Max Planck Research School for Environmental, Cellular and Molecular Microbiology in Marburg/Lahn, Germany.

References

Abell, G.C.J., Revill, A.T., Smith, C., Bissett, A.P., Volkman, J.K., and Robert, S.S. (2010). Archaeal ammonia oxidizers and *nirS*-type denitrifiers dominate sediment nitrifying and denitrifying populations in a subtropical macrotidal estuary. ISME J. *4*, 286–300.

Altschul, S.F., Madden, T.L., Schaffer, A.A., Zhang, J., Zhang, Z., Miller, W., and Lipman, D.J. (1997). Gapped BLAST and PSI-BLAST: a new generation of protein database search programs. Nucleic Acids Res. *25*, 3389–3402.

Anders, S., and Huber, W. (2010). Differential expression analysis for sequence count data. Genome Biol. *11*, R106.

Bailly, J., Fraissinet-Tachet, L., Verner, M.-C., Debaud, J.-C., Lemaire, M., Wésolowski-Louvel, M., and Marmeisse, R. (2007). Soil eukaryotic functional diversity, a metatranscriptomic approach. ISME J. *1*, 632–642.

Balzer, S., Malde, K., and Jonassen, I. (2011). Systematic exploration of error sources in pyrosequencing flowgram data. Bioinformatics *27*, i304–309.

Beissbarth, T., and Speed, T.P. (2004). GOstat: find statistically overrepresented Gene Ontologies within a group of genes. Bioinformatics *20*, 1464–1465.

Berg, I.A., Kockelkorn, D., Buckel, W., and Fuchs, G. (2007). A 3-hyroxypropionate/4-hydroxybutyrate autotrophic carbon dioxide assimilation pathway in archaea. Science *318*, 1782–1786.

Brochier-Armanet, C., Gribaldo, S., and Forterre, P. (2011). Spotlight on the thaumarchaeota. ISME J. *6*, 227–230.

Burke, C., Steinberg, P., Rusch, D., Kjelleberg, S., and Thomas, T. (2011). Bacterial community assembly based on functional genes rather than species. Proc. Natl. Acad. Sci. U.S.A. *108*, 14288–14293.

Bürgmann, H., Widmer, F., Sigler, W.V., and Bu, H. (2003). mRNA extraction and reverse transcription-PCR protocol for detection of *nifH* gene expression by *Azotobacter vinelandii*. Appl. Environ. Microbiol. *69*, 1928–1935.

Caporaso, J.G., Kuczynski, J., Stombaugh, J., Bittinger, K., Bushman, F.D., Costello, E.K., Fierer, N., Peña, A.G., Goodrich, J.K., Gordon, J.I., *et al.* (2010). Correspondence: QIIME allows analysis of high-throughput community sequencing data Intensity normalization improves color calling in SOLiD sequencing. Nature 7, 335–336.

Caspi, R., Foerster, H., Fulcher, C.A., Kaipa, P., Krummenacker, M., Latendresse, M., Paley, S., Rhee, S.Y., Shearer, A.G., Tissier, C., Walk, T.C., Zhang, P., and Karp, P.D. (2008). The MetaCyc database of metabolic pathways and enzymes and the BioCyc collection of pathway/genome databases. Nucleic Acids Res. *36*, D623–D631.

Celesnik, H., Deana, A., and Belasco, J.G. (2007). Initiation of RNA decay in *Escherichia coli* by 5′ pyrophosphate removal. Mol. Cell *27*, 79–90.

Cole, J.R., Wang, Q., Cardenas, E., Fish, J., Chai, B., Farris, R.J., Kulam-Syed-Mohideen, A.S., McGarrell, D.M., Marsh, T., Garrity, G.M., and Tiedje, J.M. (2009). The Ribosomal Database Project: improved alignments and new tools for rRNA analysis. Nucleic Acids Res. *37*, D141–145.

Curtis, T.P., Sloan, W.T., and Scannell, J.W. (2002). Estimating prokaryotic diversity and its limits. Proc. Natl. Acad. Sci. U.S.A. *99*, 10494–10499.

Curtis, T.P., Head, I.M., Lunn, M., Woodcock, S., Schloss, P.D., and Sloan, W.T. (2006). What is the extent of prokaryotic diversity? Philos. Trans. R. Soc. Lond., B. Biol. Sci. *361*, 2023–2037.

Damon, C., Vallon, L., Zimmermann, S., Haider, M.Z., Galeote, V., Dequin, S., Luis, P., Fraissinet-Tachet, L., and Marmeisse, R. (2011). A novel fungal family of oligopeptide transporters identified by functional metatranscriptomics of soil eukaryotes. ISME J. *5*, 1871–1880.

Damon, C., Lehembre, F., Oger-Desfeux, C., Luis, P., Ranger, J., Fraissinet-Tachet, L., and Marmeisse, R. (2012). Metatranscriptomics reveals the diversity of genes expressed by eukaryotes in forest soils. PloS One *7*, e28967.

Deangelis, K.M., and Firestone, M.K. (2012). Phylogenetic clustering of soil microbial communities in 16S rRNA but not 16S rRNA genes. Appl. Environ. Microbiol. *78*, 2459–2461.

DeLong, E.F., Preston, C.M., Mincer, T., Rich, V., Hallam, S.J., Frigaard, N.U., Martinez, A., Sullivan, M.B., Edwards, R., Brito, B.R., Chisholm, S.W., and Karl, D.M. (2006). Community genomics among stratified microbial assemblages in the ocean's interior. Science *311*, 496–503.

de Menezes, A., Clipson, N., and Doyle, E. (2012). Comparative metatranscriptomics reveals widespread community responses during phenanthrene degradation in soil. Environ. Microbiol. *14*, 2577–2588.

DeSantis, T.Z., Hugenholtz, P., Larsen, N., Rojas, M., Brodie, E.L., Keller, K., Huber, T., Dalevi, D., Hu, P., and Andersen, G.L. (2006). Greengenes, a chimera-checked 16S rRNA gene database and workbench compatible with ARB. Appl. Environ. Microbiol. *72*, 5069–5072.

Dumont, M.G., Pommerenke, B., Casper, P., and Conrad, R. (2011). DNA-, rRNA-, and mRNA-based stable isotope probing of aerobic methanotrophs in lake sediment. Environ. Microbiol. *13*, 1153–1167.

Felske, A., Engelen, B., Nübel, U., Backhaus, H., Felske, A., Engelen, B., Nu, U., and Backhaus, H. (1996). Direct ribosome isolation from soil to extract bacterial rRNA for community analysis. Appl. Environ. Microbiol. *62*, 4162–4167.

Field, D., Garrity, G., Gray, T., Morrison, N., Selengut, J., Sterk, P., Tatusova, T., Thomson, N., Allen, M.J., Angiuoli, S.V., *et al.* (2008). The minimum information about a genome sequence (MIGS) specification. Nat. Biotechnol. *26*, 541–547.

Fleming, J.T., Yao, W.H., and Sayler, G.S. (1998). Optimization of differential display of prokaryotic mRNA: application to pure culture and soil microcosms. Appl. Environ. Microbiol. *64*, 3698–3706.

Frias-Lopez, J., Bonheyo, G.T., and Fouke, B.W. (2004). Identification of differential gene expression in bacteria associated with coral black band disease by using RNA-arbitrarily primed PCR. Appl. Environ. Microbiol. *70*, 3687–3694.

Frias-Lopez, J., Shi, Y., Tyson, G.W., Coleman, M.L., Schuster, S.C., Chisholm, S.W., and Delong, E.F. (2008). Microbial community gene expression in ocean surface waters. Proc. Natl. Acad. Sci. U.S.A. *105*, 3805–3810.

Gans, J., Wolinsky, M., and Dunbar, J. (2005). Computational improvements reveal great bacterial diversity and high metal toxicity in soil. Science *309*, 1387–1390.

Gardner, P.P., Daub, J., Tate, J., Moore, B.L., Osuch, I.H., Griffiths-Jones, S., Finn, R.D., Nawrocki, E.P., Kolbe, D.L., Eddy, S.R., and Bateman, A. (2011). Rfam: wikipedia, clans and the 'decimal' release. Nucleic Acids Res. *39*, D141–D145.

Giannoukos, G., Ciulla, D.M., Huang, K., Haas, B.J., Izard, J., Levin, J.Z., Livny, J., Earl, A.M., Gevers, D., Ward, D.V., Nusbaum, C., Birren, B.W., and Gnirke, A. (2012). Efficient and robust RNA-seq process for cultured bacteria and complex community transcriptomes. Genome Biol. *13*, R23.

Gilbert, J.A., Field, D., Huang, Y., Edwards, R., Li, W., Gilna, P., and Joint, I. (2008). Detection of large numbers of novel sequences in the metatranscriptomes of complex marine microbial communities. PloS One *3*, e3042.

Gilbert, J.A., Meyer, F., Schriml, L., Joint, I.R., Mühling, M., and Field, D. (2010). Metagenomes and metatranscriptomes from the L4 long-term coastal monitoring station in the Western English Channel. Stand. Genomic Sci. *3*, 183–193.

Griffiths, R.I., Whiteley, A.S., Anthony, G., Donnell, O., Bailey, M.J., and Donnell, A.G.O. (2000). Rapid method for coextraction of DNA and RNA from natural environments for analysis of ribosomal DNA- and rRNA-Based microbial community composition. Appl. Environ. Microbiol. *66*, 5488–5491.

Gubry-Rangin, C., Nicol, G.W., and Prosser, J.I. (2010). Archaea rather than bacteria control nitrification in two agricultural acidic soils. FEMS Microbiol. Ecol. *74*, 566–574.

Hallam, S.J., Konstantinidis, K.T., Putnam, N., Schleper, C., Watanabe, Y., Torre, D., Richardson, P.M., Delong, E.F., Sugahara, J., and Preston, C. (2006). Genomic analysis of the uncultivated marine crenarchaeote *Cenarchaeum symbiosum*. Proc. Natl. Acad. Sci. U.S.A. *103*, 18296–18301.

Hardcastle, T.J., and Kelly, K.A. (2010). baySeq: Empirical Bayesian methods for identifying differential expression in sequence count data. BMC Bioinformatics *11*, 422.

He, S., Wurtzel, O., Singh, K., Froula, J.L., Yilmaz, S., Tringe, S.G., Wang, Z., Chen, F., Lindquist, E.A., Sorek, R., and Hugenholtz, P. (2010). Validation of two ribosomal RNA removal methods for microbial metatranscriptomics. Nat. Meth. *7*, 807–812.

Hewson, I., Poretsky, R.S., Tripp, H.J., Montoya, J.P., and Zehr, J.P. (2010). Spatial patterns and light-driven variation of microbial population gene expression in surface waters of the oligotrophic open ocean. Environ. Microbiol. *12*, 1940–1956.

Holmes, D.E., Nevin, K.P., and Lovley, D.R. (2004). *In situ* expression of *nifD* in *Geobacteraceae* in subsurface sediments. Appl. Environ. Microbiol. *70*, 7251–7259.

Huang, D.W., Sherman, B.T., and Lempicki, R. (2009). Systematic and integrative analysis of large gene lists using DAVID bioinformatics resources. Nat. Prot. *4*, 44–57.

Huang, X. (1999). CAP3: A DNA sequence assembly program. Genome Res. *9*, 868–877.

Hurt, R.A., Qiu, X., Wu, L., Roh, Y.U.L., Palumbo, A.V., Tiedje, J.M., and Zhou, J. (2001). Simultaneous recovery of RNA and DNA from soils and sediments. Appl. Environ. Microbiol. *67*, 4495–4503.

Huson, D.H., Mitra, S., Ruscheweyh, H.-J., Weber, N., and Schuster, S.C. (2011). Integrative analysis of environmental sequences using MEGAN4. Genome Res. *21,* 1552–1560.

Jansson, J.K., Neufeld, J.D., Moran, M.A., and Gilbert, J.A. (2012). Omics for understanding microbial functional dynamics. Environ. Microbiol. *14,* 1–3.

Kanehisa, M., Goto, S., Sato, Y., Furumichi, M., and Tanabe, M. (2011). KEGG for integration and interpretation of large-scale molecular data sets. Nucleic Acids Res. *40,* D109-D114.

Karner, M.B., DeLong, E.F., and Karl, D.M. (2001). Archaeal dominance in the mesopelagic zone of the Pacific Ocean. Nature *409,* 507–510.

Karsenti, E., Acinas, S.G., Bork, P., Bowler, C., De Vargas, C., Raes, J., Sullivan, M., Arendt, D., Benzoni, F., Claverie, *et al.* and the *Tara* Oceans Consortium (2011). A holistic approach to marine eco-systems biology. PLoS Biol. *9,* e1001177.

Kellner, H., Luis, P., Portetelle, D., and Vandenbol, M. (2011). Screening of a soil metatranscriptomic library by functional complementation of *Saccharomyces cerevisiae* mutants. Microbiol. Res. *166,* 360–368.

Kent, W.J. (2002). BLAT-The BLAST-like alignment tool. Genome Res. *12,* 656–664.

Klein, F., Samorski, R., Klug, G., and Evguenieva-Hackenberg, E. (2002). Atypical processing in domain III of 23S rRNA of *Rhizobium leguminosarum* ATCC 10004T at a position homologous to an rRNA fragmentation site in protozoa. J. Bacteriol. *184,* 3176–3185.

Leininger, S., Urich, T., Schloter, M., Schwark, L., Qi, J., Nicol, G.W., Prosser, J.I., Schuster, S.C., and Schleper, C. (2006). Archaea predominate among ammonia-oxidizing prokaryotes in soils. Nature *442,* 806–809.

Malde, K. (2011). Flower: extracting information from pyrosequencing data. Bioinformatics *27,* 1041–1042.

McCarren, J., Becker, J.W., Repeta, D.J., Shi, Y., Young, C.R., Malmstrom, R.R., Chisholm, S.W., and DeLong, E.F. (2010). Microbial community transcriptomes reveal microbes and metabolic pathways associated with dissolved organic matter turnover in the sea. Proc. Natl. Acad. Sci. U.S.A. *107,* 16420–16427.

McDonald, D., Price, M.N., Goodrich, J., Nawrocki, E.P., Desantis, T.Z., Probst, A., Andersen, G.L., Knight, R., and Hugenholtz, P. (2012). An improved Greengenes taxonomy with explicit ranks for ecological and evolutionary analyses of bacteria and archaea. ISME J. *6,* 610–618.

McGrath, K.C., Thomas-Hall, S.R., Cheng, C.T., Leo, L., Alexa, A., Schmidt, S., and Schenk, P.M. (2008). Isolation and analysis of mRNA from environmental microbial communities. J. Microbiol. Methods *75,* 172–176.

Mettel, C., Kim, Y., Shrestha, P.M., and Liesack, W. (2010). Extraction of mRNA from soil. Appl. Environ. Microbiol. *76,* 5995–6000.

Meyer, F., Paarmann, D., D'Souza, M., Olson, R., Glass, E.M., Kubal, M., Paczian, T., Rodriguez, A., Stevens, R., Wilke, A., Wilkening, J., and Edwards, R.A. (2008). The metagenomics RAST server – a public resource for the automatic phylogenetic and functional analysis of metagenomes. BMC Bioinformatics *9,* 386.

Meyer, H., Kaiser, C., Biasi, C., Hämmerle, R., Rusalimova, O., Lashchinsky, N., Baranyi, C., Daims, H., Barsukov, P., and Richter, A. (2006). Soil carbon and nitrogen dynamics along a latitudinal transect in Western Siberia, Russia. Biogeochemistry *81,* 239–252.

Mou, X., Vila-Costa, M., Sun, S., Zhao, W., Sharma, S., and Moran, M.A. (2011). Metatranscriptomic signature of exogenous polyamine utilization by coastal bacterioplankton. Environ. Microbiol. Rep. *3,* 798–806.

Mount, D.W. (2001). Bioinformatics: Sequence and Genome Analysis. Cold Spring Harbor Laboratory Press.

Mutter, G.L., Zahrieh, D., Liu, C., Neuberg, D., Finkelstein, D., Baker, H.E., and Warrington, J. (2004). Comparison of frozen and RNALater solid tissue storage methods for use in RNA expression microarrays. BMC Genomics *5,* 88.

Nawrocki, E.P., Kolbe, D.L., and Eddy, S.R. (2009). Infernal 1.0: inference of RNA alignments. Bioinformatics *25,* 1335–1337.

Noonberg, S.B., Scott, G.K., and Benz, C.C. (1995). Effect of pH on RNA degradation during guanidinium extraction. Biotechniques *19,* 731–733.

Ogata, H., Goto, S., Sato, K., Fujibuchi, W., Bono, H., and Kanehisa, M. (1999). KEGG: Kyoto encyclopedia of genes and genomes. Nucleic Acids Res. *27,* 29–34.

Overbeek, R., Begley, T., Butler, R.M., Choudhuri, J.V., Chuang, H.-Y., Cohoon, M., Crécy-Lagard, V., Diaz, N., Disz, T., Edwards, R., *et al.* (2005). The subsystems approach to genome annotation and its use in the project to annotate 1000 genomes. Nucleic Acids Res. *33,* 5691–5702.

Parks, D.H., and Beiko, R.G. (2010). Identifying biologically relevant differences between metagenomic communities. Bioinformatics *26,* 715–721.

Patel, R.K., and Jain, M. (2012). NGS QC Toolkit: a toolkit for quality control of next generation sequencing data. PloS One *7*, e30619.

Penaud, S., Fernandez, A., Boudebbouze, S., Ehrlich, S.D., Maguin, E., and van de Guchte, M. (2006). Induction of heavy-metal-transporting CPX-type ATPases during acid adaptation in Lactobacillus bulgaricus. Appl. Environ. Microbiol. *72*, 7445–7454.

Persoh, D., Theuerl, S., Buscot, F., and Rambold, G. (2008). Towards a universally adaptable method for quantitative extraction of high-purity nucleic acids from soil. J. Microbiol. Methods *75*, 19–24.

Pester, M., Schleper, C., and Wagner, M. (2011). The Thaumarchaeota: an emerging view of their phylogeny and ecophysiology. Curr. Opin. Microbiol. *14*, 300–306.

Poretsky, R.S., Bano, N., Buchan, A., Kleikemper, J., Pickering, M., Pate, W.M., Moran, M.A., Hollibaugh, J.T., and Lecleir, G. (2005). Analysis of microbial gene transcripts in environmental samples. Appl. Environ. Microbiol. *71*, 4121–4126.

Poretsky, R.S., Hewson, I., Sun, S., Allen, A.E., Zehr, J.P., and Moran, M.A. (2009). Comparative day/night metatranscriptomic analysis of microbial communities in the North Pacific subtropical gyre. Environ. Microbiol. *11*, 1358–1375.

Poretsky, R.S., Sun, S., Mou, X., and Moran, M.A. (2010). Transporter genes expressed by coastal bacterioplankton in response to dissolved organic carbon. Environ. Microbiol. *12*, 616–627.

Pruesse, E., Quast, C., Knittel, K., Fuchs, B.M., Ludwig, W., Peplies, J., and Glöckner, F.O. (2007). SILVA: a comprehensive online resource for quality checked and aligned ribosomal RNA sequence data compatible with ARB. Nucleic Acids Res. *35*, 7188–7196.

Radax, R., Rattei, T., Lanzen, A., Bayer, C., Rapp, H.T., Urich, T., and Schleper, C. (2012). Metatranscriptomics of the marine sponge *Geodia barretti*: tackling phylogeny and function of its microbial community. Environ. Microbiol. *14*, 1308–1324.

Rappé, M.S., Vergin, K., and Giovannoni, S.J. (2000). Phylogenetic comparisons of a coastal bacterioplankton community with its counterparts in open ocean and freshwater systems. FEMS Microbiol. Ecol. *33*, 219–232.

Rappé, M.S., Connon, S.A., Vergin, K.L., and Giovannoni, S.J. (2002). Cultivation of the ubiquitous SAR11 marine bacterioplankton clade. Nature *418*, 630–633.

Rehman, A., Lepage, P., Nolte, A., Hellmig, S., Schreiber, S., and Ott, S.J. (2010). Transcriptional activity of the dominant gut mucosal microbiota in chronic inflammatory bowel disease patients. J. Med. Microbiol. *59*, 1114–1122.

Rho, M., Tang, H., and Ye, Y. (2010). FragGeneScan: predicting genes in short and error-prone reads. Nucleic Acids Res. *38*, e191.

Ricke, P., Erkel, C., Kube, M., Liesack, W., and Reinhardt, R. (2004). Comparative analysis of the conventional and novel *pmo* (particulate methane monooxygenase) operons from *Methylocystis* strain SC2. Appl. Environ. Microbiol. *70*, 3055–3063.

Robinson, M.D., and Oshlack, A. (2010). A scaling normalization method for differential expression analysis of RNA-seq data. Genome Biol. *11*, R25.

Ronaghi, M., Karamohamed, S., Pettersson, B., Uhlén, M., and Nyrén, P. (1996). Real-time DNA sequencing using detection of pyrophosphate release. Anal. Biochem. *242*, 84–89.

Rosen, B.P. (2002). Transport and detoxification systems for transition metals, heavy metals and metalloids in eukaryotic and prokaryotic microbes. Comp. Biochem. Physiol. A. *133*, 689–693.

Rousk, J., and Bååth, E. (2007). Fungal and bacterial growth in soil with plant materials of different C/N ratios. FEMS Microbiol. Ecol. *62*, 258–267.

Rousk, J., Aldén Demoling, L., Bahr, A., and Bååth, E. (2008). Examining the fungal and bacterial niche overlap using selective inhibitors in soil. FEMS Microbiol. Ecol. *63*, 350–358.

Rousk, J., Brookes, P.C., and Bååth, E. (2009). Contrasting soil pH effects on fungal and bacterial growth suggests functional redundancy in carbon mineralization. Appl. Environ. Microbiol. *75*, 1589–1596.

Rousk, J., Brookes, P.C., and Bååth, E. (2010a). Investigating the mechanisms for the opposing pH relationships of fungal and bacterial growth in soil. Soil Biol. Biochem. *42*, 926–934.

Rousk, J., Bååth, E., Brookes, P.C., Lauber, C.L., Lozupone, C., Caporaso, J.G., Knight, R., and Fierer, N. (2010b). Soil bacterial and fungal communities across a pH gradient in an arable soil. ISME J. *4*, 1340–1351.

Selenska-Pobell, S., and Evguenieva-Hackenberg, E. (1995). Fragmentations of the large-subunit rRNA in the family *Rhizobiaceae*. Microbiology *177*, 6993–6998.

Schloss, P.D. (2009). A high-throughput DNA sequence aligner for microbial ecology studies. PloS One *4*, e8230.

Schmieder, R., and Edwards, R. (2011). Quality control and preprocessing of metagenomic datasets. Bioinformatics *27*, 863–864.

Seshadri, R., Kravitz, S.A., Smarr, L., Gilna, P., and Frazier, M. (2007). CAMERA: a community resource for metagenomics. PLoS Biol. *5*, e75.

Sessitsch, A., Gyamfi, S., Stralis-Pavese, N., Weilharter, A., and Pfeifer, U. (2002). RNA isolation from soil for bacterial community and functional analysis: evaluation of different extraction and soil conservation protocols. J. Microbiol. Methods *51*, 171–179.

Shaw, A.K., Halpern, A.L., Beeson, K., Tran, B., Venter, J.C., and Martiny, J.B.H. (2008). It's all relative: ranking the diversity of aquatic bacterial communities. Environ. Microbiol. *10*, 2200–2210.

Shi, Y., Tyson, G.W., and DeLong, E.F. (2009). Metatranscriptomics reveals unique microbial small RNAs in the ocean's water column. Nature *459*, 266–269.

Shi, Y., Tyson, G.W., Eppley, J.M., and DeLong, E.F. (2011). Integrated metatranscriptomic and metagenomic analyses of stratified microbial assemblages in the open ocean. ISME J. *5*, 999–1013.

Shi, Y., McCarren, J., and DeLong, E.F. (2012). Transcriptional responses of surface water marine microbial assemblages to deep-sea water amendment. Environ. Microbiol. *14*, 191–206.

Six, J., Frey, S.D., Thiet, R.K., and Batten, K.M. (2006). Bacterial and fungal contributions to carbon sequestration in agroecosystems. Soil Sci. Soc. Am. J. *70*, 555–569.

Shrestha, P.M., Kube, M., Reinhardt, R., and Liesack, W. (2009). Transcriptional activity of paddy soil bacterial communities. Environ. Microbiol. *11*, 960–970.

Stewart, F.J., Ottesen, E.A., and DeLong, E.F. (2010). Development and quantitative analyses of a universal rRNA-subtraction protocol for microbial metatranscriptomics. ISME J. *4*, 896–907.

Stewart, F.J., Sharma, A.K., Bryant, J.A., Eppley, J.M., and DeLong, E.F. (2011). Community transcriptomics reveals universal patterns of protein sequence conservation in natural microbial communities. Genome Biol. *12*, R26.

Stewart, F.J., Ulloa, O., and DeLong, E.F. (2012). Microbial metatranscriptomics in a permanent marine oxygen minimum zone. Environ. Microbiol. *14*, 23–40.

Subramanian, A., Tamayo, P., Mootha, V.K., Mukherjee, S., Ebert, B.L., Gillette, M.A., Paulovich, A., Pomeroy, S.L., Golub, T.R., Lander, E.S., and Mesirov, J.P. (2005). Gene set enrichment analysis: a knowledge-based approach for interpreting genome-wide expression profiles. Proc. Natl. Acad. Sci. U.S.A. *102*, 15545–15550.

Sun, S., Chen, J., Li, W., Altintas, I., Lin, A., Peltier, S., Stocks, K., Allen, E.E., Ellisman, M., Grethe, J., and Wooley, J. (2011). Community cyberinfrastructure for advanced microbial ecology research and analysis: the CAMERA resource. Nucleic Acids Res. *39*, D546–51.

Thiet, R.K., Frey, S.D., and Six, J. (2006). Do growth yield efficiencies differ between soil microbial communities differing in fungal:bacterial ratios? Reality check and methodological issues. Soil Biol. Biochem. *38*, 837–844.

Torsvik, V., Øvreås, L., and Thingstad, T.F. (2002). Prokaryotic diversity – magnitude, dynamics, and controlling factors. Science *296*, 1064–1066.

Tringe, S.G., von Mering, C., Kobayashi, A., Salamov, A.A., Chen, K., Chang, H.W., Podar, M., Short, J.M., Mathur, E.J., Detter, J.C., Bork, P., Hugenholtz, P., and Rubin, E.M. (2005). Comparative metagenomics of microbial communities. Science *308*, 554–557.

Tripp, H.J., Hewson, I., Boyarsky, S., Stuart, J.M., and Zehr, J.P. (2011). Misannotations of rRNA can now generate 90% false positive protein matches in metatranscriptomic studies. Nucleic Acids Res. *39*, 8792–8802.

Urich, T., Lanzén, A., Qi, J., Huson, D.H., Schleper, C., and Schuster, S.C. (2008). Simultaneous assessment of soil microbial community structure and function through analysis of the meta-transcriptome. PloS One *3*, e2527.

Venter, J.C., Remington, K., Heidelberg, J.F., Halpern, A.L., Rusch, D., Eisen, J.A., Wu, D., Paulsen, I., Nelson, K.E., Nelson, W., *et al.* (2004). Environmental genome shotgun sequencing of the Sargasso Sea. Science *304*, 66–74.

Vila-Costa, M., Rinta-Kanto, J.M., Sun, S., Sharma, S., Poretsky, R., and Moran, M.A. (2010). Transcriptomic analysis of a marine bacterial community enriched with dimethylsulphoniopropionate. ISME J. *4*, 1410–1420.

Walker, C.B., de la Torre, J.R., Klotz, M.G., Urakawa, H., Pinel, N., Arp, D.J., Brochier-Armanet, C., Chain, P.S.G., Chan, P.P., Gollabgir, A., *et al.* (2010). *Nitrosopumilus maritimus* genome reveals unique mechanisms for nitrification and autotrophy in globally distributed marine crenarchaea. Proc. Natl. Acad. Sci. U.S.A. *107*, 8818–8823.

Wang, Q., Garrity, G.M., Tiedje, J.M., and Cole, J.R. (2007). Naïve Bayesian classifier for rapid assignment of rRNA sequences into the new bacterial taxonomy. Appl. Environ. Microbiol. *73*, 5261–5267.

Wang, Z., Gerstein, M., and Snyder, M. (2009). RNA-Seq: a revolutionary tool for transcriptomics. Nat. Rev. Genet. *10*, 57–63.

Wang, Y., Hayatsu, M., and Fuji, T. (2012). Extraction of bacterial RNA from soil: challenges and solution. Microbes Environ. *27*, 111–121.

White, J.R., Nagarajan, N., and Pop, M. (2009). Statistical methods for detecting differentially abundant features in clinical metagenomic samples. PLoS Comp. Biol. *5*, e1000352.

Yilmaz, P., Gilbert, J.A., Knight, R., Amaral-Zettler, L., Karsch-Mizrachi, I., Cochrane, G., Nakamura, Y., Sansone, S.-A., Glöckner, F.O., and Field, D. (2011). The genomic standards consortium: bringing standards to life for microbial ecology. ISME J. *5*, 1565–1567.

Yuan, Y., Conrad, R., and Lu, Y. (2011). Transcriptional response of methanogen *mcrA* genes to oxygen exposure of rice field soil. Environ. Microbiol. Rep. *3*, 320–328.

Zeeberg, B.R., Feng, W., Wang, G., Wang, M.D., Fojo, A.T., Sunshine, M., Narasimhan, S., Kane, D.W., Reinhold, W.C., Lababidi, S., Bussey, K.J., Riss, J., Barrett, J.C., and Weinstein, J.N. (2003). GoMiner: a resource for biological interpretation of genomic and proteomic data. Genome Biol. *4*, R28.

Zhou, X., and Su, Z. (2007). EasyGO: Gene Ontology-based annotation and functional enrichment analysis tool for agronomical species. BMC Genomics *8*, 246.

Soil Proteomics

Giancarlo Renella, Laura Giagnoni, Mariarita Arenella and Paolo Nannipieri

Abstract

Proteomics is a post-genomic approach with the potential to interrogate natural complex systems such as soils. However, the great potentials of soil proteomics are currently limited by either the complexity of the soil matrix which is reactive, structured, teeming with microbial communities which are at the same time extremely diverse, in heterogeneous physiological state and normally poorly characterized. Taken together, these soil features pose problems of protein sampling, extraction and purification. This chapter, though not exhaustive, aims to illustrate the main approaches and achievements in soil proteomics and indicate some future directions for further developments soil proteomics.

Introduction

Proteome as a term was introduced by Wasinger *et al.* (1995) and Wilkins *et al.* (1996) and it considers the entire complement of proteins expressed by an organisms' genome. Therefore, proteomics is an analytical approach providing an understanding of the structure, functions and regulation of living organisms by studying the protein expression profile rather than individual proteins and it is complementary to metagenomics, metatranscriptomics and metabolomics (Tyers and Mann, 2003). The goal of proteomics is to achieve a detailed description of the active metabolic pathways in living organisms at a given time and under defined biological conditions (Patterson and Aebersold, 2003). Proteomic analysis relies on the separation and identification of proteins. Traditionally, separation of proteins is achieved by in gel electrophoresis or liquid chromatography (LC) methods, whereas separation of peptides can be performed by chromatographic methods or by peptide isoelectric focusing (Cargile *et al.*, 2004). Usually, one-dimensional gel electrophoresis is used to separate simple protein mixtures (e.g. isoforms), whereas two-dimensional electrophoresis (2-DE) are used to separate complex protein mixtures and is preliminary to protein identification. The 2-DE technique was first proposed by Macko and Stegeman (1969) and improved by Klose (1975) and O'Farrell (1975). Notwithstanding the limitations related to the number of protein spots, impossibility of monitoring low abundant, highly hydrophobic proteins and proteins with very acidic and basic IP, the 2-DE has become a standard fingerprinting technique for protein separation prior to MS analysis (Lambert *et al.*, 2005). Important improvements of the 2-DE technique have concerned the use of fluorescent stains, in the differential gel electrophoresis (DiGE) (Unlu *et al.*, 1997), with staining of proteins with different fluorophores so as to distinguish proteins of different biological samples.

Recent technical developments have led to set up high-throughput automated methods based on off-gel techniques, often refereed as multi-dimensional protein separation or 'shotgun' proteomics, which are based on LC protein separation and development of mass spectrometry and bioinformatics for data analysis. It has been reported that LC-MS can identify up to 10^4 proteins from complex protein mixtures (Link, 1999), detecting also proteins present in relatively lower abundance (Kolkman *et al.*, 2005).

Mass spectrometry can detect ionized analytes according to their mass-to-charge (m/z) ratios. The number of ions at each m/z ratio value are registered by detectors and recorded as spectra of ion intensity versus the m/z ratio. In MS, ionization can be achieved in various ways, including chemical ionization, electrospray ionization (ESI). Currently, an efficient MS analysis is carried out by matrix-assisted laser desorption ionization (MALDI), in which a low-energy laser generates gaseous macromolecules (Karas and Hillenkamp, 1988). The ESI and MALDI ionization techniques, coupled with time of flight (TOF), ion trap, quadrupole orbitrap or hybrid tandem mass spectrometers to generate ion spectra have allowed major advances in proteomic research, owing to their high resolution and sensitivity. Alternative and more recent analytical techniques, such as desorption electrospray ionization (DESI), surface-assisted laser desorption/ionization (SALDI) along with the new detectors for peptide mass fingerprinting, such as MALDI- TOF allowing the amino acid sequencing and the Fourier transform ion cyclotron resonance (FT-ICR), bring promising improvements in sensitivity and accuracy of protein mass spectrometry.

Protein identification is done by peptide mass fingerprinting or peptide fragmentation analysis, which is based on fragment ion data combined with its molecular mass. The peptide mass fingerprinting is mainly done by MALDI-TOF techniques (Karas *et al.*, 1985; Mann *et al.*, 1993), whereas the peptide fragment ion data are derived by tandem mass spectrometry (MS/MS) with MALDI-TOF/TOF or ESI-MS/MS. For peptide mass fingerprinting, the proteins are digested with trypsin (an endoprotease), producing unique peptides per protein, and the peptides molecular mass are then measured. The MS spectra are compared using database search, introducing homology matching criteria between theoretical peptide masses calculated from each sequence entry (Mann *et al.*, 1993), resulting in protein similarity scores. Matching criteria, along with MS peptide mass accuracy measurement, are extremely important for protein identification. For example, available search algorithms can be based on simple scores based on the number of common masses between the experimental and theoretical spectra, whereas other algorithms rely on the distribution of protein and peptide masses in the databases. Protein identification by the peptide fragment ion data derived by MS/MS was developed relying on the computer aided de novo peptide sequencing, which allows determining the probable amino acid sequences of a peptide, by using ion peaks obtained by MS (Sakurai *et al.*, 1984). The MS/MS technique allows the identification of peptide mass, and amino acid sequence, by evaluating the mass difference between peptides after the cleavage of the peptide bonds from the N- or C-terminus (Hernandez *et al.*, 2006). Since the 1990s, the number of available protein sequences in the databases has markedly increased, making MS/MS technique a routine method for protein identification (Hernandez *et al.*, 2006), although some limitations still exist for soil proteomics. Potentials of protein quantification by stable isotopes have been reviewed by Müller *et al.* (2008). The targeted proteomics, which is highly specific and quantitative proteomic approach for the identification and quantification of proteins through the detection of specific peptides or protein fragments by using multiple reaction monitoring in triple quadrupole MS set-up (Deutsch *et al.*, 2008), have not been applied to soil studies to our knowledge.

The various protein detection and identification techniques are discussed in Chapter 8.

Soils and their protein complement

Soils support the plant growth, host the largest microbial diversity among the terrestrial ecosystems in which soil microorganisms are responsible for a number of ecosystem services, and soils have interactive effects with climate and living organisms. Despite the progresses in taxonomical and functional analyses of the microbial communities, to date it is not possible to establish direct links between microbial diversity and specific decomposition processes (Nannipieri *et al.*, 2003).

Proteins in soil originate from plants, animals and microorganisms, either through active excretion or passive release. Generally, 96–99% of soil total N is organic and after acid hydrolysis, amino acidic N accounts for 30–50% of the total N in soil (Stevenson, 1994). It is assumed that most of the amino acids released from acid hydrolysis derive from proteins and peptides present in soils. Modern techniques, such as Curie-point-pyrolysis gas chromatography–mass spectrometry (Cp Py-GC/MS) pyrolysis-field ionization mass spectrometry (Py-FIMS), ^{15}N nuclear magnetic resonance spectroscopy (^{15}N NMR), X-ray photoelectron spectroscopy, and X-ray absorption near-edge structures spectroscopy (XANES) have shown that proteins are the most abundant organic N compounds in soil (Nannipieri and Paul, 2009). However, origin of proteins and peptides in soil and their links with the measurable soil functions (e.g. SOM decomposition, enzymatic activity) are largely unknown, despite the vast literature on the subject (Nannipieri and Paul, 2009). Several surface-reactive soil particles such as clay minerals, Fe, Al and Mn hydr(oxides), and humic substances, are involved in the protein sorption.

Protein interactions with the soil inorganic solid phases

Soil is characterized by the presence of solid inorganic or organic phases or organo-mineral reactive surfaces. Among the solid phases, clay minerals have been extensively studied for their interactions with proteins.

Early evidences of protein protection by soil clays against microbial proteolysis were shown by Ensminger and Gieseking (1942) and Pinck and Allison (1951). The extensive research reviewed by Stotzky (1986) has shown that protein sorption onto kaolinite and montmorillonite is rapid, especially at reaction conditions below the protein isoelectric point of the protein, likely due to ionization of the amino and carboxyl groups. Protein sorption by 2:1 minerals (e.g. montmorillonite) is mainly related to the protein intercalation in the interlayer expanded crystal lattice, whereas protein sorption by kaolinite is more related to surface area of the clay mineral and less to its surface charge, whereas the clay saturation cation and the protein conformation may be not major variables influencing the protein adsorption. However, as the mineral surfaces are negatively charged, the role of the polyvalent cations on the protein adsorption and the enzyme catalytic activity has been also demonstrated (Burns, 1986; Fusi *et al.*, 1989; Gianfreda and Bollag, 1996; Huang *et al.*, 1998; Naidja and Huang, 1995; Quiquampoix, 2000; Rupert *et al.*, 1987; Theng, 1974; Violante and Gianfreda, 1995).

Studies on the interactions between clay minerals and proteins do not always report reaction conditions and relative purity of proteins and clay minerals used for the experiments.

In fact, the presence of other proteins or other chemical compounds may influence the sorption of the target protein. For example, small, uncharged, or non target proteins with non polar protein domains may be also adsorbed by van der Waals interactions. The same situation is for the clay minerals, often referred as 'pure' clays, because even small amounts of impurities like other minerals such as Fe-(hydro)oxides or silica, may influence the protein adsorption by clays. In fact, silica and Fe-(hydro)oxides can interact with proteins (Norde, 1986; Stotzky, 1986), either by electrostatic forces or by entropic favourable structural rearrangements. Other variables involved include contact mode (static or not), contact time (minutes-hours), contact temperature (controlled or not) and protein-to-clays ratio.

Proteins are mainly adsorbed by clays via electrostatic interactions, mostly fitting the Freundlich isotherm model (Quiquampoix *et al.*, 1993). In addition to electrostatic interactions, van der Waals forces and hydrophobic interactions have been shown to be involved in protein sorption by clays (Hamzehi and Pflug, 1981; Quiquampoix and Ratcliffe, 1992; Staunton and Quiquampoix, 1994). Formation of protein multiple layers has also been suggested (Violante and Gianfreda, 1995). Protein adsorption by hydrophobic and ionic interactions is also accompanied by entropy gain caused by conformational changes of the protein during the adsorption (Haynes and Norde, 1995). Yu *et al.* (2000) studied the sorption of various proteins and peptides on pyrophyllite, and reported that the clay mineral denatured the adsorbed proteins due to dehydration and non specific interactions inducing torsion tension and destabilization of the tertiary structure of the adsorbed proteins. However, the general assumption that protein adsorption can reduce mobility and alter biological properties of proteins due to conformational changes, orientation on the surface (Baron *et al.*, 1999; Servagent-Noinville *et al.*, 2000) is not supported by studies on Bt toxin and glomalin, whereas protection against proteolysis was due to protein adsorption (Kleber *et al.*, 2007). Koskella and Stotzky (1997) observed that insecticidal toxins produced by *B. thuringiensis* subspp. *kurstaki* and *tenebrionis* bound to kaolinite and montomorillonite were not mineralized by mixed microbial cultures and also retained insecticidal activity, whereas the opposite occurred for the free toxins. However, due to the very broad range of protein size, conformation, electric properties of proteins, a general model of protein interactions with clay minerals is currently not available.

The fate of specific proteins such as glomalin and the insecticidal *Bacillus thuringensis* endotoxin (Bt toxin) in soil, have attracted the attention of soil scientists due to their ecological and commercial relevance (De Barjac and Frachon, 1990). These studies further illustrate the difficulty of extracting, quantifying and characterizing proteins directly extracted from soils. Glomalin is an extracellular glycoprotein produced by arbuscular mycorrhizal fungi (Wright and Upadhyaya, 1998), and usually quantified by either the Bradford colorimetric or immunological (ELISA). However, both techniques produce artefacts due to non-specific reactions with phenolic compounds and litter and humic components. Analysis of glomalin-related proteins extracted from soil by NMR has shown the presence of humic substances (Schindel *et al.*, 2007).

Similar observations were done on Bt toxin, a commercial product for agriculture use. Such studies have shown that the toxin is stabilized in soil by clay minerals and humic acids in its active form (Lee *et al.*, 2003). From an *in vitro* study Helassa *et al.* (2009) reported that sorption of the Bt Cry1Aa (released by Bt-transformed rice) was greater onto montmorillonite than kaolinite and that desorption was less with water than with alkaline buffers in the presence of surfactants such as CHAPS, Triton-X-100 and Tween 20.

Interesting information on protein production and persistence were provided by studies on litter decomposition (Criquet *et al.*, 2002). Huang *et al.* (1998) and Miltner and Zech (1999) reported that plant proteins dominated in fresh litter whereas proteins of microbial origin predominated in aged litter. In a model litter decomposition study, based on the proteomic analysis of *Pectobacterium carotovorum* and *Aspergillus nidulans*, either grown pure and co-culture or on beech litter, Schneider *et al.* (2010) showed that various proteases, pectinases and cellulases were involved in litter decomposition and that decomposition was initiated and carried out by the fungus. It cannot be excluded that these observations may also apply to soil, and a correlation between the SOM (soil organic matter) decomposition rate and proteins concentrations may exist, particularly in the early stages. In this sense, there is a need to monitor the protein synthesis and persistence in soil by the monitoring of intracellular proteins. However, information on protein turnover during SOM decomposition is still scarce.

Sorption of enzymes and other proteins, actively or passively released by soil organisms, on clay minerals and humic substances can have an ecological relevance because it reduces their availability as C and N sources for microorganisms (Calamai *et al.*, 2000; Stotzky, 1986; Violante and Gianfreda, 1995). Despite protein extraction yields are important in soil proteomics, protein desorption from soil colloids is poorly known as well as the factors controlling it, such as the effects of pH shifts induced by the different protein extraction protocols. Some information on complexity of the pH-dependent interactions is well illustrated by the variability of enzyme adsorption depending on the physico-chemical conditions. Quiquampoix *et al.* (1989) reported that the α-glucosidase of *Aspergillus niger* did not adsorb on the surface of montmorillonite above pH 6.0 whereas the bovine serum albumin (BSA) did not adsorb above pH 6.5 (Quiquampoix and Ratcliffe, 1992). Quiquampoix and Mousain (2005) showed that acid phosphomonoesterases released by ectomycorrhizal fungi could have different behaviour towards clay minerals: sorption or total repulsion for *Suillus mediterraneensis* or *Pisolithus tinctorius*, respectively. The importance of pH-induced conformational changes on the extent of adsorption of specific proteins has been reviewed by Quiquampoix (2008).

Studies on the formation of mixed lipid–protein layers at solid phase interfaces can be also important because they can help to understand the nature of interactions in such this type of interface which likely occur in soil (Bos and Nylander, 1996; Cornell, 1982). Such interactions can significantly change hydrophobicity or hydrophility and conformation of proteins (Bos and Nylander, 1996; Cornell, 1982).

Protein interactions with the soil organic matter (SOM)

The SOM is composed of molecules such as structural and functional sugars, lipids, nucleic acids, proteins and secondary metabolites, produced and released by living and dead cells at various stages of decomposition. The soil organic matter (SOM) exists as discrete organic phases associated or not with mineral surfaces, or dissolved into the soil solution. The particulate SOM has various particle sizes, variable solubility and richness of functional groups and variable decomposition stages (Schmidt *et al.*, 2011).

Humic substances are natural organic substances ubiquitous in nature and represent a variable proportion of the SOM, and have been traditionally described as polymeric structures with high molecular weight, of brown to black colour, formed by condensation reactions, rich in carboxylic and phenolic groups and therefore present in the soil environment as negatively charged branched polyelectrolytes (Stevenson, 1994). The protein surface is highly heterogeneous with positive and negative charges, groups with hydrogen bonding abilities, as well as non polar regions. Due to such complexity, each protein can interact with humic molecules in several modes, including ionic interactions (both repulsive and attractive), hydrogen bonding, hydrophobic interactions, hydration forces, acid–base interactions and van der Waals forces. According to Cheshire and Hayes (1990) and Schnitzer (1986), interactions between proteins and polysaccharides involve mainly oxygen or hydroxyl functions of the proteins, through hydrogen bonding for neutral and negatively charged polysaccharides.

It has been hypothesized that in soil humic substances are supramolecular associations of self-assembling heterogeneous molecules, originating from the decomposition of biological macromolecules, with their supramolecular conformation stabilized by weak dispersive forces such as hydrophobic interactions and H-bonds Piccolo (2002). Therefore, fulvic and humic acids are operationally defined SOM fractions of varying molecular weight and properties, and likely formed by the supramolecular aggregation of small organic molecules, rich in functional groups, held together by weak molecular interactions (Piccolo, 2002). The molecular association of humic substances was confirmed by Kelleher and Simpson (2006) through NMR studies, demonstrating that HS fractions could contain intact and degraded known biopolymers, apparently confirming that soil humic substances originate from complex degradation mixtures of biological molecules.

While protein–clay interactions have been extensively studied, interactions between proteins and humic substances are considerably less characterized, even if they have been long postulated (Bremner, 1951; Hsu and Hatcher, 2005, 2006; Knicker and Hatcher, 1997; Jenkinson and Tinsley, 1960; Simonart *et al.*, 1967; Swaby and Ladd, 1964; Zang *et al.*, 2000). Usually, proteins, including enzymes which are ideal for testing the interactions with humic substances because effects can be also evaluated by changes in the enzymes' kinetic properties following adsorption by humic substances, are not resistant against proteolysis (Nannipieri *et al.*, 1996). For example, Ceccanti *et al.* (1989) reported the extraction of humo–enzyme complexes from soils retained their catalytic activity, whereas Solaiman *et al.* (2007) reported degradation after contact with HS. From a model study, Tan *et al.* (2009) reported synthetic humic acids were able to complex lysozyme, and such complexation altered the catalytic activity (Zang *et al.*, 2000).

However, proteins can be entrapped by humic substances as shown by Tomaszewski *et al.* (2011). In addition, complexation of humic acids and specific domains of the human serum albumin have been reported by Ding *et al.* (2011).

There is increasing evidence that interactions between proteins and humic substances are mainly hydrophobic, and thus proteins can interact with humic substances also at pH values above the protein isoelectric point, when both proteins and partner humic substances are negatively charged (De Kruif *et al.*, 2004). Thermodynamically favourable hydrophobic interactions following electrostatic complexation between lysozyme and synthetic humic acids were reported by Tan *et al.* (2008). Such hydrophobic interactions may be also due to dehydration which may destabilize and partially disassembly humic substances in contact with the protein, as hypothesized by Tomaszewski *et al.* (2011). Hydrophobic interactions may also reduce the affinity of the positively charged protein for the negatively charged soil mineral phases, as reported by Tan *et al.* (2009) for lysozyme. However, our knowledge on the protein SOM interactions comes from oversimplified *in vitro* experiments, and divalent cations (e.g. Mg^{2+}, Ca^{2+}) in soil solution can form bridge bonds between negatively charged protein and SOM functional groups, as reported by Yuan and Zydney (2000).

Soil proteomics

Early papers on soil proteomics have focused on factors influencing the protein extraction from soil (Nannipieri, 2006) so as to set up for soil proteomic studies giving high yields of extracted proteins. Both direct and indirect protein extraction protocols have been used, with the latter being based on the protein extraction from microorganisms isolated from soil in the absence of clays and humic substances. A review of direct methods has been published by Keller and Hettich (2009), and four protein extraction methods have been compared by Keiblinger *et al.* (2012). Methods for direct protein extraction from soils are based on the use of various lysis buffers, protease inhibitor cocktails, NaOH, surfactants (e.g. SDS), phenol and chelating agents (e.g. EDTA), with or without sonication, autoclaving and freeze–thawing cycles to lyse microbial cells (Table 6.1).

Singleton *et al.* (2003) observed that soil pollution by Cd decreased the content of extracted proteins and increased synthesis of low molecular weight (< 21 kDa) proteins; shifts in the microbial proteome of trace element polluted soils were also reported by Maron *et al.* (2006), who used a protein indirect extraction method, and showed microbial synthesis of proteins with molecular weights ranging from 20 to 50 kDa possibly involved in heavy metal resistance.

Chen *et al.* (2009) reported that sequential extractions of soil with citrate and SDS buffers followed by phenol extraction improved the protein yields from various soils.

Schulze *et al.* (2005) presented the first comprehensive environmental proteomic approach by extracting proteins from the soil dissolved organic matter and from clay minerals of a forest soil. After protein purification by gel filtration, to remove humic acids and phenolic compounds, and protein separation by SDS-PAGE, proteins were digested with trypsin, and the tryptic peptides separated by nanoflow liquid chromatography and analysed by mass spectrometry. Proteins involved in ribosomal transcription, membrane proteins and enzymes of plants, animals and soil microorganism were detected. To date the study by Schulze *et al.* (2005) is the only soil proteomic study where enzymes involved in SOM mineralization, such as cellulases and lactases, were identified as associated to soil solid phases. Benndorf *et al.* (2007, 2009) characterized the regulation of a 1,2-dioxygenase and other proteins involved in microbial degradation of dichlorophenoxy acetic acid and benzene in contaminated soils, sediments and aquifers.

Despite these studies have provided pioneering information extraction and characterization of proteins from soil and presence of specific proteins in soil, they miss the main target of proteomics, as they do not provide information on the intracellular protein expression profile of soil organisms. Only the most recent papers quoted in Table 6.1 clearly addressed the problem of extracting representative (i.e. functional) proteins. Indeed, the extracellular stabilized proteins are not related to microbial responses to specific stimuli and should not be considered for characterizing the proteomic responses of soil microbial communities and related soil functions, which mostly depend on the activity of newly expressed microbial proteins (Nannipieri, 2006). Protein extraction protocols should prevent alteration of the microbial proteome and protein losses by hydrolysis or irreversible denaturation. On the contrary, the indirect extraction methods allow bacterial isolation from soil colloids, thus

Table 6.1 Soil and environmental (meta)proteomic studies in chronological order, with main extraction conditions and outcome

Reference	Aim	Analytical conditions	Outcome	Comment
Ogunseitan (1993)	Extraction and characterization of whole soil proteins	Two extraction methods: (1) samples in boiling lysis buffer (50mM Tris-HCl pH 6.8, 100mM dithiothreitol, 2% SDS, 10% glycerol, and 0.2% bromophenol blue) (2) incubation at 0°C in a lysis buffer (50mM Tris-HCl pH 7.6, 1mM EDTA, 10% sucrose, 1mM dithiothreitol, 300µg/ml lysozyme, 0.1% polyoxyethylene 20 acetyl ether), followed by four freeze–thaw cycles. Proteins characterization by SDS-PAGE	The freeze–thaw method yielded more proteins (20–50µg/g) Protein size obtained by both methods ranged from less than 14kDa to greater than 97kDa	The first paper on soil and sediment metaproteomic. Both the tested extraction methods were suggested as suitable for soil and environmental metaproteomics Interferences by phenolic compounds on soil-extracted proteins that make impossible a quantitative comparison between extraction methods were demonstrated in later papers (Roberts and Jones, 2008)
Wright and Upadhyaya (1996, 1998)	Arbuscular mycorrhizal glycoprotein (glomalin) extraction from 12 soils	Autoclaving+extraction by citrate buffer or malate buffer (pH 7.0 or pH 8.0). Protein characterization by ELISA	Protein characterization by ELISA confirmed that the extracted protein was glomalin	In a later paper on Schindler *et al*. (2007) showed the possible formation of artefacts
Ogunseitan (1997)	Study of extracellular enzymes produced by microbial communities of freshwater and wastewater sludge and from the respective culturable microbial communities and by a *P. putida* strain	Proteins from environmental matrices, culturable microorganisms and *P. putida* were extracted by pulsed sonication on ice from 3minutes. Soluble proteins recovery by centrifugation (25minutes 25,000 *g*). Protein immobilization onto nitrocellulose filters (0.2µm) pre-soaked by the appropriate buffer and detection of enzyme activity by chromogen substrates and native-PAGE	The protein extraction protocol permitted the detection of catechol oxidase, nitrate reductase, peroxidase, and xanthine dehydrogenase activities and the differentiation of extracellular activities between sludge and freshwater samples	A similar conservative approach has not been sufficiently tested in soil proteomics The comparison with control experiments based on the measurement of enzyme activity of a *P. putida* strain, shows the detection limits in soil proteomics. The xanthine dehydrogenase activity was undetectable at population density below 10^3 CFU, and below 3.3ng of total protein. The importance of sample amount for environmental proteomics was discussed later by Thompson *et al* (2008)

Craig and Collins (2000)	Immuno-detection of BSA adsorbed to siliceous minerals (DACIA) using three different buffer (Urea, PBS and HF)	The lysis buffer contained 10mM PBS (pH 7.4), 0.87% NaCl, 8M urea (pH 8.5) and 4M HF. Samples were incubated at 4°C under shaking for 24h, followed by a washing step with 0.1M carbonate-bicarbonate (pH 9.6)	Using DACIA immunological test, HF resulted a suitable extractant for mineral-sorbed proteins	The use of BSA as test protein is questionable due to its inherent instability. Irreversible protein denaturation due to acidic pH values of the extractant should be considered for soil proteomic studies
Craig and Collins (2002)	Characterization of BSA in the presence of quartz and illite and using three different buffers	The lysis buffer contained: 2% SDS (pH 6.4), 6M guanidine-HCl (pH 5.6), 8M urea, 10mM PBS, 0.149M NaCl pH (7.4), 5% ammonia (pH 11.5), 10% K_2-EDTA (pH 10.0). Protein characterization by DACIA and spectrophometric analysis	Proteins were tightly bound to mineral surfaces via short–range bonds. The extraction efficiency was not satisfactory.	Use of immunological test (DACIA) for identifying mineral bound protein, after digestion of the mineral phase with HF was proposed.
Singleton *et al*. (2003)	Metaproteomic study of Cd-polluted soils	Bead-beating or freezing-thawing cycles, Tris-HCl buffer+protease inhibitors, sucrose, dithiothreitol and EDTA. Protein characterization by SDS-PAGE	Soil Cd pollution decreased the content of extracted proteins and induced the synthesis of some low molecular weight proteins	Low number of protein bands on SDS-PAGE in spite of the theoretical soil metaproteome and the used harsh method for disaggregating soil matrix. No specific proteins could be identified
Murase *et al*. (2003)	Extraction of extracellular proteins from a greenhouse soil	The lysis buffer contained 67mM phosphate buffer at pH values 6.0, 7.0 and 7.7, and pH 7.7 + NaN_3. Filtration by filter paper and 0.2μm cellulose acetate filters. TCA and ethanol were used for protein precipitation and characterization by SDS-PAGE	The best results were obtained using buffer at pH 6.0 but a low amount of protein was detected on gel. Impurities were present in protein SDS-PAGE.	The amount of protein extracted was very low in relation with the theoretical soil protein content, likely due to the weakness of the used extractant. Information on other proteins were not provided
Wilmes and Bond (2004)	Indirect proteomic analysis of microbial communities extracted from activated sludge	Microbial communities extraction by centrifugation. Protein extraction by 7M urea, 2M thiourea, 4% (w/v) CHAPS, 10mM Tris-1mM EDTA, 50mM DTT, 25mM PefablocSC and 2mM Pefabloc protector. Protein precipitation by TCA. Protein fingerprinting by 2-DE, MALDI-TOF	Specific functional proteins were identified	The used de novo sequencing approach helped to identify the functional proteins of the extracted microbial communities

Table 6.1 (Continued)

Reference	Aim	Analytical conditions	Outcome	Comment
Schulze *et al*. (2005)	Off gel proteomic fingerprint of dissolved organic matter (DOM)	Flotation of DOM with Na polytungstate+washing with H_2O, 10% HF+gel filtration using Sepharose 4B. Protein analysis by SDS-PAGE and LC/MS/MS	One hundred forty-eight proteins isolated by SDS-PAGE. Graphics showed the different percentage of identified proteins belonging to organisms from different kingdoms	No gel image was reported. No identified protein list with related identification parameters was shown
Benndorf *et al*. (2007)	Enrichment of soil with *C. necator* JMP134, *Rhodoferax* sp. P230 and *S. herbicidovorans* B488.	Protein extraction by 0.1M NaOH+phenol, protein precipitation by 0.1M CH_3COONH_4 in CH_3OH. Protein extraction and characterization by SDS-PAGE,2D-E and MS/MS	Four proteins of *C. necator* JMP134 were identified by SDS-PAGE. Proteomic analysis of groundwater by SDS-PAGE and 2-DE showed 20 bands and ca. 100 spots, respectively. Twenty-nine proteins of 19 SDS-PAGE bands and 26 proteins of 50 2-DE spots were identified by MS.	Low protein recovery particularly from soil and low number of identified proteins. SDS-PAGE gel showed impurities on gel.
Maron *et al*. (2008)	Protein fingerprint of soil microbial communities of three soils coupled to soil microbial communities characterization	Extraction of soil microorganisms by centrifugation density gradient, resuspension in NaCl and further separation by Nycodenz density gradient. Protein extraction by ultrasonic treatment and lysis buffer. Protein characterization by SDS-PAGE and microbial characterization by ARISA	This was the first attempt to combine between genetic and proteomic approaches for the study of soil functionality	Large number of protein bands detected on SDS-PAGE but no proteins were identified
Benndorf *et al*. (2009)	Proteomic study of river sediments and lava granules	Sonication for 10minutes, extraction with 20mM Tris-HCl pH 7.5. 2-DE analysis		No protein identification
Chen *et al*. (2009)	Comparison between different direct soil protein extraction methods	Sequential extraction with citrate+SDS buffers, followed by phenol extraction. SDS-PAGE and 2-DE protein characterization	The protocol with SDS buffer was more efficient	No protein identification

Williams *et al*. (2010)	Metaproteomic and metagenomics study of soil microbial communities in toluene-amended soil, and toluene-amended soil inoculated with microbial cultures	Protein extraction of soil microorganisms, separated from soil by density gradient centrifugation (Nycodenz), using the lysis buffer (0.5M TRIS-HCl (pH 8.7), 0.9M sucrose, 0.05M EDTA, 0.1MKCl, and 2% 2-mercaptoethanol). Phenol treatment (0.5 mL phenol for several phase inversion separation). Protein precipitation by 0.1M CH3COONH4/1% β-mercaptoethanol in methanol and 80% acetone. Proteins analysis by SDS-PAGE – MALDI-TOF/TOF MS, using NCBInr in MASCOT	Fourty-seven proteins were identified; several proteins were common between toluene-amended soil communities and toluene impacted bacteria, mostly stress related proteins, analysed by fatty acid methyl ester analysis profiles. The 16S rRNA gene analysis indicated high dominance with 80% of the OTUs related to the *Bacillus* genus.	One of the few integrated approaches for the study of soil microbial diversity and functional responses to specific stressors. The used proteomic approach highlighted the type of stress imposed by toluene to soil microbial communities
Chourey *et al*. (2010)	Off gel soil proteomic analysis by direct and indirect extraction *P. putida* was added to soil as internal standard strain	Three extraction protocols: (1) SDS-TCA lysis buffer (5% SDS, 50mM Tris-HCl, pH 8.5, 0.15M NaCl, 0.1mM EDTA, 1mM $MgCl_2$, 50mM DTT). Protein precipitation by 100% TCA and acetone. Protein pellet in guanidine buffer (6M guanidine-HCl, 10mM DTT in Tris-$CaCl_2$ buffer at 60°C for 1h. Proteins trypsin digestion and (2) Guanidine-HCl lysis buffer. Protein digestion by double treatment with trypsin, protein reduction using 20mM DTT (3) Isolation of soil microorganisms by centrifugation with cold PBS 10,000*g* for 10minutes at 4°C. Protein extraction by SDS-TCA or guanidine lysis buffers. Protein characterization by 2D-LC-MS/MS Protein identification using SEQUEST.	1043 identified protein sof *P. putida* in liquid culture, 925 identified proteins of *P. putida* extracted from soil The approach showed to be suitable for Gram-negative and Gram-positive bacteria, as direct soil protein resulted in identification of more than 500 unique proteins with no apparent bias in terms of protein size, localization, functions model strains grown in pure cultures	First soil metaproteomic study using 2D-LC-MS/MS Relatively low protein identification rates from soil and reduced protein identification of the inoculated *P. putida* strain. Large amount of unknown proteins from using the direct extraction approach

Table 6.1 (Continued)

Reference	Aim	Analytical conditions	Outcome	Comment
Bastida *et al*. (2010)	Microbial proteomic responses to soil pollution with hydrocarbons	Soil microbial communities were extracted by phosphate buffer, inoculated and cultured onto LB medium; then they were lysed with 20mM Tris-HCl (pH 7.5) plus 0.2g/l SDS and also using the method proposed by Benndorf *et al*. (2007). Protein analysis by SDS-PAGE and LC/MS analysis	Nineteen proteins belonging to *B. cereus*, *B. thuringiensis*, *B. Anthracis* were identified	Relatively low number of proteins were identified, and SDS-PAGE gel image were not shown. The main conclusion was that *Bacillus* sp. dominated the soil microbial communities of hydrocarbon polluted soil could be biased by the intermediated microbial cultural step
Wang *et al*. (2011)	Metaproteomic analysis of crop soil Soil microbial composition analysis by TRLFP	Two protein extraction protocols: (1) SDS 1.25% w:v, 1M Tris-HCl, pH 6.8, 20mM DTT+ buffered (pH 8) phenol (2) 0.25M citrate buffer (pH 8)+buffered (pH 8) phenol. Proteins were precipitated by CH_3COONH_4 in methanol. Protein analysis by SDS-PAGE, 2DE and MS/MS	Two-hundred and eighty protein randomly selected were analysed by MS: 122 protein were identified. Correspondence between protein MS/MS results and TRLFP results	First report matching soil proteomic and genomic data. The extraction protocol with SDS buffer was more efficient than that with citrate buffer. Several proteins were still not identified
Keiblinger *et al*. (2012)	Comparison of four different soil protein extraction method	Pre-treatment with PVPP. Four protein extraction protocols were compared: (1) 50mM Tris, 1% SDS pH 7.5 and 10% TCA (2) NaOH, phenol buffer and 0.1M ammonium acetate (3) 50mM Tris, 1% SDS pH 7.5 and phenol pH 8.0 (4) pre-treatment with TCA and Methanol. Then 50mM Tris, 1% SDS pH 7.5 and phenol (pH 8.0).	The method with SDS and phenol were the two most efficient for the studied soils	The protein extraction efficiency from soil samples is severely hampered by the complex matrix. The SDS-phenol was showed to be the best extractant

almost eliminating the presence of organic and inorganic contaminants, but the validity of such methods has not been thoroughly tested yet, and changes in the bacterial physiological status during the extraction may introduce bias in the evaluation of soil microbial proteome.

Information by model studies

In order to better understand the main factors influencing the proteomic analysis of soil microorganisms, Nannipieri *et al.* (2009) proposed to use model studies based on the analysis of the proteomic responses of selected soil borne microorganisms with known genome and proteome, inoculated into soils. Since then, some model studies have been published (Table 6.2), and have allowed to evaluate the effects of selected soil properties on microbial proteomic responses and the potentials of current soil proteomic approaches.

Taylor and Williams (2009) reported a poor recovery of proteins of *S. maltophila* after its inoculation into soil, and Williams *et al.* (2010) detected the presence of 187 proteins of the same strain in soil extracts, but only 47 proteins could be identified. Chourey *et al.* (2010) reported variable protein yields of proteins from *P. putida* and *A. Chlorophenolicus* inoculated into soil using different extraction protocols. Giagnoni *et al.* (2011, 2012) reported a poor recovery of proteins of *C. metallidurans* CH34 inoculated into inorganic soil solid phases with different reactivity and a decreased recovery of proteins from an artificial soil containing quartz sand, clays, goethite and humic acids, by prolonging the incubation time.

In this context, Luo *et al.* (2007) demonstrated the bacterial starvation in extracted soil organic matter (SESOM) by determining the differences in the proteome of *Bacillus cereus* grown either in Luria Bertani culture medium or in SESOM. Maron *et al.* (2008) showed that soil borne bacterial communities inoculated into sterilized modified their protein expression profile according to the physico-chemical properties of the recipient soil.

A detailed protocol for preparing artificial soil microcosms for proteomic model studies was proposed by Brözel *et al.* (2011). This protocol takes into consideration major factors affecting protein extraction from soil and soil factors which can influence the proteomic data.

Information from low-diversity biotopes

To detect and identify all proteins produced by complex environmental microbial communities, with uneven species distribution, broad range of protein expression levels for various microbial species, and the large genetic heterogeneity within microbial communities is to date technically impossible. However, the potentials of proteomics to link genetic diversity and activities of microbial communities and understand their impact on ecosystem function has been well proven in some extreme biotopes such as biosolids and mine spoils (Schneider e Riedel, 2010). In one of the first environmental metaproteomic studies, Ehlers and Cloete (1999) demonstrated the functional similarity of 21 activated sludge systems differing for design and phosphorous removal. Wilmes and Bond (2004) characterized the functional responses of microbial communities in an activated sludge selected for their polyphosphate accumulating capacity, showing the presence of proteins from an uncultured *Rhodocyclus* strain. Tyson *et al.* (2004) and Ram *et al.* (2005) provided a deep insight into the metabolic pathways of the dominant *Leptospirillum* and *Ferrobacillum* strains by an integrated metagenome/metaproteomic approach for the characterization of microbial communities of an acid mine drainage. Their studies elucidated the survival strategies of the two strains in this extreme environment, and detected a novel protein responsible for Fe oxidation. Similar

Table 6.2 Sample of soil and environmental model proteomic studies in chronological order, with main extraction conditions and outcome

Reference	Aim	Analytical conditions	Outcome	Comment
Luo *et al.* (2007)	Proteomic study of *B. cereus* grown in soil organic matter extracts	The *B.cereus* proteome was analysed after growth in soil organic matter and in LB medium until mid exponential phase were analysed. Cells were resuspended in urea CHAPS buffer and lysed by freeze–thaw cycles and pulsed sonication Protein characterization by 2-DE and MALDI-TOF MS (*B. cereus* sequence database)	The 2-DE analysis revealed 234 up-regulated proteins and 201 down regulated proteins of *B. cereus* grown in soil organic matter extracts. Thirty-five proteins were overexpressed and eight proteins were underexpressed after *B. cereus* growth in soil organic matter, all involved in the fundamental cellular metabolism	The first model approach for microbial proteomics related to a specific soil component. The study revealed that the studied organic matter was relatively poor in nutrients
Taylor and Williams (2009)	Comparison of direct and indirect protein extraction method from soil after inoculation with *S. maltophilia* and BSA to soil as protein internal standard	Direct method: soil proteins extracted directly from soil Indirect method: microorganisms extracted from soil by density gradient centrifugation Protein extraction for both methods: 0.5M Tris–HCl (pH 8.7), 0.9M sucrose, 0.05M EDTA, 0.1M KCl, and 2% 2-mercaptoethanol, phenol. Protein precipitation: 0.1M ammonium acetate and 1% β-mercaptoethanol in methanol and 80% acetone. Protein characterization by SDS-PAGE.	SDS-PAGE of proteins using indirect method was better than direct method BSA as control was evident	Direct extraction of protein from soil was affected by organic matter and clay particles Low amount of protein and unclear SDS-PAGE patterns from direct protein extraction from soil were obtained and no protein identification was provided

Schneider *et al.* (2010)	Proteomic analysis of bacterial and fungal species involved in litter decomposition. *P. carotovorum* and *A. nidulans* were used as bacterial and fungal model organisms	Enzymes in the secretome were extracted by ultrafiltration (10kDa cut-off membrane) and analysed by SDS-PAGE Proteins were subjected to in-gel tryptic digestion and analysed by MS/MS Protein identification was done by using the MASCOT algorithm	The *P. carotovorum* grew better in co-culture with the fungus. The bacterium showed a limited litter decomposition capability but profited by the fungal degradation products	Proteomic analysis revealed the differences in the regulation and in the production levels of degradative enzymes in the secretome of the studied bacterial and fungal strains and their ecological role in litter decomposition
Giagnoni *et al.* (2011)	Proteomic analysis of *C. metallidurans* in artificial soils containing quartz sand, kaolinite, montomorillonite, goethite and humic acids, singly or mixed for the evaluation of the effects of selected soil solid phases on the bacterial proteome	The bacterial proteome was extracted by PBS-SDS lysis buffer containing nuclease and protease inhibitor cocktail. Proteome analysis was done by 2-DE and MALDI-TOF (MASCOT algorithm in NCBInr)	The presence of highly reactive clays affected the analysis of the bacterial proteome as less proteins were recovered in the artificial soils as compared to the sand and kaolinites substrates. No proteins could be recovered by microcosms containing sole montmorillonite	Microbial proteome analysis in soils is influenced by the reactivity of the soil solid phases, by reducing the protein extraction efficiency.
Giagnoni *et al.* (2012)	Proteomic analysis of *C. metallidurans* CH34 incubated in an artificial soil at different contact times	The bacterial proteome was extracted by PBS-SDS lysis buffer containing nuclease and protease inhibitor cocktail. Proteome analysis was done by 2-DE and MS (MASCOT algorithm in NCBInr)	The number of protein spots in 2-DE was reduced during the incubation time. Apparently, a large fraction of the bacterial proteome was stabilized by soil solid phases after the bacterial death	The paper underlined the importance of protein extracellular stabilization for soil proteomics. Relatively low rates of protein identification was reported, apparently due to interactions between proteins and humic substances

proteomic high-resolution studies have been successfully conducted on other specific environmental samples with relatively low microbial diversity such as plant phyllosphere (Delmotte *et al.*, 2009).

Specificity of soil proteomics

A peculiar aspect of the past research in soil proteomics is that some of the milestone concepts developed and tested in decades of research in soil microbiology and biochemistry were apparently neglected. Here we focus on four key concepts related to soil microbial activity and ecology that, if adopted could have led to a more soil oriented and knowledge-based development of soil proteomics. In particular, we discuss specific aspects of the high soil microbial diversity, diffuse dormancy and biofilm formation by soil microorganisms, the complexity of the rhizosphere, and on the identification of proteins modified by interactions with soil surface-reactive particles.

High biological diversity

The theoretical complexity of the soil metaproteome can be exemplified by the fact that relatively simple prokaryotic and eukaryotic unicellular organisms such as *E. coli* and *S. cerevisiae* possess 4.3×10^3 and 6.1×10^3 open reading frames, respectively, coding for proteins with molecular weights between 1.7 and 559 kDa, with a number of amino acids in the order of 10^6. By considering the high microbial diversity of most soils, it is clear that no techniques are currently available for resolving the soil metaproteome. Moreover, some microbial species may be ubiquitous in soil whereas other may only inhabit specific soil niches such as aggregates and rhizosphere. Furthermore, microbial activity is generally localized in hotspots characterized by sufficient availability of nutrients and water. Such specific soil features require specific soil protein sampling procedures as compared to DNA extraction from soil, because it is likely that soil microorganisms of the same species may differ for their metabolic activity and therefore for their responses at proteomic and transcriptomic levels.

One of the major limitations of the past soil proteomic studies is the lack of information on the microbial community composition, and on the presence of key microbial species with annotated proteomes in the analysed soils. Although the soil metagenomic studies were launched since the 2000s (Rondon *et al.*, 2000), joint soil genomic-proteomic research, which can give valid indications on the presence and expression of genes, as discussed Chapter 8, were not conducted. In early proteomic studies, soil microbial communities were implicitly assumed to be evenly distributed and active, without considering the uneven distribution of soil microbial communities, with microbes localized in hot spots such as rhizosphere (Brimecombe *et al.*, 2001), as mentioned above. Another important neglected factor is the typical species dominance within the presence of rare species within soil microbial communities, which are difficult to be detected despite the marked improvement of the sequencing techniques, as discussed in Chapter 10.

Another current limitation in soil proteomics is the relative poor information of proteomic databases, which is not comparable with the increasing genomic and transcriptomic databases. A comparison of some available genomic and proteomic information in the NCBI databases is reported in Table 6.3.

Both bacterial and animal databases are more developed (King *et al.*, 2006), than fungal and plant proteomic databases, because the latter have only become active research areas in

Table 6.3 Example of genomic and proteomic information available in NBCI and Expasy SwissProt database for living organisms of various kingdoms (October 2012), of specific interest for soil proteomics.

	Genomes (NBCI)	**Proteins** (NBCI)	**Proteomes** (SwissProt)
	Total	**Entries**	**Total**
Eukaryote	4.30×10^3	1.10×10^7	3.05×10^2
Opisthokonta	3.47×10^3	7.70×10^6	2.37×10^2
Metazoa	4.28×10^3	5.52×10^6	8.50×10^1
Mesozoa	None	8.70×10^{-1}	none
Eumetazoa	3.01×10^3	5.23×10^6	8.30×10^1
Bilateria	3.02×10^2	5.18×10^6	8.20×10^1
Nematoda	7.1×10^1	2.97×10^5	9
Arthropoda	4.85×10^2	1.66×10^6	3
Anellida	9	17.67×10^3	none
Fungi	3.56×10^2	2.49×10^6	1.49×10^2
Dikarya	3.17×10^2	2.28×10^6	1.40×10^2
Ascomycota	2.50×10^2	1.90×10^6	1.18×10^2
Basidiomycota	6.70×10^1	5.08×10^5	2.20×10^1
Green plants (Viridiplantae)	4.55×10^2	2.23×10^6	2.00×10^1
Green algae (Chlorophyta)	9.8×10^1	1.68×10^5	7
Amoebozoa	2.20×10^1	1.20×10^5	6
Bacteria (eubacteria)	5.50×10^3	4.37×10^5	1.58×10^3
Proteobacteria	2.47×10^3	2.29×10^5	7.30×10^2
Alphaproteobacteria	5.63×10^2	2.81×10^6	1.67×10^2
Betaproteobacteria	3.76×10^2	2.25×10^6	1.12×10^2
Gammaproteobacteria	1.33×10^3	9.90×10^6	3.46×10^2
Firmicutes	1.47×10^3	7.84×10^6	3.40×10^2
Cyanobacteria (blue-green algae)	1.45×10^2	6.56×10^6	4.10×10^1
Fusobacteria	3.6×10^1	1.59×10^5	5
Actinobacteria	4.83×10^2	3.77×10^6	1.73×10^2
Nitrospirae	4	2.87×10^4	2
Archaea	2.43×10^2	1.25×10^3	1.23×10^2
Viruses	3.96×10^3	1.17×10^6	1.17×10^3
Plastids	1.05×10^5 (nucleotides)	4.83×10^3	None
Mitochondria	9	9.32×10^5	1

recent years (Jorrin *et al.*, 2006; Kim *et al.*, 2007). A key aspect in development of soil proteomics, is that the annotation of proteomes of environmental strains, is lower than that of human- or animal-related microbiome (Table. 6.3). In addition, the current rate of protein and proteome annotation of protozoa and earthworms in the databases is also low, and does not allow the proteomic study of soil fauna. Therefore, the present poor database does allow a limited proteomic study of soil microorganisms and soil fauna.

Even if the proteomic databases can be used for the identification of specific functional proteins and for the identification of soil microorganisms, as one protein is theoretically sufficient for organism identification via databases (Aebersold and Mann, 2003), soil metagenomics should theoretically precede soil proteomic to first reconstruction of the main metabolic pathways present and potentially active in the studied soils (Torsvik *et al.*, 2002). Proteomics provide information on functions actually expressed by soil microbial communities and their reflections at ecosystem level. From this point of view, the strain resolution level approach followed by Luo *et al.* (2007) for the study of the microbial community of a mine drainage environment, where the genomic data were used to identify proteins from dominant community members coupled with multidimensional proteomics, appears as the most promising to be applied to complex microbial communities, like those of soils (Fig. 6.1). The proteomic responses of known microorganisms with annotated proteome may be useful to increase the yields of whole soil protein extraction and provide new information on the relationship between microbial diversity and soil functions. Quantitative (sub)proteomic approaches based on stable isotope metabolic tagging using ^{13}C-glucose or, $^{15}NH_4^+$ or ^{13}C-labelled amino acids (Gygi *et al.*,1999; Ong and Mann, 2006), followed by separation and identification of the labelled proteins by LC-MS/MS, can be also helpful in detecting synthesized proteins, although this method is expensive and labour intensive. Stable isotope probing may also reveal low-abundance proteins and distinguish between low expression rates from low recovery. Targeted proteomics may be a suitable technique for detecting and quantifying proteins in soil since it can detect specific peptides or protein fragments of relevant interest for soil studies (e.g. stress biomarker protein, nutrient

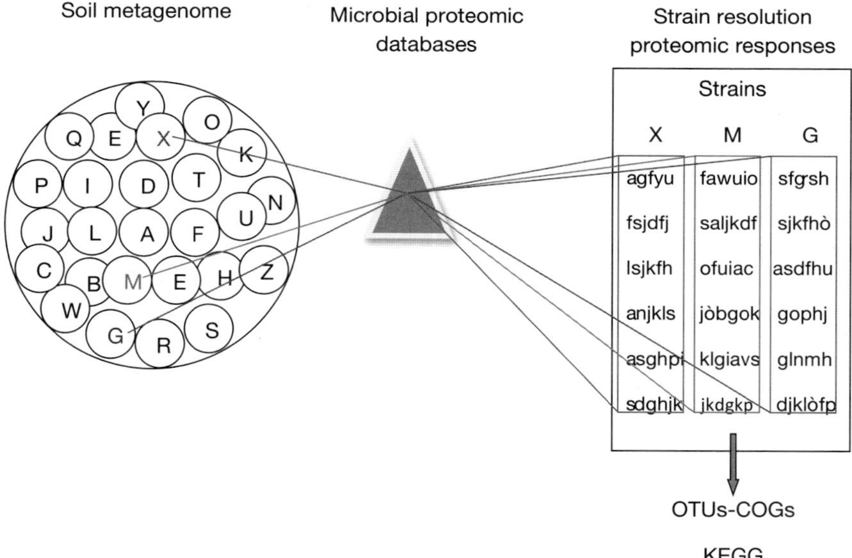

Figure 6.1 Strain resolution approach for soil proteomics. A colour version of this figure is available in the plate section at the back of the book.

hydrolysing enzymes). In addition, targeted proteomics is a quantitative approach as target stable isotope-labelled specific peptides can be prepared and added as internal standards to soil extracts (Deutsch *et al.*, 2008).

Identification of soil extracted proteins

Another present drawback of soil proteomics is the relatively low rate of protein identification, despite of the large progresses in protein detection technologies. For example, Wang *et al.* (2011) reported that one-third of the protein spots could not be identified by MS in their rhizosphere metaproteomic study. Wu *et al.* (2011) reported that only six bacterial proteins and one fungal protein were differentially expressed in their *R. glutinosa* rhizosphere proteomic study. Low rates and quality of protein identification is also common to other environments. Kan *et al.* (2005) reported that of the 140 proteins extracted and detected by the Chesapeake Bay marine microbial communities, only few of them were identified with high quality scores (i.e. two or more peptides). Even if soils display a strong proteolytic activity (Renella *et al.*, 2002), proteolysis as well as chemical denaturation may be only partially responsible for the low rate of protein recovery and identification, also by considering that lysis buffers contain protease inhibitors.

As mentioned above, proteins may react and be stabilized by soil colloids maintaining or modifying their native conformation (Miller *et al.*, 2006). Therefore, it is reasonable to hypothesize that the low protein identification rates may also be due to protein interactions with low molecular weight organic ligands (LMWOLs) occurring before or during protein extraction from soil. Such interactions may induce pitfalls in the current protein identification by MS analysis, which is based on the probability score for ranking possible peptide sequences that best fits an observed tandem mass spectra, automatically computed by the proteomic databases (e.g. Mascot). The basic assumption behind this model is that if a peptide sequence produces an established number of fragment ions, and a sufficient number of MS peaks match the predicted ones, then the 'random-chance probability is computed as to have a 5% chance of yielding correct scores. Therefore, the use of the probability score makes search engine results easier to be interpreted. However, other scores based on other mathematical procedures can be used [e.g. SEQUEST's cross-correlation score (XCorr), Mowse score] but these bioinformatics procedures require more basic knowledge (Gasteiger *et al.*, 2005). Differently from the SEQUEST, the Mascot database uses a probability based algorithm for the assignment of MS/MS spectra to peptides, and can be coupled with statistical models such as ProteinProphet and PeptideProphet, allowing a more robust and database-independent assessment of the validity of the protein identification (Keller *et al.*, 2002). While it is common to generate thousands of MS/MS spectra, the major challenge in soil proteomics is the correct peptide assignment, due to the low quality of spectra. However, in all models, all the mass-to-charge ratios (m/z) peaks are mathematically assumed to be 'independent', but this is not always the case in MS data. Therefore, the results depend on clean spectra and well-fitting data, with problems in assigning peak probability to ion fragments having low quality spectra. Although different MS manufacturers may require site specific optimal set up, uniformity in the MS spectra analysis may be useful, especially for soil proteomics due to the high diversity of protein sources and potentials of protein extracellular modification or re-arrangements. Schulze *et al.* (2005) identified proteins extracted from soil and identified by off-gel LC-MS/MS analysis. Mass spectra were searched against NBCI database, using the following search parameters: maximum of one missed trypsin cleavage, cysteine carbamidomethylation, methionine oxidation, and a maximum 0.2 Da error tolerance in both the MS and MS/MS data (40 ppm after dynamic recalibration). This setup allowed the identification of only 75 proteins, of which less of than 50% was bacterial and about 30% of proteins were identified by a single tryptic peptide.

Chourey *et al.* (2010) in their soil metaproteomic study identified 716 redundant and 333 non redundant proteins belonging to characterized soil microorganisms, based on off gel nano-LC-MS/MS, and spectra search using SEQUEST with the following parameters: system parent mass tolerance 3.0, fragment ion tolerance 0.5, up to four missed cleavages allowed; the determination was only based on fully tryptic peptides.

Wang *et al.* (2011) in their rhizosphere metaproteomic study analysed 287 2-DE spots by MALDI-TOF/TOF followed by MS/MS, and relative spectra were searched against NCBI (National Centre for Biotechnology Information) in SwissProt (http://www.matrixscience.com/search_form_select.html), applying the following parameters: one missed cleavage site, carbamidomethyl as fixed modification of cysteine and oxidation of methionine as a variable modification, MS tolerance of 100 ppm, MS/MS tolerance of 0.6 Da, and exclusions of known contaminant ions (e.g. keratin). This is an important criterion because keratins are proteins identified by MS of biological samples and their abundance can overwhelm the analytical LC-MS capacity leading to missing peptides present in low concentrations in the extracted proteome (Ding *et al.*, 2003). In the work by Chourey *et al.* (2010) common contaminants such as trypsin and keratin were included.

For soil induced modifications, potential protein modification sites should be identified and protein treatments under different conditions. Technics comparable to those commonly adopted in the study of post-translational protein modifications (e.g. phosphorylation, glycosylation), such as enrichment of phosphorylated proteins and peptides and MS set up for protein identification (Mann *et al.*, 2002) may be used in soil proteomics. In this view, the use of MS signal analysis (e.g. TPP) and the analysis of collision-induced dissociation (CID) may lead to the identification of modified ions which may serve as 'reporter ions' for preferential HS–protein interactions.

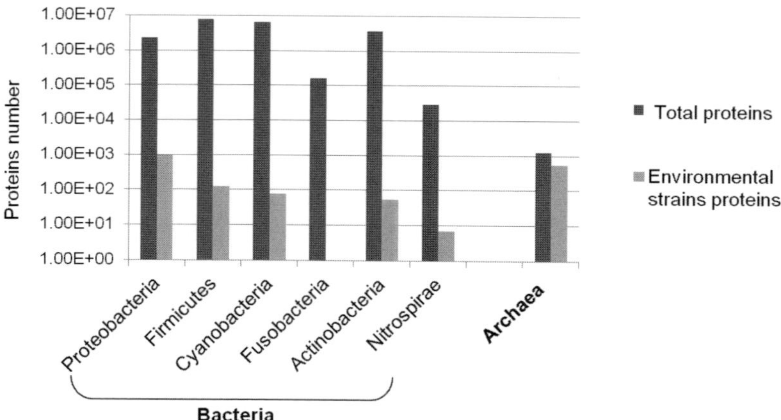

Figure 6.2 Bacterial classes with protein annotation in Expasy SwissProt database (October 2012). A colour version of this figure is available in the plate section at the back of the book.

Another important aspect related to the optimal protein identification issue is the possibilities to perform metabolic pathway analysis for soils. Nowadays, pathway approach can be done using databases such as UniProtKB/Swiss-Prot integrating information from several pathway databases (e.g. KEGG, UniPathway, Reactome). It is possible to define the role of a given protein in 1012 pathways currently available in UniPathway (Release 2012_04), with 272 biochemically defined pathways, 1010 enzymatic-reactions, and 494 sub-pathways and (Morgat *et al.*, 2012). Current uncertainties in protein identification hampers the reconstruction of proteomic-based soil metabolic pathways because identification of active metabolic networks and the protein–protein interactions require the extraction and optimal identification of the largest possible number of proteins, for a reliable analysis of soil functions from a pathway perspective. For example, Wu *et al.* (2011) identified 26 plant proteins from a rhizosphere metaproteome experiment, and recognized proteins of eight KEGG categories: carbohydrate metabolism and energy (38%), amino acid metabolism (23%), stress/defence response (11%), and proteins from the glycan metabolism, protein metabolism and signal transduction all representing 4% of the total.

Future soil proteomic studies, should be based on testing different protein identification methodologies to better understand which is the most suitable for reliable soil proteomic data. The ever increasing genome annotation in the databases has allowed the reconstruction of gene homology and speciation, and the creation of the clusters of orthologues and paralogues, which is critical for reliable prediction of gene function in newly sequenced genomes. This aspect is important for clustering orthologous genes (COG) in microbial soil communities and pathway analysis, based on the recovery and identification of whole soil proteins. According to the supramolecular theory of the soil humic substances the presence of such low molecular weight organic molecules may influence detection of proteins in soil proteomic studies by electrospray ionization mass spectrometry (ESI-MS). For example, the detection of a model tripeptide was drastically reduced after contact with hydrophobic low molecular weight organic ligands (LMWOLs) analysed by ESI-MS, due to differences in ionization of the mixture (Nebbioso *et al.*, 2010). Therefore, interactions between proteins and SOM may influence not only the protein extraction, from soil but also their purification and identification by MS. An important drawback of soil proteomic studies published in the period 1993–2006, is the absence of the used MS parameters, making the evaluation and comparisons of the reported proteins quite problematic.

Another factor that may affect protein coverage % in MS after interactions with humic fractions might be the different efficiency in producing peptides by tryptic hydrolysis, which being carried out in highly standardized conditions, has not been tested when proteins are modified by their association with LMWOLs.

Another unsolved problem is the protein quantification in soil extracts with the available routine methods, as polyphenolic compounds can interfere with the colorimetric protein quantification (Roberts and Jones, 2008). This problem prevents the use of quantitative proteomic methods such as the two-dimensional gel electrophoresis for differential analysis in proteomics (2-DiGE), which relies on the protein labelling using different fluorescent cyanine dyes and on the quantitative evaluation of fluorescence differences on the same 2-D gel, but also other quantitative proteomics such as the iTRAQ shotgun approach. This proteomic technique has the potential of discriminating the newly expressed by the analysis of proteins time course experiments, independently on inherent biological variation or differences in protein recovery from soils. However, this technique relies on the precise protein quantification among samples to be compared, and therefore the limitations of the current methods for protein quantification and uncertainty on protein recovery, which depends on protein size, abundance and reactivity towards the soil solid phases, does not meet the established procedures for the 2-D DiGE analysis.

Therefore, an additional issue to address in the next proteomic studies will be a critical evaluation of factors affecting the identification of proteins extracted from soils, and particularly the definition of potential soil organic ligands with high affinity for various protein types.

The rhizosphere

Despite of the importance of the rhizosphere for crop production (Bestel-Corre *et al.*, 2004), only few studies have focused on the plant–soil–microbe interactions at the proteomic level (Cheng *et al.*, 2010). For example, proteins involved in the symbiotic interactions between leguminous plant roots and rhizobia have been reported (Morris and Djordjevic, 2001), but the outcome of this type of *in vitro* studies cannot be scaled up to real soils. A proteomic study of soils under monoculture of *R. glutinosa* was carried out by Wu *et al.* (2011), and changes in microbial activity as evaluated by the microbial identified proteins, were due to accumulation of specific (e.g. diterpenoids) plant metabolites released in the rhizosphere.

Wang *et al* (2011) set up an extraction method suitable for metaproteome extraction of rhizosphere of rice plants, and obtained ca. 1000 2-DE spots, of which ca. 100 plant and microbial proteins were identified by MS. Interestingly, the identified proteins were mainly involved in energy balance, secondary metabolite synthesis and signal transduction.

Further progresses in rhizosphere proteomics will surely come from the increasing information on plant genome sequencing, as showed by some of the pioneering papers on the *A. Thaliana* microbe-induced proteomic responses (Peck *et al.*, 2001; Ndimba *et al.*, 2003; Bestel-Corre *et al.*, 2004).

Another important contribution to soil proteomic may derive from studies on the plant N nutrition. It has been generally accepted that protein and peptide N is available to plants only after mineralization to NH_4^+ by rhizosphere microbial community or mycorrhizae. However, plant may take up amino acids (Chapin *et al.*, 1993; Nasholm *et al.*, 2000; Nannipieri and Paul, 2009) and small peptides (Rentsch *et al.*, 2007). According to Paungfoo-Lonhienne *et al.* (2008), protein N can be taken up by plants either after protein hydrolysis on the root surface or by intact protein endocytosis. Secretion of proteases by roots of various plant species has been reported (Godlewski and Adamczyk, 2007); most of the proteases are released in the rhizosphere against soil borne plant pathogens and/or as chemical signals. Therefore, advances in soil proteomics may be helpful for a better understanding of the N cycling in soil, such as elucidating the underlying mechanisms of the 'microbial loop' concept (Bonkowski, 2004).

The sleeping beauty paradox, viable but non-culturable (VBNC) status and biofilm formation

It is well established that most (more than 90%) soil microorganisms are non-cultivable (Schloss and Handelsman, 2005). Soil microbial populations are characterized by diffuse dormancy (Grey and Williams, 1971) owing to low nutrient and H_2O availability. The concept of the sleeping beauty paradox for describing this specific physiological aspect of soil microbial communities was introduced by Lavelle *et al.* (1994) to describe the large nutrient flushes occurring upon H_2O or nutrient availability, and this may also explain why most of soil microorganisms are in a viable but non culturable (VBNC) status. However, soil conditions leading to changes in microbial viability and culturability status and metabolic pathways involved in the physiological transition phases in *in vitro* studies, showed that specific proteins (e.g. phasins) and stress-related enzymes are produced by metabolically resting cells (Xu *et al.*, 1982; Nystrom, 2005; Rodriguez-Verdugo *et al.*, 2012). The presence of such specific proteins in soil has not been reported or overlooked. Because the VBNC phenomenon was also observed in some human pathogenic bacteria such as *E. coli* (Oliver, 2005), the detection of such specific proteins may be informative about the potential virulence of dangerous human pathogens, permanently or transiently present in the soil environment, and about soil factors inducing the metabolic switch mechanisms.

Despite this, these concepts have been proposed and also proven since decades, and some bacterial marker protein of the VBNC status have been reported while none of the soil proteomics studies so far published has reported the detection of such protein markers.

Soil microorganisms may form biofilms adhering to the surface of plant roots and soil solid phases (Grossman and Lynn, 1967; Harris, 1972; Burmølle *et al.*, 2007), by releasing polymeric substances such as polysaccharides, proteins and nucleic acids (Hoiby *et al.*, 2010). Soil microorganisms forming biofilms, express different phenotypes (Sauer *et al.*, 2002) from cells living in planktonic status, and share ecological interactions which are still largely unknown (Branda *et al.*, 2006). Proteomic studies may be useful by detecting strain-specific proteins markers of biofilm formation. Vilain *et al.* (2006) showed that three oligopeptide permeases (OppA) enolase, glutamate dehydrogenase and the β-chain of the ATP synthase, were involved in the formation of biofilm by the soil-borne bacterium *B. cereus* grown on soil organic matter extracts, along with extracellular DNA. In addition, *in vitro* studies have shown that the expression of this *opp* operon is activated in various bacteria by the availability of organic C (0.05 to 1 mM) and inorganic N (0.2 mM to 20 mM) (Wang *et al.*, 2002; Lee *et al.*, 2004). Proteomic experiments with single, dual- or multi model species, incubated in the presence of nutrients, may reveal the mode in which some microbial species promote or prevent the biofilm formation in the soil environment.

Conclusions

Proteomic approach holds the potential to interrogate complex natural soils and improve our knowledge on soil functionality by assessing the gene expression through the detection of gene products and reconstruction of active metabolic pathways. However, despite two decades of research soil proteomics can be considered still in its infancy compared to other proteomic areas, mainly due to the large complexity of soil proteome, the reactivity of the soil constituents which likely cause low recovery and unknown modifications of newly expressed proteins, and relatively poor information on soil relevant species available in public databases.

Important progress in soil proteomics can be made through model studies and strain resolution approach, and significant improvements of soil proteomic can come by coupling soil (meta)genomic and proteomic studies through the assessment of the microbial dominance and the possibilities to work at the strain resolution level as shown in Fig. 6.1. This integrated approach may rely on the regulation responses of specific proteins and also support the detection of proteins identified with low confidence parameters. Even more important results will be achieved by coupling soil metaproteomics with other post-genomic approaches such as transcriptomics and metabolomics.

The current main limitations in soil proteomics to overcome appear to be in the set-up of soil protein extraction capable at the same time of sampling the whole soil proteins and minimize their degradation and permanent modifications due to affinity interactions with organic and inorganic soil solid phases and also in the genome annotation in the accessible databases.

Acknowledgements

The authors thank the Ente Cassa di Risparmio di Firenze for the financial support to the soil proteomic studies carried out at the University of Florence.

References

Aebersold, R., and Mann, M. (2003). Mass spectrometry-based proteomics. Nature *422*, 198–207.

Baron, M.H., Revault, M., Servagent-Noinville, S., Abadie, J., and Quiquampoix, H. (1999). Chymotrypsin Adsorption on Montmorillonite: Enzymatic Activity and Kinetic FTIR Structural Analysis. J. Coll. Inter. Sci. *214*, 319–332.

Bastida, F., Nicolas, C., Moreno, J.L., Hernandez, T., and Garcia, C. (2010). Tracing Changes in the Microbial Community of a Hydrocarbon-Polluted Soil by Culture-Dependent Proteomics. Pedosphere *20*, 479–485.

Benndorf, D., Balcke, G.U., Harms, H., and von Bergen, M. (2007). Functional metaproteome analysis of protein extracts from contaminated soil and groundwater. ISME J. *1*, 224–234.

Benndorf, D., Vogt, C., Jehmlich, N., Schmidt, Y., Thomas, H., Woffendin, G., Shevchenko, A., Richnow, H.H., and von Bergen, M. (2009). Improving protein extraction and separation methods for investigating the metaproteome of anaerobic benzene communities within sediments. Biodegradation *20*, 737–750.

Bestel-Corre, G., Dumas-Gaudot, E., and Gianinazzi, S. (2004). Proteomics as a tool to monitor plant–microbe endosymbioses in the rhizosphere. Mycorrhiza *14*, 1–10.

Bonkowski, M. (2004). Protozoa and plant growth: the microbial loop in soil revisited. New Phytol. *162*, 617–631.

Bos, M.A., and Nylander, T. (1996). Interaction between α-lactoglobulin and phospholipids at the air/water interface. Langmuir *12*, 2791–2797.

Branda, S.S., Chu, F., Kearns, D.B., Losick, R., and Kolter, R. (2006). A major protein component of the Bacillus subtilis biofilm matrix. Mol. Microbiol. *59*, 1229–1238.

Bremner, J.M. (1951). A review of recent work on soil organic matter. J. Soil Sci. *2*, 67–82.

Brimecombe, M.J., De Leij, F.A., and Lynch, J.M. (2001). The effect ofroot exudates on rhizosphere microbial populations. In The Rhizosphere: Biochemistry and Organic Substances at the Soil-Plant Interface, R. Pinton, Z. Varanini, and P. Nannipieri, eds. (New York, USA: Marcel Dekker), pp. 95–140.

Brözel, V.S., Luo, Y., and Vilain, S.(2011). Studying the life cycle of aerobic endospore-forming bacteria in soil. In Endospore Forming Soil Bacteria, L.A. Logan, and P. De Vos, eds. (Berlin Heidelberg, Germany: Springer-Verlag), pp. 115–134.

Burmølle, M., Hansen, L.H., and Sørensen, S.J. (2007). Establishment and early succession of a multispecies biofilm composed of soil bacteria. Microb. Ecol. *54*, 352–362.

Burns, R.G. 1986. Interaction of enzymes with soil mineral and organic colloids. In Interactions of Soil Minerals with Natural Organics and Microbes, P.M. Huang, and M. Schnitzer, eds. (Madison, USA: Soil Science Society of America), pp. 429–452.

Calamai, L., Lozzi, I., Ristori, G.G., Fusi, P., and Stotzky, G. (2000). Interaction of catalase with montmorillonite homoionic to cations with different hydrophobicity: effect on bound enzyme activity. Soil Biol. Biochem. *32*, 815–823.

Chapin, F.S. III, Moilanen, L., and Kielland, K. (1993). Preferential use of organic nitrogen for growth by a non-mycorrhizal arctic sedge. Nature *361*, 150–153.

Cargile, B.J., Talley, D.L., and Stephenson, J.L. Jr. (2004). Immobilized pH gradients as a first dimension in shotgun proteomics and analysis of the accuracy of pI predictability of peptides. Electrophoresis 25, 936–945.

Ceccanti, B., Calcinai, M., Bonmati-Pont, M., Ciardi, C., and Tansitano, R. (1989). Molecular size distribution of soil humic substances with ionic strength. Sci. Tot. Envir. *81*, 471–479.

Chen, S., Rillig, M.C., and Wang, W. (2009). Improving soil protein extraction for metaproteome analysis and glomalin related soil protein detection. Proteomics *9*, 4970–4973.

Cheng, Z., McConkey, B.J., and Glick, B.R. (2010). Proteomic studies of plant–bacterial interactions. Soil Biol. Biochem. *42*, 1673–1684.

Cheshire, M.V., and Hayes, M.H.B. (1990). Compositions, origins, structures, and reactivities of soil polysaccharides. In Soil Colloids and Their Associations in Aggregates, M.F. De Boodt, M.H.B. Hayes, and A. Herbillon, eds. (New York, USA and London, UK: Plenum), pp. 307–336.

Chourey, K., Jansson, J., VerBerkmoes, N., Shah, M., Chavarria, K.L., Tom, L.M., Brodie, E.L., and Hettich, R.L. (2010). Direct cellular lysis/protein extraction protocol for soil metaproteomics. J. Prot. Res. *9*, 6615–6622.

Craig, O.E., and Collins, M.J.(2000). Digestion and Capture Immunoassay (DACIA): An improved method for the immunological detection of mineral bound protein using hydrofluoric acid and direct capture. J. Immun. Meth. *236*, 89–97.

Craig, O.E., and Collins, M.J. (2002). An improved method for the immunological detection of mineral bound protein using hydrofluoric acid and direct capture. J. Arch. Sci. *29*, 1077–1082.

Cornell, D.G. (1982). Lipid–protein interactions in monolayers: Egg yolk phosphatidic acid and β-lactoglobulin. J. Coll. Inter. Sci. *88*, 536.

Courtois, S., Frostegård, A., Göransson, P., Depret, G., Jeannin, P., and Simonet, P. (2001).Quantification of bacterial subgroups in soil: comparison of DNA extracted directly from soil or from cells previously released by density gradient centrifugation. Environ. Microbiol. *3*, 431–439.

Criquet, S., Farnet, A.M., and Ferre, E. (2002). Protein measurement in forest litter. Biol. Fertil. Soils *35*, 307–313.

De Barjac, H., and Frachon, E. (1990). Classification of *Bacillus thuringiensis* strains. Entomophaga *35*, 233–240.

De Kruif, C.G., Weinbreck, F., and DeVries, R. (2004). Complex coacervation of proteins and anionic polysaccharides. Curr. Op. Coll. Inter. Sci. *9*, 340–349.

Delmotte, N., Knief, C., Chaffron, S., Innerebner, G., Roschitzki, B., Schlapbach, R., von Mering, C., and Vorholt, J.A. (2009). Community proteogenomics reveals insights into the physiology of phyllosphere bacteria. Proc. Natl Acad. Sci. U.S.A. *106*, 16428–16433.

Deutsch, E.W., Lam, H., and Aebersold, R. (2008). PeptideAtlas: a resource for target selection for emerging targeted proteomics workflows. EMBO Reports *9*, 429–434.

Ding, Q., Xiao, L., Xiong, S., Jia, Y., Que, H., Guo, Y., and Liu, S. (2003). Unmatched masses in peptide mass fingerprints caused by cross-contamination: an updated statistical result. Proteomics *3*, 1313–1317.

Ding, F., Diao, J.-X., Yang, X.-L., and Sun, Y. (2011). Structural analysis and binding domain of albumin complexes with natural dietary supplement humic acid. J. Luminesc. *131*, 2244–2251.

Ehlers, M.M., and Cloete, T.E. (1999). Comparing the protein profiles of 21 different activated sludge systems after SDS-PAGE. Wat. Res. *33*, 1181–1186.

Ensminger, L.E., and Gieseking, J.E. (1942). Resistance of clayadsorbed proteins to proteolytic hydrolysis. Soil Sci. *53*, 205–209.

Fusi, P., Ristori, G.G., Calamai, L., and Stotzky, G. (1989). Adsorption and binding of protein on 'clean' (homoionic) and dirty (coated with Fe oxyhydroxides) montmorillonite, illite and kaolinite. Soil Biol. Biochem. *21*, 911–920.

Gasteiger, E., Hoogland, C., Gattiker, A., Duvaud, S., Wilkins, M.R., Appel, R.D., and Bairoch, A. (2005). Protein Identification and Analysis Tools on the ExPASy Server. In The Proteomics Protocols Handbook, J.M. Walker, ed. (Totowa, NJ: Humana Press), pp. 571–607.

Giagnoni, L., Magherini,F., Landi, L., Taghavi, S., Modesti, A., Bini, L., Nannipieri, P., van der Lelie, D., and Renella, G. (2011). Extraction of microbial proteome from soil: Potential and limitations assessed through a model study. Eur. J. Soil Sci. *62*, 74–81.

Giagnoni, L., Magherini,F., Landi, L., Taghavi, S., van der Lelie, D., Puglia, M., Bianchi, L., Bini, L., Nannipieri, P., Renella, G., and Modesti, A. (2012). Soil solid phases effects on the proteomic analysis of *Cupriavidus metallidurans* CH34. Biol. Fertil. Soils *48*, 425–433.

Gianfreda, L., and Bollag, J.M. (1996). Influence of natural and anthropogenic factors on enzyme activity. In Soil Biochemistry, 9, G. Stotzky, and J.-M. Bollag, eds. (New York, USA: Marcel Dekker) pp. 123–193.

Godlewski, M., and Adamczyk, B. (2007). The ability of plants to secrete proteases by roots. Pl. Physiol. Biochem. *45*, 657–664.

Gray, T.R.G., and Williams, S.T.(1971). Soil Microorganisms (Edinburgh,UK: Oliver and Boyd).

Grossman, R.B., and Lynn W.C. 1967. Gel-like films that may form at the air–water interface in soils. Proc. Soil Sci. Soc. Am. *31*, 259–262.

Gygi, S.P., Rochon, Y., Franza, B.R., and Aebersold, R. (1999). Correlation between protein and mRNA abundance in yeast. Mol. Cell Biol. *19*, 1720–1730.

Habermann, B., Oegema, J., Sunyaev,S., and Shevchenko, A. (2004). The power and the limitations of cross-species protein identification by mass spectrometry-driven sequence similarity searches. Mol. Cell. Proteom. *3*, 238–249.

Hamzehi, E., and Pflug, W. (1981). Sorption and binding mechanism of polysaccharide cleaving soil enzymes by clay minerals. Zeit. Pflanzen. Bodenkd, *144*, 505–513.

Harris, P.J. (1972). Microorganisms in surface films from soil crumbs. Soil Biol. Biochem. *4*, 105–106.

Haynes, C.A., and Norde, W. (1995). Structures and stabilities of adsorbed proteins. J. Coll. Inter. Sci. *169*, 313–328.

Helassa, N., Quiquampoix, H., Noinville, S., Szponarski, W., and Staunton, S. (2009). Adsorption and desorption of monomeric Bt (*Bacillus thuringiensis*) Cry1Aa toxin on montmorillonite and kaolinite. Soil Biol. Biochem. *41*, 498–504.

Hernandez, P., Müller, M., and Appel, R.D. (2006). Automated protein identification by tandem mass spectrometry: issues and strategies. Mass Spect. Rev. *25*, 235–54.

Hoiby, N., Bjarnsholt, T., Givskov, M., Molin, S., and Ciofu, O. (2010). Antibiotic resistance of bacterial biofilms. Int. J. Antimicrob. Ag. *35*, 322–332.

Hsu, P.H., and Hatcher, P.G. (2005). New evidence for covalent coupling of peptides to humic acids based on 2D NMR spectroscopy: A means for preservation. Geochim. Cosmochim. Acta *69*, 4521–4533.

Hsu, P.H., and Hatcher, P.G. (2006). Covalent coupling of peptides to humic acids: Structural effects investigated using 2D NMR spectroscopy. Org. Geochem. *37*, 1694–1704.

Huang, Q., Jiang, M., and Li, X. (1998). Effects of iron and aluminum oxides and kaolinite on adsorption and activity of invertase. Pedosphere *8*, 251–260.

Huang, Q., Zhao, Z., and Chen, W. (2003). Effects of several low molecular weight organic acids and phosphate on the adsorption of acid phosphatase by soil colloids and minerals. Chemosphere *52*, 571–579.

Karas, M., Bachmann, D., and Hillenkamp, F. (1985). Influence of the Wavelength in High-Irradiance Ultraviolet Laser Desorption Mass Spectrometry of Organic Molecules. Anal. Chem. *57*, 2935–2939.

Karas, M., and Hillenkamp, F. (1988). Laser desorption ionization of proteins with molecular mass exceeding 10000 daltons. Anal. Chem. *60*, 2299–2301.

Kan, J., Hanson, T.E., Ginter, J.M., Wang, K., and Chen, F. (2005). Metaproteomic analysis of the Chesapeake Bay microbial communties. Saline Syst. *1*, 7–14.

Keller, A., Nesvizhskii, A.I., Kolker, E., and Aebersold, R. (2002). Empirical stattistical model to estimate the accuracy of peptide identifications made by MS/MS database search. Anal. Chem. *74*, 5383–5392.

Keller, M., and Hettich, R. (2009). Environmental proteomics: a paradigm shift in characterizing microbial activities at the molecular level. Microbiol. Mol. Biol. Rev. *73*, 62–70.

Kelleher, B.P., and Simpson, A.J. (2006). Humic substances in soils: are they really chemically distinct? Env. Sci. Technol. *40*, 4605–4611.

Keiblinger, K.M., Wilhartitz, I.C., Schneider, T., Roschitzki, B., Schmid, E., Eberl, L., Riedel, K., and Zechmeister-Boltenstern, S. (2012). Soil metaproteomics e Comparative evaluation of protein extraction protocols. Soil Biol. Biochem. *54*, 14–24.

Kim, Y., Nandakumar, M.P., and Marten, M.R. (2007). Proteomics of filamentous fungi. Trends in Biotechnol. *25*, 395–400.

King, N.L., Deutsch, E.W., Ranish, J.A., Nesvizhskii, A.I., Eddes, J.S., Mallick, P., Eng. J., Desiere, F., Flory, M., Martin, D.B., Kim, B., Lee, H., Raught, B., and Aebersold, R. (2006). Analysis of the *Saccharomyces cerevisiae* proteome with PeptideAtlas. Genome Biol. *7*, R106.

Kleber, M., Sollins, P., and Sutton, R. (2007). A conceptual model of organo–mineral interactions in soils: self assembly of organic molecular fragments into zonal structures on mineral surfaces. Biogeochemistry *85*, 9–24.

Klose, J. (1975). Protein mapping by combined isoelectric focusing and electrophoresis of mouse tissues. A novel approach to testing for induced point mutations in mammals. Humangenetik *26*, 231–243.

Knicker, H., and Hatcher, P.G. (1997). Survival of protein in an organicrich sediment. Possible protection by encapsulation in organic matter. Naturwissenschaften. *84*, 231–234.

Kolkman, A., Dirksen, E.H., Slijper, M., and Heck, A.J. (2005). Double Standards in Quantitative Proteomics. Direct Comparative Assessment of Difference in Gel Electrophoresis and Metabolic Stable Isotope Labeling. Mol. Cell Proteom. *4*, 255–266.

Koskella, J., and Stotzky, G. (1997). Microbial utilization of free and clay-bound insecticidal Toxins from *Bacillus thuringiensis* and their retention of insecticidal activity after incubation with microbes. Appl. Environ. Microbiol. *63*, 3561–3568.

Jenkinson, D.S., and Tinsley, J.A. (1960). Comparison of the ligno-protein isolated from a mineral soil and from a straw compost. Royal Dublin Soc. Sci. Proc. *10*, 141–147.

Jenny, H. (1941). Factors of soil formation: a system of quantitative pedology (NewYork, USA: McGraw-Hill Book Company).

Jorrin, J.V., Maldonado, A.M., and Castillejo, M.A. (2006). Plant proteome analysis: a 2006 update. Proteomics 7, 2947–2962.

Lambert, J.P., Ethier, M., Smith, J.C., and Figeys, D. (2005). Review Proteomics: from gel based to gel free. Anal. Chem. *77*, 3771–3787.

Lavelle, P., Lattaud, C., Trigo, D., and Barois, I. (1994). Mutualism and biodiversity in soil. Pl. Soil *170*, 23–33.

Lee, L., Saxena, D., and Stostky, G. (2003). Activity of free and clay-bound insecticidal proteins from *Bacillus thuringiensis* subsp. *israelensis* against the mosquito *Culex pipiens*. Appl. Environ. Microbiol. *69*, 4111–4115.

Lee, E.M., Ahn, S.H., Park, J.-H., Lee, J.H., Ahn, S.C., and Kong, I.S. (2004). Identification of oligopeptide permease (*opp*) gene cluster in Vibrio fluvialis and characterization of biofilm production by oppA knockout mutation. FEMS Microbiol. Lett. *240*, 21–30.

Link, A.J. (1999). Autoradiography of 2-D gels. Meth. Mol. Biol. *112*, 285–290.

Luo, Y., Vilain, S., Voigt, B., Albrecht, D., Hecker, M., and Brözel, V.S. (2007). Proteomic analysis of *Bacillus cereus*, growing in liquid soil organic matter. FEMS Microbiol. Lett. *271*, 40–47.

Macko, V., and Stegemann, H. (1969). Mapping of potato proteins by combined electrofocusing and electrophoresis identification of varieties. Hoppe-Seyler's Zeitsch. Physiol. Chemie *350*, 917–919.

Mann, M., Højrup, P., and Roepstorff, P. (1993). Use of mass spectrometric molecular weight information to identify proteins in sequence databases. Biol. Mass Spec. *22*, 338–345.

Mann, M., Ong, S.-E., Grønborg, M., Steen, H., Jensen, O.N., and Pandey, A. (2002). Analysis of protein phosphorylation using mass spectrometry: deciphering the phosphoproteome. Tr. Biotechnol. *20*, 261–268.

Maron, P.A., Schimann, H., Ranjard, L., Brothier, E., Domenach, A.M., Lensi, R., and Nazaret, S. (2006). Evaluation of quantitative and qualitative recovery of bacterial communities from different soil types by density gradient centrifugation. Eur. J. Soil Biol. *42*, 65–73.

Maron, P.-A., Maitre, M., Mercier, A., Lejon, D.P.H., Nowak, V., and Ranjard, L. (2008). Protein and DNA fingerprinting of a soil bacterial community inoculated into three different sterile soil. Res. Microbiol. *159*, 231–236.

Miller, I., Crawford, J., and Gianazza, E. (2006). Protein stains for proteomic applications: which, when, why? Proteomics *6*, 5385–408.

Miltner, A., and Zech, W. (1999). Microbial degradation and resynthesis of proteins during incubation of beech leaf litter in the presence of mineral phases. Biol. Fertil. Soils *30*, 48–51.

Morgat, A., Coissac, E., Coudert, E., Axelsen, K.B., Keller, G., Bairoch, A., Bridge, A., Bougueleret, L., Xenarios, I., and Viari, A. (2012). UniPathway: a resource for the exploration and annotation of metabolic pathways. Nucl. Acids Res. *40*, D761-D769.

Morris, A.C., and Djordjievic, M.A. (2001). Proteome analysis of cultivar–specific interactions between *Rhizobium leguminosarum* bv *trifolii* and subterranean clover cultivar Woogenellup. Electrophoresis *22*, 586–598.

Müller, L.N., Brusniak, M.Y., Mani, D.R., and Aebersold, R. (2008). An assessment of software solutions for the analysis of mass spectrometry based quantitative proteomics data. J. Prot. Res. *7*, 51–61.

Naidja, A., and Huang, P.M. (1995). Deamination of aspartic acid by aspartase–Ca–montmorillonite complex. J. Mol. Catal. A: Chemical *106*, 255–265.

Nannipieri, P. (2006). Role of stabilised enzymes in microbial ecology and enzyme extraction from soil with potential applications in soil proteomics. In Nucleic Acids and Proteins in Soil, P. Nannipieri and K. Smalla, eds. (Heidelberg, Germany: Springer), pp. 75–94.

Nannipieri, P., Sequi, P., and Fusi, P. (1996). Humus and enzyme activity. In Humic Substances in Terrestrial Ecosystems, A. Piccolo, ed. (New York, USA: Elsevier), pp 293–328.

Nannipieri, P., Ascher, J., Ceccherini, M.T., Landi, L., Pietramellara, G., and Renella, G. (2003). Microbial diversity and soil functions. Eur. J. Soil Sci. *54*, 655–670.

Nannipieri, P., and Paul, E. (2009). The chemical and functional characterization of soil N and its biotic components. Soil Biol. Biochem. *41*, 2357–2369.

Nannipieri, P. (2009). Proteomic analysis of Cupriavidus metallidurans CH34:a model for soil proteomic studies. In Proceedings of the 10th International Symposium on Bacterial Genetics and Ecology (Bageco-10). Uppsala, Sweden, June 15–19, 2009, p. 67.

Nasholm, T., Huss-Danell, K., and Hogberg, P. (2000). Uptake of organic nitrogen in the field by four agriculturally important plant species. Ecology *81*, 1155–1161.

Ndimba, B.K., Chivasa, S., Hamilton, J.M., Simon, W.J., and Slabas, A.R. (2003). Proteomic analysis of changes in the extracellular matrix of *Arabodopsis* cell suspension cultures induced by fungal elicitors. Proteomics 3, 1047–1059.

Nebbioso, A., Piccolo, A., and Spiteller, M. (2010). Limitations of electrospray ionization in the analysis of a heterogeneous mixture of naturally occurring hydrophilic and hydrophobic compounds. Rapid Comm. Mass Spec. *24*, 3163–3170.

Norde, W. (1986). Adsorption of proteins from solution at the solid–liquid interface. Adv. Coll. Inter. Sci. *25*, 267–340.

Norde, W., and Lykema, G. (1978). The adsorption of human plasma albumin and bovine pancreas ribonuclease at negatively charged polystyrene surfaces. J. Coll. Inter. Sci. *66*, 257–265.

Nystrom, T. (2005). Bacterial senescence, programmed death, and premeditated sterility. ASM News *71*, 363–369.

O'Farrell, P.H. (1974). High resolution two-dimensional electrophoresis of proteins. J. Biol. Chem. *250*, 4007–4021.

Ogunseitan, O.A. (1993). Direct extraction of proteins from environmental samples. J. Microbiol. Methods *17*, 273–281.

Ogunseitan, O.A. (1997). Direct extraction of catalytic proteins from natural microbial communities. J. Microbiol. Methods *28*, 55–63.

Oliver, J.D. (2005). The viable but non-culturable state in bacteria. J. Microbiol. *43*, 93–100.

Ong, S.E., and Mann, M. (2005). Mass spectrometry-based proteomics turns quantitative. Nat. Chem. Biol. *1*, 252–262.

Ong, S.E., and Mann, M. (2006). A practical recipe for stable isotope labeling by amino acids in cell culture (SILAC). Nat. Prot. *1*, 2650–2660.

Paungfoo-Lonhienne, C., Lonhienne, T.G.A., Rentsch, D., Robinson, N., Christie, M., Webb, R.I., Gamage, H.K., Carroll, B.J., Schenk, P.M., and Schmidt, S. (2008). Plants can use protein as a nitrogen source without assistance from other organisms. Proc. Natl. Acad. Sci. U.S.A. *105*, 4524–4529.

Patterson, S.D., and Aebersold, R.H. (2003). Proteomics: the first decade and beyond. Nat. Genetics 33, 311–323.

Peck, S., Nuhse, C., Hess, T.S., Iglesias, T., Meins, A., and Boller, T. (2001). Direct proteomics identifies a plant-specific protein rapidly phosphorilated in response to baterial and fungal elicitors. Pl. Cell *13*, 1467–1475.

Piccolo, A. (2002). The Supramolecular structure of humic substances. A novel understanding of humus chemistry and implications in soil science. Adv. Agr. *75*, 57–134.

Pinck, L.A., and Allison, F.E. (1951). Resistance of a protein montmorillonite complex t decomposition by soil microorganisms. Science *114*, 130–131.

Quiquampoix, H., Chassin, P., and Ratcliffe, R.G. (1989). Enzyme activity and cation exchange as tools for the study of the conformation of proteins adsorbed on mineral surfaces. Prog. Coll. Pol. Sci. *79*, 59–63.

Quiquampoix, H., and Ratcliffe, R.G. (1992). A ^{31}P NMR study of the adsorption of bovine serum albumin on montmorillonite using phosphate and the paramagnetic cation Mn^{2+}: modification of conformation with pH. J. Coll. Inter. Sci. *148*, 343–352.

Quiquampoix, H., Staunton, S., Baron, M.H., and Ratcliffe, R.G. (1993). Interpretation of the pH dependence of protein adsorption on clay mineral surfaces and its relevance to the understanding of extracellular enzyme activity in soil. Coll. Surf. A: Physicochemical and Engineering Aspects 75, 85–93.

Quiquampoix, H. (2000). Mechanisms of protein adsorption on surfaces and consequences for extracellular enzyme activity in soil. In Soil Biochemistry vol. 10, J.M. Bollag, and G. Stotzky, eds. (NewYork, USA: Marcel Dekker), pp. 171–206.

Quiquampoix, H., Servagent-Noinville, S., and Baron, M. (2002). Enzyme adsorption on soil mineral surfaces and consequences for the catalytic activity. In Enzymes in the Environment, R.G. Burns, R.P. Dick, eds. (NewYork, USA: Marcel Dekker), pp. 285–306.

Quiquampoix, H., and Mousain, D. (2005). Enzymatic hydrolysis of organic phosphorus. In Organic Phosphorus in the Environment, B.L. Turner, E. Frossard, and D.S. Baldwin, eds. (Wallingford, UK: CABI Publishing), pp 89–112.

Quiquampoix, H., and Burns, R.G. (2007). Interactions between proteins and soil mineral surfaces: environmental and health consequences. Elements *3*, 401–406.

Quiquampoix, H. 2008. Enzymes and proteins, interactions with soil constituent surfaces. In Encyclopedia of Soil Science, W. Chesworth, ed. (Dordrecht, Berlin, Heidelberg, Germany;New York, USA: Springer), pp. 210–216.

Ram, R.J., VerBerkmoes, N.C., Thelen, M.P., Tyson, G.W., Baker, B.J., Blake, R.C., Shah, M., Hettich, R.L., and Banfield, J.L. (2005). Community proteomics of a natural microbial biofilm. Science *308*, 1915–1920.

Renella, G., Landi, L., and Nannipieri, P. (2002). Hydrolase activity during and after the chloroform-fumigation of soils as affected by protease activity. Soil Biol. Biochem. *34*, 51–60.

Rentsch, D., Schmidt, S., and Tegeder, M. (2007). Transporters for uptake and allocation of organic nitrogen compounds in plants. FEBS Lett. *581*, 2281–2289.

Roberts, P., and Jones, D.L. (2008). Critical evaluation of methods for determining total protein in soil solution. Soil Biol. Biochem. *40*, 1485–1495.

Rodriguez-Verdugo, A., Souza, V., Eguiarte, L.E., and Escalante, A.E. (2012). Diversity across seasons of culturable *Pseudomonas* from a desiccation lagoon in cuatro cienegas, Mexico. Int. J. Microbiol. *2012*, article ID 201389, p. 10.

Rondon, M.R., August, P.R., Bettermann, A.D., Brady, S.F., Grossman, T.H., Liles, Loiacono, K.A., Lynch, B.A., MacNeil,.I.A., Minor, C., Tiong, C.L., Gilman, M., Osburne, M.S., Clardy, J., Handelsman, J., and Goodman, R.M. (2000). Cloning the soil metagenome: a strategy for accessing the genetic and functional diversity of uncultured microorganisms. Appl. Environ. Microbiol. *66*, 2541–2547.

Rupert, J.P., Granquist, W.T., and Pinnavaia, T.J. (1987). Catalytic Properties of Clay Minerals. In Chemistry of Clays and Clay Minerals, Newman, A.C.D., ed. (London, UK: Mineralogical Society), pp. 275–318.

Sakurai, T., Matsuo, T., Matsuda, H., and Katakuse, I. (1984). Mathematical tools in analytical mass spectrometry. Biom. Mass Spect. *11*, 396–402.

Sauer, K., Camper, A.K., Ehrlich, G.D., Costerton, J.W., and Davies, D.G. (2002). *Pseudomonas aeruginosa* displays multiple phenotypes during development as a biofilm. J. Bacteriol. *184*, 1140–1154.

Schloss, P.D., and Handelsman, J. (2005). Metagenomics for studying unculturable microorganisms: cutting the Gordian knot. Genome Biol. *6*, 229.

Schindel, F.V., Mercer, E.J., and Rice, J.A. (2007). Chemical characteristics of glomalin-related soil protein (GRSP) extracted from soils of varying organic matter content. Soil Biol. Biochem. *39*, 320–329.

Schmidt, M.W., Torn, M.S., Abiven, S., Dittmar, T., Guggenberger, G., Janssens, I.A., Kleber, M., Kögel-Knabner, I., Lehmann, J., Manning, D.A., Nannipieri, P., Rasse, D.P., Weiner, S., and Trumbore, S.E. (2011). Persistence of soil organic matter as an ecosystem property. Nature *478*, 49–56.

Schneider, T., and Riedel, K. 2010. Environmental proteomics: Analysis of structure and function of microbial communities. Proteomics *10*, 785–798.

Schneider, T., Gerrits, B., Gassmann, R., Schmid, E., Gessner, M.O., Richter, A., Battin, T., Eberl, L., and Riedel, K. (2010). Proteome analysis of fungal and bacterial involvement in leaf litter decomposition. Proteomics *10*, 1819–1830.

Schnitzer, M. (1986). Binding of humic substances by soil mineral colloids. In Interactions of Soil Minerals with Natural Organics and Microbes, P.M. Huang, and M. Schnitzer, eds. (Madison, USA: SSSA Special Publication 17, Soil Science Society of America), pp.78–102.

Schulze, W.X., Gleixner, G., Kaiser, K., Guggenberger, G., Mann, M., and Schulze, E.D. (2005). A proteomic fingerprint of dissolved organic carbon and of soil particles. Oecologia *142*, 335–343.

Servagent-Noinville, S., Revault, M., Quiquampoix, H., and Baron, M.H. (2000). Conformational Changes of Bovine Serum Albumin Induced by Adsorption on Different Clay Surfaces: FTIR Analysis. J. Coll. Inter. Sci. *221*, 273–283.

Simonart, P., Batistic, L., and Mayaudon, J. (1967). Isolation of protein from humic acid extracted from soil. Pl. Soil *27*, 153–161.

Singleton, I., Merrington, G., Colvan, S., and Delahunty, J.S. (2003). The potential of soil protein-based methods to indicate metal contamination. Appl. Soil Ecol. *23*, 25–32.

Solaiman, Z., Marschner, P., Wang, D.M., and Rengel, Z. (2007). Growth, P uptake and rhizosphere properties of wheat and canola genotypes in an alkaline soil with low P availability. Biol. Fertil. Soils *44*, 143–153.

Steinberg, P.D., and Rillig, M.C. (2003). Differential decomposition of arbuscular mycorrhizal fungal hyphae and glomalin. Soil Biol. Biochem. *35*, 191–194.

Staunton, S., and Quiquampoix, H. (1994). Adsorption and conformation of bovine serum album on montmorillonite: modification of the balance between hydrophobic and electrostatic interactions by protein methylation and pH variation. J. Coll. Inter. Sci. *166*, 89–97.

Stevenson, F.J. (1994). Humus Chemistry: genesis, composition, reactions, 2nd edition (New York, USA: Wiley and Sons), p. 49.

Stotzky, G. (1986). Influence of soil mineral colloids on metabolic processes, growth, adhesion and ecology of microbes and viruses. In Interactions of Soil Minerals with Natural Organics and Microbes, P.M. Huang, M. Schnitzer, eds. (Madison, USA: SSSA Special Publication 17, Soil Science Society of America,), pp. 305–428.

Swaby, R.J., and Ladd, J.N. (1964). Chemical nature, microbial resistance and origin of soil humus. Chem. Abstr. *60*, 4727b.

Tan, W.F., Koopal, L.K., Weng, L.P., van Riemsdijk, W.H., and Norde, W. (2008). Humic acid protein complexation. Geochim. Cosmochim. Acta *72*, 2090–2099.

Tan, W.F., Koopal, L., and Norde, W. (2009). Interaction between humic acid and lysozyme, studied by dynamic light scattering and isothermal titration calorimetry. Env. Sci. Technol. *43*, 591–596.

Taylor, E.B., and Williams, M.A. (2009). Microbial protein in soil: influence of extraction method and C amendment on extraction and recovery. Microb. Ecol. *59*, 390–399.

Theng, B.K.G. (1974). The chemistry of clay-organic reactions (New York, USA: John Wiley and Sons).

Thompson, M.R., Chourey, K., Froelich, J.M.. Erickson, B.K., VerBerkmoes, N.C., and Hettich, R.L. (2008). Experimental approach for deep proteome measurements from small-scale microbial biomass samples. Anal. Chem. *80*, 9517–9525.

Tomaszewski, J.E., Schwarzenbach, R.P., and Sander, M. (2011). Protein Encapsulation by Humic Substances. Env. Sci. Technol. *45*, 6003–6010.

Torsvik, V., Øvreås, L., and Thingstad, T.F. (2002). Prokaryotic diversity – magnitude, dynamics and controlling factors. Science *296*, 1064–1066.

Tyson, G.W., Chapman, J., Hugenholtz, P., Allen, E.E., Ram, R.J., Richardson, P.M., Solovyev, V.V., Rubin, E.M., Rokhsar, D.S., and Banfield, J.F.(2004). Community structure and metabolism through reconstruction of microbial genomes from the environment. Nature *428*, 37–43.

Tyers, M., and Mann, M. (2003). From genomics to proteomics. Nature *422*, 193–197.

Unlu, M., Morgan, M.E., and Minden, J.S. (1997). Difference gel electrophoresis: a single gel method for detecting changes in protein extracts. Electrophoresis *18*, 2071–2077.

Vilain, S., Luo, Y., Hildreth, M.B., and Brözel, V.S. (2006). Analysis of the life cycle of the soil saprophyte *Bacillus cereus* in liquid soil extract and in soil. Appl. Environ. Microbiol. *72*, 4970–4977.

Violante, A., and Gianfreda, L. (1995). Adsorption of phosphate on variable charge minerals: competition effect of organic ligands. In Environmental Impact of Soil Component Interactions. In: Metals, Other Inorganic and Microbial Activities, vol. II, P.M. Huang, J. Berthelin, J.M. Bollag, W.B. McGill, A.L. Page, eds. (Boca Raton, USA: CRC Press), pp. 29–38.

Wang, X.-G., Lin, B., Kidder, J.M., Telford, S., and Hu, L.T. (2002). Effects of environmental changes on expression of the oligopeptide permease (*opp*) genes of *Borrelia burgdorferi*. J. Bacteriol. *184*, 6198–6206.

Wang, H.B., Zhang, Z.X., Li, H., He, H.B., Fang, C.X., Zhang, A.J., Li, Q.S., Chen, R.S., Guo, X.K., Lin, H.F., Wu, L.K., Lin, S., Chen, T., Lin, R.Y., Peng, X.X., and Lin, W.X. 2011. Characterization of metaproteomics in crop rhizospheric soil. J. Prot. Res. *10*, 932–940.

Wasinger, V.C., Cordwell, S.J., Cerpa-Poljak, A., Yan, J.X., Gooley, A.A., Wilkins, M.R., Duncan, M.W., Harris, R., Williams, K.L., and Humphery-Smith, I. (1995). Progress with gene product mapping of the Mollicutes: Mycoplasma genitalium. Electrophoresis *16*, 1090–1094.

Wilkins, M.R., Sanchez, J.C., Gooley, A.A., Appel, R.D., Humphery-Smith, I., Hochstrasser, D.F., and Williams, K.L. (1996). Progress with proteome projects: why all proteins expressed by a genome should be identified and how to do it. Biotech. Gen. Eng. Rev. *13*,19–50.

Williams, M.A., Taylor, E.B., and Mula, H.P. (2010). Metaproteomic characterization of a soil microbial community following carbon amendment. Soil Biol. Biochem. *42*, 1148–1156.

Wilmes, P., and Bond, P.L. (2004). Metaproteomics: studying functional gene expression in microbial ecosystems. Trend Microbiol. *14*, 92–97.

Wright, S.F., and Upadhyaya, A. (1996). Extraction of an abundant and unusual protein from soil and comparison with hyphal protein of arbuscular mycorrhizal fungi. Soil Sci. *161*, 575–585.

Wright, S.F., and Upadhyaya, A. (1998). A survey of soils for aggregate stability and glomalin, a glycoprotein produced by hyphae of arbuscular mycorrhizal fungi. Pl. Soil *198*, 97–107.

Wu L., Wang, H., Zhang, Z., Lin, R., Zhang, Z., and Lin, W. (2011). Comparative metaproteomic analysis on consecutively *Rehmannia glutinosa*-monocultured rhizosphere soil. PlosOne *6*, e20611.

Xu, H.S., Roberts, N., Singleton, F.L., Attwell, R.W., Grimes, D.J., and Colwell, R.R. (1982). Survival and viability of nonculturable *Escherichia coli* and *Vibrio cholerae* in the estuarine and marine environment. Micr. Ecol. *8*, 313–323.

Yu, C.H., Norman, M.A., Newton, S.Q., Miller, D.M., Teppen, B.J., and Schäfer, L. (2000). Molecular dynamics simulations of the adsorption of proteins on clay mineral surfaces. J. Mol. Str. *556*, 95–103.

Yuan, W., and Zydney, L. (2000). Humic acid fouling during ultrafiltration Env. Sci. Tech. *34*, 5043–5050.

Zang, X., van Heemst, J.D.H., Dria, K.J., and Hatche, P.G. (2000). Encapsulation of protein in humic acid from a histosol as an explanation for the occurrence of organic nitrogen in soil and sediment Org. Geochem. *31*, 679–695.

7 Soil Volatile Organic Compounds as Tracers for Microbial Activities in Soils

Heribert Insam

Abstract

Soil volatilomics is a scientific discipline that deals with the multitude of volatile organic compounds produced, stored or degraded in soils. These compounds may be of plant or microbial origin, or they may enter or leave the atmosphere through diffusion and convection processes. The present article focuses on soil volatile organic compounds of which many are produced by microorganisms. The challenges and chances of using such VOCs for characterizing soil microbial communities are addressed, and several examples on microbe-to-microbe and microbe–plant interactions involving VOCs are given. Also, some problems concerning the measurement of these VOCs are addressed.

Introduction

Genomics, metagenomics, transcriptomics and whatever method and term is used to determine microbial community composition and activity on a genome basis, they have all in common that first of all DNA has to be extracted, amplified and sequenced or analysed by a fingerprint method. Extraction steps apply also to other methods for community analysis, like phospholipid fatty acid (PLFA) analyses, or quantification of specific cell-wall compounds such as murein or ergosterol for bacteria and fungi, respectively. As described in Chapter 10, extractions of nucleic acids and proteins from soil are not only a problem of yields but also completeness. Often, extraction methods bias against or towards certain members of communities, be it DNA (Bakken and Frostegård, 2006; Delmont *et al.*, 2011), or in analogy proteins or certain cell wall components (Nannipieri, 2006; Giagnoni *et al.*, 2010). Not enough, extracted compounds may interact with other components, such as clay, organic matter or co-extracts, again inducing undesired biases (Chapter 10).

If some compounds were available that would emerge from the sample by themselves and that could be used for microbial fingerprinting the researcher would have an exciting tool. Any extraction step could be skipped, and the evolved compounds be analysed without an extraction bias: no shaking or soaking or boiling or freezing-and-thawing or judging which of the weak or strong extractants should possibly be used. Volatilomics would thus offer a new window into the black box of the soil biota (Insam and Seewald, 2010).

A special group of such volatile substances, which are classified by chemical properties like a vapour pressure exceeding 0.1 mbar (at 20°C) and a boiling point below 240°C (at standard pressure) are called volatile organic compounds (VOCs). VOCs can be found in widespread environments, including soils, composts, litter and sediments and their origin may be from various sources, plants, animals or microbes. A significant proportion of the VOCs released from soil are presumably of microbial origin (Leff and Fierer, 2008) produced of different metabolic pathways of the secondary metabolism of microorganisms. They may be metabolic waste products or used as signal substances or for other purposes. Such microbial (m)VOCs include alcohols, ketones, aldehydes, terpenes, amines and sulphur compounds (Dott *et al.*, 2004). They can be detected before any visible sign of microbial growth appears. These mVOCs are commonly used to specify and characterise fungi, especially moulds (Bjurman *et al.*, 1997). Specific VOCs can be used to indicate fungal, bacterial and other food spoilage (Börjesson *et al.*, 1990, 1992; Keshri *et al.*, 1998; Gao and Martin, 2002; Mayr *et al.*, 2003), to characterize odour contamination in composting processes (Smet *et al.*, 1999) and during storage of organic household wastes (Mayrhofer *et al.*, 2006) and to detect mould growth in buildings (Wilkins *et al.*, 2000; Fischer and Dott, 2003). Soil VOCs have been used for studying the effect of organic fertilization with composts on the soil microbial community (Seewald *et al.*, 2010), and they are a considerable component of any ecosystem-level VOC flux study (e.g. Brilli *et al.*, 2011, Ruuskanen *et al.*, 2011).

Different technologies are available to detect and quantify VOCs, including gas chromatography/mass spectrometry (Börjesson *et al.*, 1992; Smet *et al.*, 1999), electronic noses (Rajamäki *et al.*, 2005) or the assignment of sniffing dogs to detect moulds in buildings. These methods for the characterisation of VOCs in soils are expensive or time consuming (Alvarado and Rose, 2004) and do usually not allow on-line measurements. With proton transfer reaction mass spectrometry (PTR-MS), there is a new and very sensitive technique for online VOC measurements available (Lindinger *et al.*, 1998). The PTR-MS has been used for the detection of VOCs in food quality control (Mayr *et al.*, 2003), decomposition of organic wastes (Mayrhofer *et al.*, 2006), for plant physiological studies (Mayrhofer *et al.*, 2004), grassland (Ruuskanen *et al.*, 2011) and forest (Grabmer *et al.*, 2004) ecology and for several other environmental studies. The advantage of PTR-MS is the capability of on-line measurement of several volatile organic compounds with a detection limit of less than 0.1 ppbv.

This review is destined to give an overview on current knowledge on volatile organic compounds present in soils that might be used for fingerprinting purposes. These VOCs might be produced by microorganisms in their soil habitat, they might be produced by plant roots or enter the soil from the atmosphere by diffusion. Eventually, they may be degraded to different degrees by the soil microflora. It has, however, always to be kept in mind that soils may act as a source or sink of VOCs, depending on the vegetation, season, temperature and moisture regime and on several other factors.

Soil smells?

Textbook knowledge about microbially produced odorous compounds emitted from soils often begins and ends with geosmin (*trans*-1, 10-dimethyl-*trans*-9-decalol) produced by *Streptomyces coelicolor* (Schöller *et al.*, 2002; Jiang *et al.*, 2007). Geosmin is the typical scent of fresh forest or garden soil many compost producers would love to have in their products, indicating both compost maturity (Li *et al.*, 2004) and attracting the consumer who is looking for an earthy-smell product. Nutritional and physical factors influencing geosmin production have been examined (Dionigi *et al.*, 1996). However, it is not only this one molecule that makes up the scent of soil. Schöller *et al.* (2002) characterized a total of 120 VOCs from 26 *Streptomyces* spp. on agar plate cultures, which will be discussed in more detail below. For soils, Leff and Fierer (2008) were able to show that VOC production is closely related to microbial respiration and biomass. In an extensive review, Insam and Seewald (2010) published a table summarizing more than 70 papers related to mVOC production (Table 7.1).

Volatiles produced by microorganisms

From cultivation studies it is known that most microorganisms produce VOCs. Some VOCs may be used for taxonomic purposes since they are produced by specific phylogenetic groups or species (Larsen and Frisvad, 1995). Despite the multitude of compounds that are produced by microorganisms, the reasons why they do so are in most cases unresolved. While communication and defence have been suggested, the proof of these assumptions is rare (Schulz and Dickschat, 2007); in many cases it may be assumed that certain volatile emissions are not purposeful but are just an expression of certain biochemical pathways where intermediate products are being 'lost'. For example, Dickschat *et al.* (2005) report of members of the *Cytophaga–Flavobacterium–Bacteroides* (CFB) group that use valine, and others that use isoleucine, for ketone biosynthesis. Like bacteria, also fungi can produce a multitude of volatiles, some of which are common to many species, while others seem to be unique for certain species (Schnürer *et al.*, 1999): These authors showed that monitoring fungal metabolites can be used to detect fungal infestation in general or specifically.

The best-studied group of VOC-emitting microorganisms definitely are the actinobacteria (e.g. Schöller *et al.*, 2002; Wilkins, 1996). Apart from geosmin, detailed knowledge is available for a 2-methylisoborneol emitting *Streptomyces* sp., known for its detrimental effect on freshwater quality (Wood *et al.*, 1985, 2001). For surface water, however, the current notion is that geosmin producers are cynobacteria (Durrer *et al.*, 1999). Particularly malodorous VOC emissions come from household biowastes and may be a nuisance during collection. Emissions from biodegradable household waste are comprised of compound classes such as amines, carboxylic acids, alcohols, aldehydes, ketones, esters, ethers, isoprenoids, aromatic and sulphur-containing compounds (Wilkins, 1996; Smet *et al.*, 1999). Bacteria are known to emit predominantly sulphur compounds, hydrocarbons (methane and isoprene), and alcohols (Wilkins, 1996; Schöller *et al.*, 1997). Wheatley (2002) demonstrated the importance of VOCs for the mediation of bacterial and fungal interactions

Individual microbial species produce a typical pattern of VOCs depending on environmental conditions. However, small variations in nutrient composition may change the type and the amount of the individual VOCs emitted. VOCs are frequently used as infochemicals to mediate positive, negative or neutral interactions between microorganisms. In a study on degrading household waste, Mayrhofer *et al.* (2006) found high numbers of *Lactococcus lactis*, and at the same time high amounts of certain VOCs.

Kai *et al.* (2006) show how little is known about VOC emissions from small bacterial VOCs emitted from bacterial antagonists negatively influenced the mycelial growth of *Rhizoctonia solani*, a soil-borne phytopathogen. Inhibitions of up to 99% were observed with strains of *Stenotrophomonas maltophilia*, *Serratia plymuthica*, *Stenotrophomonas rhizophila*, *Serratia odorifera*, *Pseudomonas trivialis*, and *Bacillus subtilis*. Commonly known general antagonists like *Pseudomonas fluorescens* and *Burkholderia cepacia* achieved 30% growth reduction. The VOC profiles comprised 1 to almost 30 mostly unidentified compounds,

Table 7.1 Summary of the literature on VOC production

Source	Organisms investigated	Habitat	VOCs found
Asensio *et al.* (2007a,b)	Microbial community	Mediterranean soil	Diverse VOCs
Bjurman *et al.* (1997)	*Penicillium* spp.	Pine wood	1-Octene-3-ol, 2-heptanone, 4-allylanisole, 3-methyl-1-butanol
Farag *et al.* (2006)	*Bacillus amyloliquefaciens*, *B. subtilis*, *E. coli*, *Pseudomonas fluorescens*	Rhizosphere	Diverse VOCs
Fernando *et al.* (2005)	*Pseudomonas fluorescens*, *P. corrugata*, *P. chlororaphis*, *P. aurantiaca*	Canola stem, soy bean	Fungistatic VOCs
Kai *et al.* (2006, 2008, 2009)	*Stenotrophomonas maltophilia*, *St. rhizophila*, *Serratia plymuthica*, *S. odorifera*, *S. plymuthica*, *Pseudomonas trivialis*, *Ps. fluorescens*, *Bacillus subtilis*, *Burkholderia cepacia*, *Staphylococcus epidermidis*	Rhizosphere isolates	Diverse VOCs
Leff and Fierer (2008)	Microbial community	Soil and litter	Diverse VOCs (17 identified)
Liu *et al.* (2008)	*Bacillus subtilis*	Soil isolate	Diverse VOCs
Mackie and Wheatley (1998)	diverse bacteria	Soil isolate	No VOC identified
Mattheis and Roberts (1992)	*Penicillium expansum*	Czapek agar	Geosmin
McNeal and Herbert (2009)	Microbial community	Hyperthermic, hypersaline soils	diverse VOCs (72 identified)
Ryu *et al.* (2002)	*Pseudomonas fluorescens*, *Bacillus pumilus*, *B. pasteurii*, *B. subtilis*, *B. amyloliquefaciens*, *Serratia marescens*, *Enterobacter cloacae*	Rhizosphere isolates of *Arabidopsis thaliana*	No VOC identified
Schöller *et al.* (2002)	Actinobacteria (26 *Streptomyces* spp.)	Yeast starch agar	120 VOCs, mainly terpenoide
Seewald *et al.* (2010)	Microbial community	Soil treated with different composts	Diverse VOCs (seven tentatively identified)
Serrano and Gallego (2006)	Microbial community	Different Mediterranean soils	Diverse VOCs (25 identified)
Spilvallo *et al.* (2007)	Truffle (3 *Tuber* spp.)	Fruiting bodies	Diverse VOCs (119 identified)
Stahl and Parkin (1976)	actinobacteria, bacteria, fungi	soil	geosmin, 2methylisoborneol, other VOCs
Wheatley *et al.* (1997)	*Trichoderma* spp.	decaying wood	diverse VOCs (45 identified)
Wheatley *et al.* (1996)	microbial community	soil cropped to potatoes	diverse VOCs (35 identified)
Zhang *et al.* (2007)	*Bacillus subtilis*	rhizosphere	no VOC identified
Zou *et al.* (2007)	328 soil bacteria, Alcaligenaceae, Bacillales, Micrococcaceae, Rhizobiaceae, Xanthomonadaceae	soil bacterial isolates	diverse VOCs

From Insam and Seewald, 2010 (abbreviated).

most VOCs were species-specific, but overlapping patterns were found for *Serratia* spp. and *Pseudomonas* spp.

Xu *et al.* (2004) studied fungistasis in soils and found a close relationship of soil fungistasis and fungistasis attributed to VOCs. These authors also showed that trimethylamine, benzaldehyde, and N,N-dimethyloctylamine, among other VOCs, had strong antifungal activities even at low levels (4–12 mg/l). In a study on degrading household waste, Mayrhofer *et al.* (2006) found high numbers of *Lactococcus lactis,* and at the same time high amounts of certain VOCs. The aim of that study had been to relate the abundance of specific VOCs to bacterial species or groups, however, this was not found to be possible. Thus malodour emissions could not be assigned to specific microorganisms.

Volatiles from plant roots

Despite recent advances, little is known about the contribution of plant roots to VOC emissions. Five to 20% of all photosynthetically fixed carbon is released by plant roots, creating a carbon-rich environment for numerous rhizoplane and rhizosphere microorganisms, including plant pathogens and symbiotic microbes (Steeghs *et al.,* 2004). These authors found that the major VOCs released were either simple metabolites, ethanol, acetaldehyde, acetic acid, ethyl acetate, 2-butanone, 2,3-butanedione, and acetone, or the monoterpene, 1,8-cineole. Some of the compounds were found to be produced constitutively, while other VOCs were induced specifically as a result of interactions between microbes and insects and *Arabidopsis* roots. Environmental studies on VOCs first focussed on plant Plants are considered to be the main contributor to soil VOC emission. In numerous investigations, VOCs emitted by plant roots (and associated mycorrhiza) or seedlings, were identified and suggestions for their functions were proposed. A very detailed insight on the studies conducted in this field is given by Linton and Wright (1993). Plant VOC production is beyond the aims of this review.

Microbial volatiles affecting plant growth

Fungistatic VOCs are of special interest since they are known to be involved in soil disease suppressiveness (Fuchs *et al.,* 2004). The repression of mould and pathogen growth in household biowaste by fungistatic VOCs can also help in reducing the health risk for the consumer and waste collection personnel. For example, *Actinobacteria* and bacteria of the genera *Bacillus* and *Pseudomonas* are known to produce fungistatic VOCs such as 1-octen-3-ol, dimethyldisulphide, trimethylamine, mono- and sesquiterpenes (Wilkins, 1996; Schöller *et al.,* 1997; Chitarra, 2003). Organic acids in compost can have phytotoxic or plant growth suppressive properties and the highest emissions occur mostly during the first 20 days of composting. As a result, organic acid emissions can be used as indicator substances to determine compost maturity (Brinton, 1998).

Several fungal and bacterial VOCs positively impact other individuals (Stotzky and Schenck, 1976; Wheatley, 2002; Farag *et al.,* 2006); e.g. bacterial and rhizobacterial VOCs from several strains of *Bacillus subtilis, B. amyloliquefaciens* and *Enterobacter cloacae* promote the growth of *Arabidopsis thaliana* (Ryu *et al.,* 2003; Zhang *et al.,* 2007). Kloepper *et al.* (2004) showed that *Bacillus* sp. derived VOCs induced systemic resistance in tomato, bell pepper, muskmelon, watermelon, sugar beet, tobacco and loblolly pine, amongst others.

On the other hand, VOCs are known to act as inhibitors, like the manifold of VOCs antagonistic to microorganisms that are produced from several plants (Dorman and Deans, 2000). The VOCs prevent them from direct bacterial or fungal attack and thus contribute to soil disease suppressiveness (Fuchs *et al.*, 2004). A multitude of antifungal VOCs are further produced by bacteria (Mackie and Wheatley, 1999; Wheatley, 2002; Bruce *et al.*, 2003; Chuankun *et al.*, 2004, Zou *et al.*, 2007; Liu *et al.*, 2008). The repression of phytopathogens in soils through VOCs emitted by microorganisms or (transgenic) plants is regarded as a future alternative to conventional bactericides and fungicides. Examples for fungistatic VOCs are 1-octen-3-ol, mono- and sesquiterpenes, nonanal acid, trimethylamine and dimethyldisulphide which are produced by actinobacteria and by the bacterial genera *Bacillus* and *Pseudomonas* (Wilkins, 1996; Schöller *et al.*, 1997; Chitarra, 2003). VOCs generated by bacteria isolated from canola and soybean inhibit sclerotia and ascospore germination and mycelial growth of *Sclerotinia sclerotiorum*. This has been observed under laboratory as well as under field conditions. Fernando *et al.* (2005) demonstrated that Benzothiazole, cyclohexanol, *n*-decanal, dimethyl trisulphide, 2-ethyl 1-hexanol, and nonanal completely inhibit mycelial growth or sclerotia formation.

On the other hand, growth of some plants like *Arabidopsis thaliana* may negatively be affected by VOCs produced by soil fungi, e.g. truffle (*Tuber melanosporum*) volatiles that induce an oxidative burst in *A. thaliana* leaf parenchyma tissue (Splivallo *et al.*, 2007).

Degradation of VOCs

While abiotic VOC degradation in the atmosphere is well investigated little is known about abiotic degradation processes in soils (Willson and Jones, 1996). This knowledge gap may be attributed to the inability to sterilize soils without directly impacting soil VOCs. Anyway, Atkinson and Arey (2003) give a very detailed overview on physical–chemical degradation of VOCs in the atmosphere. They identified photolysis, and spontaneous reactions with OH radicals (under presence of light), NO_3 radicals (under absence of light), O_3 (under presence of light) and Cl atoms (at coastlines) as the main abiotic processes responsible for VOC disintegration. Since light is not able to enter soil more than some millimetres (Ciani *et al.*, 2005), photolysis and other light driven reactions (with OH radicals and O_3) are restricted to a thin surface zone (Konstantinou *et al.*, 2001) where OH radicals and O_3 are present through microbial production or diffusion from the atmosphere. The degradation of VOC through NO_3 radicals seems to be a rather plausible way of VOC degradation in soils. Nitrate is formed in soils under aerobic conditions (nitrification) and the reaction between VOCs and NO_3 radicals does not require light. In a similar way hydrogen peroxide could also react with VOCs. However, clay minerals, humic acids or phospholipids may compete with VOCs for these radicals, showing the interdependence of soil properties with VOC emissions.

In soils both anaerobic and aerobic microhabitats can be found, and very often end products of anaerobic metabolism serve as nutrients for aerobic microorganisms and vice versa. Many anaerobic microorganisms found in soils are able to degrade VOCs like formic or acetic acid (Guyot and Brauman, 1986). This metabolic pathway is well known from syntrophic methanogenic archaea. In methanogenic environments, saturated fatty acids, unsaturated fatty acids, alcohols and hydrocarbons are degraded by the action of syntrophic communities. These syntrophic communities consist of an acetogen and a methanogen, who cannot grow alone on a certain organic compound, but when present together they can

(Plugge *et al.*, 2010). The degradation of many substrates is thermodynamically unfavourable under anoxic conditions if the product concentrations are at standard concentrations (1 M concentration, or 10^5 Pa for gases). The role of methanogens is to consume such products to low partial pressure (10^{-4}–10^{-5} atm) to allow energy gain. The diffusion distances for metabolite transfer should be as short as possible. In their natural habitat, syntrophic propionate degrading bacteria such as *Syntrophobacter fumaroxidans* form microcolonies with methanogens, e.g. *Methanospirillum hungateii* (Harmsen *et al.*, 1998).

In the gas phase, VOCs are inaccessible to microorganisms thus solubilisation in the aqueous phase is a prerequisite for microbial degradation. VOC water solubility is a decisive factor as is the adsorption to humic acid or clay mineral surfaces (polar and apolar interactions). The bioconversion of VOC-pollutants to metabolic end- and intermediate products (VOCs), biomass or carbon dioxide and water is the second step. Soils may be regarded perfect natural biofilters as they provide a multitude of species and microbial consortia (capable for different organic compound degrading pathways), environmental conditions (from anaerobic to aerobic) and a variety of different VOC adsorbents (water, humic acids, clay minerals).

Retention, emission and measurement

Apart from substrate availability, several environmental factors determine VOC emission, among them temperature, pH, O_2 availability and moisture content (Brinton, 1998). In soils, most important is the action of adsorbents like clay minerals or humic acids. Ruiz *et al.* (1998) investigated the adsorption of *n*-hexane, *n*-heptane, *n*-octane, toluene, xylene, ethylbenzene, and methyl ethyl ketone and of water vapour on sand, clay, and limestone. They found considerable differences between the adsorption levels of the three soil minerals. Polar compounds were more strongly adsorbed than aliphatic and aromatic compounds. Serrano and Gallego (2006) investigated the effect of clay minerals (7.0–69.7%) and C_{org} (0.2–3.5%) contents in acid and alkaline (pH 5.3–8.8) soils on the sorption of 25 VOCs. All compounds were sorbed more strongly in alkaline than in acid soils. In alkaline soils VOC sorption increased with C_{org} while it decreased in acid soils. This indicates that VOCs are often positively charged and interact with the negatively charged organic matter and clay mineral particles. Thus, upon a decrease of pH, the number of positive charges on soil particles increase and VOCs are being more easily released. In acid soil, clay minerals (bentonite) played an important role in the sorption of VOCs. The sorption/desorption behaviour of VOCs must be seen in the context of soil texture and particle architecture (Aochi and Farmer, 2005) who showed that soil physically impeded the flow of both liquid and gaseous phase rather than chemical adsorption.

As for microbial accessibility, the solubility in water, and hence the water content, does have a significant effect on the VOC emissions. The water affects the VOC adsorption by decreasing the retention of these compounds to a greater extent for aromatic and aliphatic compounds than for the polar compound and by linearizing the isotherms (Ruiz *et al.*, 1998).

Methods of VOC measurement

Most of the published papers describe the analysis of VOCs with the aid of head space or thermal desorption gas chromatography in combination with detection methods like mass spectrometry (MS), flame ionization detector (FID), flame photometric detector (FPD), infrared analyser (IA) or photoionization detector (PID) (Chung, 2006). Soil microbial VOC output determined with the aid of GC-MS has been discussed in many publications (e.g. Leff and Fierer, 2008). Proton-transfer-reaction mass-spectrometry (PTR-MS) is discussed elsewhere (Hansel *et al.*, 1995; Lindinger *et al.*, 1998) and has so far been used for environmental and medical applications but not often for soil related applications (Seewald *et al.*, 2010).

Besides mass spectrometric approaches also Microresp® tubes detecting single volatile compounds (Kaufmann *et al.*, 2005) have been used as well as metal oxide-based olfactory sensors (electronic noses, e.g. Rajamäki *et al.*, 2005) or nanoparticle-structured sensing array materials (Han *et al.*, 2005).

Soils are living systems, and the numerous biotic and abiotic factors ruling VOC emissions have been mentioned above. Soil sampling, transport and storage may have a considerable effect on the results, as do the environmental conditions (temperature, moisture, aeration) during the analysis. It makes a difference if we are interested in the current abundance of VOCs or in the current VOC production or degradation. Interrupting the plant–microbe interactions upon sampling will have an immediate effect.

In a long-term field experiment where for several years different organic fertilizers and composts were used, the composition of the microbial community, its functional diversity and the VOC patterns were assessed (Innerebner *et al.*, 2006; Ros *et al.*, 2006a,b). It was found that the different organic amendments left a significant imprint on the VOC emission

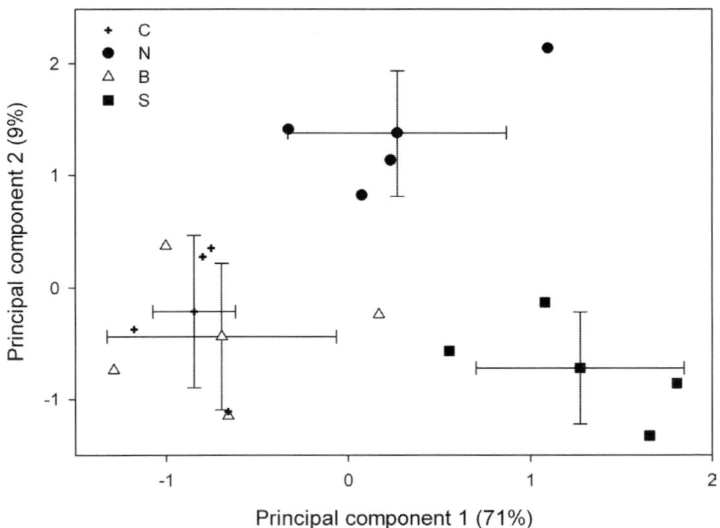

Figure 7.1 VOC emission patterns of differently treated soils (1) unfertilized control (C), (2) mineral fertilizer (NH_4NO_3, corresponding to 80kg N ha^{-1}) (N), (3) combined amendment of mineral fertilizer (80kg N/ha) plus one of two different composts corresponding to a total of 175kg N ha^{-1}: urban organic waste compost (so called 'Bio-waste' compost, B), and Sewage sludge compost (S). Visualization by a multi dimensional scaling (MDS) diagram. Principal components of all significant masses are shown as dimensions 1 and 2 (cumulatively, explaining 80% of the variance). Group means and standard deviations are given (from Seewald *et al.*, 2010).

patterns (Seewald *et al.*, 2010, Fig. 7.1), showing that soil volatilomics may offer the possibility of distinguishing between differently treated soils, just as genomic approaches do. However, cause–effect relationships are difficult to reveal.

Microbial mass products

This review has a focus on VOCs, however, a term soil volatilomics should definitely also encompass other gases that are mostly found in much higher concentrations. Such microbial 'mass products' like carbon dioxide, methane, nitrous oxide, dinitrogen, ammonia, dihydrogen sulphide are important indicators of soil fertility and indicators of major cycles of C, N and S. However, these compounds are beyond the focus of this review.

Conclusions

It sounds easy and straightforward to use volatilomics for characterizing microbial communities in soils, or for characterizing soils in general. The main advantage over other approaches like metagenomic or metabolomic ones would be the avoidance of extraction steps. The complicated nature of VOCs, however, makes any results highly dependent on the conditions during measurement. VOCs can be found in the gas and aqueous phase, they bind on organic and clay molecules. These are highly susceptible to environmental conditions, and thus any changes in moisture availability, temperature, diffusion gradients or pH add to the data variability. Albeit volatilomics may give additional information, and tell about soils acting as sinks or sources of particular compounds, it will be extremely difficult to use them as indicators of microbial community composition or activity.

References

Alvarado, J.S., and Rose, C. (2004). Static headspace analysis of volatile organic compounds in soil and vegetation samples for site characterization. Talanta *62*, 17–23.

Aochi, Y.O., and Farmer, W.J. (2005). Impact of soil microstructure on the molecular transport dynamics of 1,2-dichloroethane. Geoderma *127*, 137–153.

Asensio, D., Penuelas, J., Filella, I., and Llusià, J. (2007a). On-line screening of soil VOCs exchange responses to moisture, temperature and root presence. Plant Soil *291*, 294–261.

Asensio, D., Penuelas, J., Llusià, J., Ogaya, R., and Filella, I. (2007b). Interannual and interseasonal soil CO_2 efflux and VOC exchange rates in a Mediterranean holm oak forest in response to experimantal drought. Soil Biol. Biochem. *39*, 2471–2484.

Atkinson, R., and Arey, J. (2003). Atmospheric degradation of volatile organic compounds. Chem. Rev. Vol. *103*, 4605–4638.

Bakken, L.R., and Frostegård, Å. (2006). Nucleic acid extraction from soil. In Nucleic Acid and Proteins in Soil, P. Nannipieri, and K. Smalla, eds. (Berlin, Germany: Springer), pp. 49–73.

Bjurman, J., Nordstrand, E., and Kristensson, J. (1997). Growth-phase-related production of potential volatile-organic tracer compounds by moulds on wood. Indoor Air *7*, 2–7.

Börjesson, T., Stollman, U., and Schnürer, J. (1990). Volatile Metabolites and Other Indicators of *Penicillium aurantiogriseum* Growth on Different Substrates, Appl. Env. Microbiol. *56*, 3705–3710.

Börjesson, T., Stollman, U., and Schnürer, J. (1992). Volatile metabolites produced by six fungal species compared with other indicators of fungal growth on cereal grains. Appl. Envir. Microbiol. *58*, 2599–2605.

Brilli, F., Hörtnagl, L., Hammerle, A., Haslwanter, A., Hansel, A., Loreto, F., and Wohlfahrt, G. (2011). Leaf and ecosystem response to soil water availability in mountain grasslands Agr. For. Meteor. *151*, 1731–1740.

Brinton, W.F.(1998). Volatile organic acids in compost, Production and odorant aspects. Compost Sci. Util. *6*, 75–82.

Chitarra, G.S., Abee, T., Rombouts, F.M., Posthumus, M.A., and Dijksterhuis, J. (2004). Germination of *Penicillium paneum* conidia is regulated by 1-octen-3-ol, a volatile self-inhibitor. Appl. Environ. Microbiol. *70*, 2823–2829.

Chuankun, X., Minghe, M., Leming, Z., and Keqin, Z. (2004). Soil volatile fungistasis and volatile fungistatic compounds. Soil Biol. Biochem. *36*, 1997–2004.

Chung, Y.-C. (2006). Evaluation of gas removal and bacterial community diversity in a biofilter developed to treat composting exhaust gases. J. Hazard. Mat. *144*, 377–385..

Ciani, A., Goss, K.-U., and Schwarzenbach, R.P. (2005). Light penetration in soil and particulate minerals. Eur. J. Soil Sci. *56*, 561–574.

Delmont, T.O., Robe, P., Cecillon, S., Clark, I.M., Constancias, F., Simonet, P., Hirsch, P.R., and Vogel, T.M. (2011). Accessing the soil metagenome for studies of microbial diversity. Appl. Env. Microbiol. *77*, 1315–1324.

Dickschat, J.S., Helmke, E., and Schulz, S. (2005). Volatile Organic Compounds from Arctic Bacteria of the *Cytophaga-Flavobacterium-Bacteroides* Group, A Retrobiosynthetic Approach in Chemotaxonomic Investigations. Chem. Biodiv. *2*, 318–353.

Dionigi, C.P., Ahten, T.S., and Wartelle, L.H. (1996). Effects of several metals on spore, biomass, and geosmin production by*Streptomyces tendae* and*Penicillium expansum*. J. Ind. Microb. Biotech. *17*, 84–88.

Dorman, H.J.D., and Deans, S.G. (2000). Antimicrobial agents from plants, antibacterial activity of plant volatile oils. J. Appl. Microb. *88*, 308–316.

Dott, W., Thißen, R., Müller. T., Wiesmüller, G.A., and Fischer, G. (2004). Belastung der Arbeitnehmer bei Schimmelpilzsanierungsarbeiten in Innenräumen. Literaturstudie. Tiefbau-Berufsgenossenschaft, München, pp. 1–71.

Durrer, M., Zimmermann, U., and Jüttner, F. (1999). Dissolved and particle-bound geosmin in a mesotrophic lake (lake Zürich), spatial and seasonal distribution and the effect of grazers. Water Res. *33*, 3628–3636..

Farag, M.A., Ryu, C.-M., Sumner, L.W., and Paré, P.W. (2006). GC–MS SPME profiling of rhizobacterial volatiles reveals prospective inducers of growth promotion and induced systemic resistance in plants. Phytochem. *67*, 2262–2268.

Fernando, W.G.D., Ramarathnam, R., Krishnamoorthy, A.S., and Savchuk, S.C. (2005). Identification and use of potential bacterial organic antifungal volatiles. Soil Biol. Biochem. *37*, 955–964..

Fischer, G., and Dott, W. (2003). Relevance of airborne fungi and their secondary metabolites for environmental, occupational and indoor hygiene. Minireview. Arch. Microbiol. *179*, 75–82..

Fuchs, J.G., Bieri, M., and Chardonnes, M. (2004). Auswirkungen von Komposten und Gärgut auf die Umwelt, die Bodenfruchtbarkeit sowie die Pflanzengesundheit. Zusammenfassende Übersicht der aktuellen Literautur. Forschungsinstitut für biologischen Landbau. FiBL-Report. Frick. Schweiz.

Gao, P., and Martin, J. (2002). Volatile Metabolites Produced by Three Strains of Stachybotrys chartarum Cultivated on Rice and Gypsum Board. Appl. Occup. Environ. Hyg. *17*, 430–436.

Giagnoni, L., Magherini, F., Landi, L., Taghavi, S., Modesti, A., Bini, L., Nannipieri, P., Vanderlelie, D., and Renella, G. (2010). Extraction of microbial proteome from soil, potential and limitations assessed through a model study. Eur J. Soil Sci. *62*, 74–81.

Grabmer, W., Graus, M., Lindinger, C., Wisthaler, A., Rappenglück, B., Steinbrecher, R., and Hansel, A. (2004). Disjunct eddy covariance measurements of monoterpene fluxes from a Norway spruce forest using PTR-MS. Int. J. Mass Spectr. *239*, 111–115.

Guyot, J.-P., and Brauman, A. (1986). Methane Production from Formate by Syntrophic Association of *Methanobacterium bryantii* and *Desulphovibrio vulgaris*. Appl. Environ. Microb. *52*, 1436–1437..

Han, L., Shi, X., Wu, W., Kirk, F.L., Luo, J., Wang, L., Mott, D., Cousineau, L., Lim, S.I.I., Lu, S., and Zhong, C.-J. (2005). Nanoparticle-structured sensing array materials and pattern recognition for VOC detection. Sensors Actuators *106*, 431–441.

Hansel, A., Jordan, A., Holzinger, R., Prazeller, P., Vogel, and Lindinger, W. (1995). Proton transfer reaction mass spectrometry, on-line trace gas analysis at ppb level. Int. J. Mass Spect. Ion Proc., *149*, 609–619.

Harmsen, H.J.M., VanKuijk, B.L.M., Plugge, C.M., Akkermans, A.D.L., DeVos, W.M., and Stams, A.J.M. (1998). *Syntrophobacter fumaroxidans* sp. nov., a syntrophic propionate-degrading sulphate-reducing bacterium. Int. J. Syst. Bacteriol. *48*, 1383–1387.

Innerebner, G., Knapp, B., Vasara, T., Romantschuk, M., and Insam, H. (2006). Traceability of ammonia-oxidizing bacteria in compost-treated soils. Soil Biol. Biochem. *38*, 1092–1100.

Insam, H., and Seewald, M.S.A. (2010). Volatile organic compounds (VOCs) in soils. Biol. Fert. Soils *46*,199–213.

Jiang, J., He, X., and Cane, D.E. (2007). Biosynthesis of the earthy odorant geosmin by a bifunctional *Streptomyces coelicolor* enzyme. Nature Chem. Biol. *3*, 711–715.

Kai, M., Effmert, U., Berg, G., and Piechulla, B. (2006). Volatiles of bacterial antagonists inhibit mycelial growth of the plant pathogen *Rhizotonia solani*. Arch. Microbiol. *187*, 351–360..

Kai, M., Haustein, M., Molina, F., Petri, A., Scholz, B., and Piechulla, B. (2009). Bacterial volatiles and their action potential. Appl. Microbiol. Biotech. *81*,1001–1012.

Kai, M., Vespermann, A., and Piechulla, B. (2008). The growth of fungi and *Arabidopsis thaliana* is infuenced by bacterial volatiles. Plant Signaling Behav. 3, 482–484.

Kaufmann, K., Chapman, S.J., Campbell, C.D., Harms, H., and Höhener, P. (2006). Miniaturized test system for soil respiration induced by volatile pollutants. Environ. Poll. *140*, 269–278.

Keshri, G., Magan, N., and Voysey, P. (1998). Use of an electronic nose for the early detection and differentiation between spoilage fungi. Lett. Appl. Microbiol. *27*, 261–264.

Kloepper, J.W., Ryu, C.-M., Zhang, S.(2004). Induced systemic resistance and promotion of plant growth by *Bacillus* spp. Phytopathology *94*, 1259–1266.

Konstantinou, I.K., Zarkadis, A.K., and Albanis, T.A. (2001). Photodegradation of selected herbicides in various natural waters and soils under environmental conditions. J. Envir. Qual. *30*, 121–130.

Larsen, T.O., and Frisvad, J.C. (1995). Characterization of volatile metabolites from 47 *Penicillium* taxa. Mycol. Res. *99*, 1153–1166.

Leff, J.W., and Fierer, N. (2008). Volatile organic compound (VOC) emissions from soil and litter samples. Soil Biol. Biochem. *40*, 1629–1636.

Li, H.F., Imai, T., Ukita, M., Sekine, M., and Higuchi, T. (2004). Compost stability assessment using a secondary metabolite, geosmin. Environ, Technol. *25*,1305–1312.

Lindinger, W., Hansel, A., and Jordan, A. (1998). Proton transfer reaction-mass spectrometry (PTR-MS), on-line monitoring of volatile organic compounds at pptv levels. Chem. Soc. Rev. *27*, 347–354.

Linton, C.J., and Wright, S.J.L. (1993). Volatile organic compounds, microbiological aspects and some technological implications. J. Appl. Bacteriol. *75*, 1–12.

Liu, W., Mu, W., Zhu, B., and Liu, F. (2008). Antifungal activities and components of VOCs produced by *Bacillus subtilis* G_8. Curr. Res. Bacteriol. *1*, 28–34.

Mackie, A.E., and Wheatley, R.E. (1999). Effects and incidence of volatile organic compound interactions between soil bacterial and fungel isolates. Soil Biol. Biochem. *31*, 375–385.

Mattheis, J.P., and Roberts, R.G. (1992). Identification of geosmin as a volatile metabolite of *Penicillium expansum*. Appl. Environ. Microbiol. *58*, 3170–3172.

Mayr, D., Margesin, R., Klingsbichel, E., Hartungen, E., Jenewein, D., Schinner, F., and Märk, T.D. (2003). Rapid Detection of Meat Spoilage by Measuring Volatile Organic Compounds by Using Proton Transfer Reaction Mass Spectrometry, Appl. Env. Microbiol. *69*, 4697–4705.

Mayrhofer, S., Heizmann, U., Magel, E., Eiblmeier, M., Müller, A., Rennenberg, H., Hampp, R., Schnitzler, J.P., and Kreuzwieser, J. (2004). Carbon balance in leaves of young poplar trees. Plant Biol. *6*, 730–739.

Mayrhofer, S., Mikoviny, T., Waldhuber, S., Wagner, A., Innerebner, G., Leuenberger, J., Franke-Whittle, I.H., Märk, T., Hansel, A., and Insam, H. (2006). Correlation of microbial community and VOC emission in household biowaste. Env. Microbiol. *8*, 1960–1974.

McNeal, K.S., and Herbert, B.E. (2009). Volatile Organic Metabolites as Indicators of Soil Microbial Activity and Community Composition Shifts. Soil Sci. Soc. Am. J. *73*, 579–588.

Nannipieri, P. (2006). Nucleic acid extraction from soil. In Nucleic Acid and Proteins in Soil, P. Nannipieri, and K. Smalla, eds. (Berlin, Germany: Springer), pp. 49–73.

Plugge, C.M., van Lier, J.B., and Stams, J.M. (2010). Syntrophic Communities in Methane Formation from High Strength Wastewaters. In Microbes at Work. From Wastes to Resources, H. Insam, I.H. Franke-Whittle, M. Goberna, eds. (Heidelberg, Germany: Springer), pp. 59–77.

Rajamäki, T., Arnold, M., Venelampi, O., Vikman, M., Räsänen, J., and Itävaara, M. (2005). An Electronic Nose and Indicator Volatiles for Monitoring of the Composting Process. Water Air Soil Poll. *162*, 71–87.

Ros, M., Klammer, S., Knapp, B., Aichberger, K., and Insam, H. (2006a). Long term effects of compost amendment of soil on functional and structural diversity and microbial activity. Soil Use Manage. *22*, 209–218.

Ros, M., Pascual, J.A., Garcia, C., Hernandez, M.T., and Insam, H. (2006b). Hydrolase activities, microbial biomass and bacterial community in a soil after long-term amendment with different composts. Soil Biol. Biochem. *38*, 3443–3452.

Ruiz, J., Bilbao, R., and Murillo, M.B. (1998). Adsorption of different VOC onto soil minerals from the gas phase, influence of mineral, type of VOC, and air humidity. Environ. Sci. Technol. *32*, 1079–1084.

Ruuskanen, T.M., Müller, M., Schnitzhofer, R., Karl, T., Graus, M., Bamberger, I., Hörtnagl, L., Brilli, F., Wohlfahrt, G., and Hansel, A. (2011). Eddy covariance VOC emission and deposition fluxes above grassland using PTR-TOF, Atm. Chem. Phys. *11*, 611–625.

Ryu, C.-M., Farag, M.A., Hu, C.-H., Reddy, M.S., Wei, H.-X., Paré, P.W., and Kloepper, J.W. (2003). Bacterial volatiles promote growth in Arabidopsis. PNAS *100*, 4927–4932.

Schnürer, J., Olsson, J., and Börjesson, T. (1999). Fungal volatiles as indicators of food and feeds spoilage. Fungal Gen. Biol. *27*, 209–217.

Schöller, C.E.G., Gürtler, H., Pedersen, R., Molin, S., and Wilkins, K. (2002). Volatile metabolites from Actinomycetes. J. Agric. Food Chem. *50*, 2615–2621.

Schöller, C., Molin, S., and Wilkins, S. (1997). Volatile metabolites from some Gram-negative bacteria. Chemosphere *35*, 1487–1495.

Schulz, S., and Dickschat, J.S. (2007). Bacterial volatiles, the smell of small organisms. Nat. Prod. Rep. *24*, 814–842.

Seewald, M., Bonfanti, M., Singer, W., Knapp, B.A., Hansel, A., Franke-Whittle, I.H., and Insam, H. (2010). Substrate-induced volatile organic compound emissions from compost-amended soils. Biol. Fert. Soils *46*, 371–382.

Serrano, A., and Gallego, M. (2006). Sorption study of 25 volatile organic compounds in several Mediterranean soils using headspace-gas chromatography–mass spectrometry. J. Chromat. *1118*, 261–270.

Smet, E., Van Langenhove, H., De Bo, I. (1999). The emission of volatile compounds during the aerobic and the combined anaerobic/aerobic composting of biowaste. Atm. Env. *33*, 1295–1303.

Spilvallo, R., Novero, M., Bertea, C.M., Bossi, S., and Bonfante, P. (2007). Truffle volatiles inhibit growth and induce an oxidative burst in *Arabidopsis thaliana*. New Phytol. *175*, 417–424.

Stahl, P.D., and Parkin, T.B. (1976). Microbial production of volatile organic compounds in soil microcosms. Soil Sci. Soc. Am. *60*, 821–828.

Steeghs, M., Bais, H.P., de Gouw, J., Goldan, P., Kuster,W., Northway, M., Fall, R., and Vivanco, J.M. (2004). Proton-transfer-reaction mass spectrometry as a new tool for real time analysis of root-secreted volatile organic compounds in *Arabidopsis*. Plant Physiol. *135*, 47–58.

Stotzky, G., and Schenck, S. (1976). Volatile organic compounds and microorganisms. CRC Crit. Rev. *4*, 333–382.

Wheatley, R.E. (2002).The consequences of volatile organic compound mediated bacterial and fungal interactions. Antonie van Leeuwenhoek *81*, 357–364.

Wheatley, R.E., Millar, S.E., and Griffiths, D.W. (1996). The production of volatile organic compounds during nitrogen transformation in soils. Plant Soil *181*, 163–167.

Wheatley, R., Hackett, C., Bruce, A., and Kundzewicz, A. (1997). Effect of substrate composition on production of volatile organic compounds from *Trichoderma* spp. Inhibitory to wood decay fungi. Int. Biodet. Biodegrad. *39*, 199–205.

Wilkins, K., Larsen, K., and Simkus, M. (2000). Volatile metabolites from mold growth on building materials and synthetic media. Chemosphere *41*, 437–446.

Wilkins, K.. (1996). Volatile metabolites from actinomycetes. Chemosphere *32*, 1427–1434.

Willson, S.C., and Jones, K.C. (1996). The fate and behavior of volatile aromatic hydrocarbons in sewage sludge-amended soil. ASTM Special Technical Publication *1261*, 119–123.

Wood, S., Williams, S.T., and White, W.R. (2001). Microbes as as source of earthy flavours in potable water – a review. International Biodet. Biodegrad. *48*, 26–40.

Wood S., Williams, S.T., and White, W.R. (1985). Potential sites of geosmin production by streptomycetes in and around reservoirs. J. Appl. Microbiol. *58*, 319–326.

Xu, Chuankun, Mo, Minghe, Zhang, Leming, Zhang, Keqin (2004). Soil volatile fungistasis and volatile fungistatic compounds. Soil Biol. Biochem. *36*, 1997–2004.

Zhang, H., Kim, M.S., Krishnamachari, V., Payton, P., Sun, Y., Grimson, M., Farag, M.A., Ryu, C.-M., Allan, R., Melo, I.S., and Paré, P.W. (2007). Rhizobacterial volatile emissions regulate auxin homeostasis and cell expansion in *Arabidopsis*. Int. J. Plant Biol. *4*, 839–851.

Zou, C.-S., Mo, M.-H., Gu, Y.-Q., Zhou, J.-P., and Zhang, K.-Q. (2007). Possible contributions of volatile-producing bacteria to soil fungistasis. Soil Biol. Biochem. *39*, 2371–2379.

Proteogenomics: A New Integrative Approach for a Better Description of Protein Diversity Found in Soil Microflora

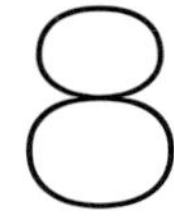

Céline Bland and Jean Armengaud

Abstract

Proteogenomics is a relatively recent field at the junction of genomics and proteomics which consists of refining the annotation of the genome of model organisms with the help of high-throughput proteomic data. To get a comprehensive view on how a given microorganism functions, elucidating its genome is a prerequisite. Since the first complete genome of a cellular organism was sequenced, that of *Haemophilus influenzae* in 1995, an impressive catalogue of genomes has been reported. Because automatic annotation software are not yet sufficiently confident, the annotation process should be complemented with experimental data. Alongside the development of high-throughput sequencing techniques, important innovations in tandem mass spectrometry and proteomic approaches have led to the possibility of analysing thousands of proteins from a given sample. Proteogenomics has proved to be helpful in discovering new genes that were forgotten by automatic annotation software, identifying the true translational initiation codon of coding domain sequences and characterizing maturation events at the protein level, such as signal peptide processing. Consequently, proteogenomics is now proposed at the earliest stage of a genome sequencing project as exemplified by the *Deinococcus deserti* genome, for which unexpected results, such as the reversal of gene sequences in different bacteria or the use of non-canonical start codons for translation in *Deinococcus* species, are only some of the numerous corrections obtained by proteogenomics. Because an important issue is the identification of the correct translational start codons, we have pointed out the need for developing N-terminal-oriented strategies to reveal experimentally the precise sites of translation initiation. Today, a better description of the protein universe found in soil microflora can be achieved if proteogenomics is performed on a given set of representative models from this environment.

Introduction

Establishing high-throughput and confident genome sequencing on model microorganisms has led to considerable development of new scientific fields over the last fifteen years: genomics, comparative genomics, structural genomics, transcriptomics and proteomics are the main examples. Because of the success of this genome-centric approach, the last decade was called the 'post-genomic era'. Basically, it consists of the deep analysis of all the biological components of a given model with a specific approach depending on the nature of these components: the transcriptome comprises the whole set of mRNAs produced by a cell in a given condition, the proteome is the list of all the proteins, the metabolome is the whole catalogue of metabolites. The integration of all these data has led to the emergence of systems biology which is aimed at modelling the functioning of the whole system based on these high-throughput data (Armengaud, 2010). Today, a new era is emerging with the high-throughput analysis of more complex biological samples. This new era was pioneered with the metagenomic analysis of samples harvested from the Sargasso Sea (Yooseph *et al.*, 2007, 2010). The analysis consisted of sequencing as much as possible of environmental DNA extracted from the microflora of the open sea in order to better understand the diversity present in such ecosystems. Impressive data were obtained with the identification of more than 6 million proteins from organisms that, for most of them, will not grow in the laboratory using current cultivation techniques. Scientists are logically extending the pioneering concepts of metagenomics towards other fields. Recently, some interesting studies related to metaproteomics have been published (Siggins *et al.*, 2012; Wilmes and Bond, 2006). The integration of high-throughput data harvested with several approaches is also gaining ground. Currently, more and more studies are presenting data obtained with various omics techniques (Armengaud, 2009). For example, the genome of the *Ruegeria pomeroyi* SS3 bacterium has been reannotated with a combination of proteomic and transcriptomic data (Christie-Oleza *et al.*, 2012a). Another mixed transcriptomics and proteomics approach was focused on methanol metabolism by the *Methylotenera mobilis* JLW8 bacterium (Beck *et al.*, 2011). The work done on *Geobacter sulfurreducens* is also worth noting both for the technical point of view with several high-throughput approaches carried out as well as the scientific insights into the regulatory complexity of a bacterial genome (Qiu *et al.*, 2010).

Here, we are reviewing the current state-of-the-art regarding the proteomic tools and approaches, the explosion of genomic sequences for thousands of microorganisms, the integration of both approaches giving rise to proteogenomics, a new science aimed at better annotating genomes with proteome high-throughput data to hand. We also present the specific approaches currently available to study the protein diversity of soil microflora, taking the work we have recently done on the Saharan bacterium *Deinococcus deserti* as the main example. Because the integration of multi-omic approaches is rich of information, the readers should take into consideration the other chapters of this book related to metagenomics, transcriptomics and proteomics (Chapters 3, 5 and 6).

The current proteomic tools and approaches

Basic principles for mass spectrometry of peptides and proteins

One of the initial goals of proteomics was the identification of a limited set of proteins in a sample. The other goals were to get the most important coverage of the sequence of these proteins (to facilitate localizing the corresponding gene onto the genome for further study) and identify all possible post-translational modifications. This has been achieved first by chemical Edman sequencing of a few residues of either the N-terminus or an internal sequence of a purified protein after cleavage with a protease. However, the technique is inefficient when a blocked N-terminus is present and quite difficult when the product is not sufficiently purified. Another approach for protein identification relied on the use of antibodies to perform Western blots, but non-specific binding and the availability of antibodies to all proteins limited its application in proteomics. As genome sequence information has accumulated, the pattern further shifted from sequencing part of the sequence of a purified protein to the identification of a set of proteins resolved on 2D-gels (Lin *et al.*, 2003). Edman-sequencing was replaced in the early 1990s by mass spectrometry because the latter technique requires less sample and is more informative. Peptide mass fingerprint was the most convenient strategy for protein identification. With the tremendous progress in mass spectrometry tools and the development of new approaches aimed at fractionating the protein sample, it quickly turned out to be possible to identify all the main protein components of a cell or tissue sample from one living cell model in a given condition. Mass spectrometry is now so performant in terms of accuracy, precision and speed of acquisition that a wealth of structural information can be quickly obtained on proteins: identification, sequence coverage, post-translational modifications and quantification.

The predominance of mass spectrometry originates from the fundamental improvements in ionization sources and mass analysers combined with advances in data processing such as the ability to correlate mass spectrometry data of peptides and proteins to the numerous sequences stored in databases. At the end of the 1980s, two main developments regarding the soft ionization techniques that create intact gas-phase ions from biomolecules considerably improved the analysis of proteins. Michael Karas, Franz Hillenkamp and their colleagues proposed the matrix-assisted laser desorption ionization (MALDI) where peptides are first dissolved in a matrix with a UV-absorbing compound before being included in the matrix crystals as the solvent dries. Pulses of UV laser light are used to vaporize small amounts of the matrix and the included peptide ions are carried into the gas phase. After ionization, the peptide ions can be analysed, i.e. their mass over charge ratio can be accurately determined with mass analysers coupled to ion detectors (Hillenkamp *et al.*, 1991; Karas and Hillenkamp, 1988). On the other hand, Koichi Tanaka who obtained in 2002 the Nobel Prize in Chemistry, was able to ionize high molecular weight biomolecules as large as the 34 kDa protein carboxypeptidase-A by choosing the right matrix and adapting the laser used for ionization of biomolecules. At the same time, John B. Fenn, who also obtained in 2002 the Nobel Prize in Chemistry, developed with his collaborators the electrospray ionization (ESI) technique. In this process, ions are created by application of an electrical potential to a flowing liquid leading the liquid to charge and subsequently spray. The electrospray creates very small droplets of solvent containing the analytes. Solvent is removed as the charged droplets enter the mass spectrometer by heat or by energetic collisions with a gas for example, and multiply charged ions are formed (Fenn *et al.*, 1989; Whitehouse *et al.*, 1985). These two soft ionization approaches, namely MALDI and ESI, have been used successfully for the analysis of peptides and proteins by mass spectrometry. SELDI, surface-enhanced laser desorption ionization, is a MALDI-derived strategy for solid-phase extraction of specific biomolecules via chemical and/or biochemical modified probe grafted on a surface. In practice, SELDI has been overcome by other approaches for proteins and peptides (Tang *et al.*, 2004).

Determination of protein sequences by tandem mass spectrometry

As ions exit the ion source, they are conducted within a mass analyser which is responsible for separating ions by their mass-to-charge (m/z) ratios. Mass analysers use electric and/or magnetic fields to manipulate ions in a mass-dependent manner. Several are available to accommodate the large mass range of peptides or proteins and the pulsed nature of the ionization technique: time-of-flights (TOF), quadrupoles and ion traps (Cotter *et al.*, 1994;

McLuckey *et al.*, 1994) are the main analysers currently in use in proteomic platforms. MALDI produces predominantly singly charged ions and because of the pulsed laser radiation, ions are created in bunches or packets. TOF mass analysers, which accelerate a packet of ions with a set of electric potentials and differentiate them by the time they take to travel a flight tube, are well-suited for analysing ions created in an intermittent fashion. On the other hand, ESI uses a steady stream of solvent to produce a continuous beam of ions. Coupling ESI to liquid chromatography in order to separate peptides on a reverse phase column prior the mass spectrometry measurements is readily feasible (Ducret *et al.*, 1998). Another feature of ESI is the production of multiply charged ions lowering the m/z values for high molecular weight compounds and allowing measurement of m/z values on mass spectrometers with limited m/z ranges. Multiple protonation of peptides and proteins also promotes easier amide bond fragmentation when the ions are activated for dissociation. All of these characteristics made ESI a widespread technique for proteomic research. Moreover, hybrid instruments (TOF/TOF, Q-TOF, LTQ-Orbitrap, etc.) coupled several mass analysers in order to perform tandem mass-spectrometry (MS/MS) to obtain as much data as possible specific to an individual peptide (McLafferty, 1981). Ionized peptides are first measured as intact fragment ions in the first mass analyser. They are further selected based on their m/z ratio and then pass into a collision cell where they undergo high-energy collisions with an inert gas such as helium, causing fragmentation of the charged peptides. This fragmentation of ions is referred to as collision-induced dissociation (CID) (McCormack *et al.*, 1993). One of the greatest strengths of tandem mass spectrometry for protein identification is the ability to sequence peptides directly from mixtures of peptides. To avoid laborious and time-consuming de novo sequencing of a given protein, (i.e. direct identification of the amino-acid chain from a mass spectrum), advantages from the direct relationship between mass spectrometry data and amino acid sequence are taken (Dongre *et al.*, 1997). Under the low-energy conditions employed for CID, peptide ions are fragmented into predictable patterns. Because peptide molecular ions fragment preferentially at certain points along the peptide backbone, theoretical spectra can be predicted for all the possible peptidic sequences derived from the protein databases. Computer algorithms have been developed to match the experimental spectra to these predicted spectra. By this means, they use the CID fragmentation patterns recorded for the peptides to determine their sequences (Yates, 1998). Many scoring algorithms have been devised to suggest which peptide sequence best matches a given spectrum. As a result, matching several tandem mass spectra to sequences in the same protein provides a high level of confidence in the identification of the proteins that are effectively present in a mixture.

The dawn of the shotgun era

Mass spectrometry allows the direct identification of the individual protein components of a given sample. However, if the mixture of peptides resulting from the proteolysis of the sample is highly complex, it is advantageous to introduce a separation step prior to the analysis in order to limit the number of peptides the mass spectrometer sees at a given time over the whole analysis. Current approaches and tools in proteomics are summarized in Fig. 8.1. The method most commonly used to reduce sample complexity prior to introduction into the mass spectrometer is the separation of proteins by gel electrophoresis followed by excision of the individual protein spots from the gel and in-gel digestion with a protease. One-dimensional (1D) SDS-PAGE gels, which separate proteins based on their molecular

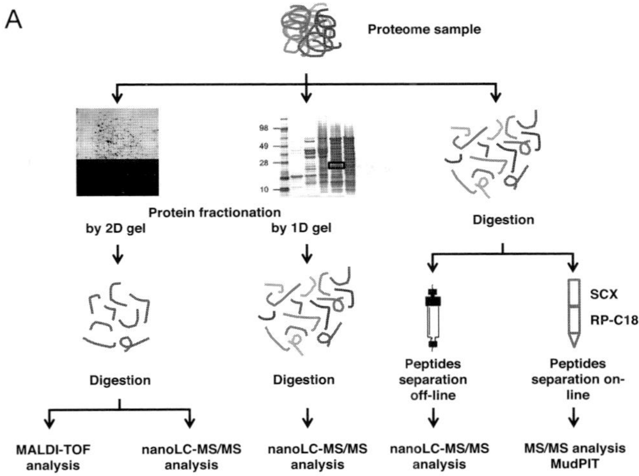

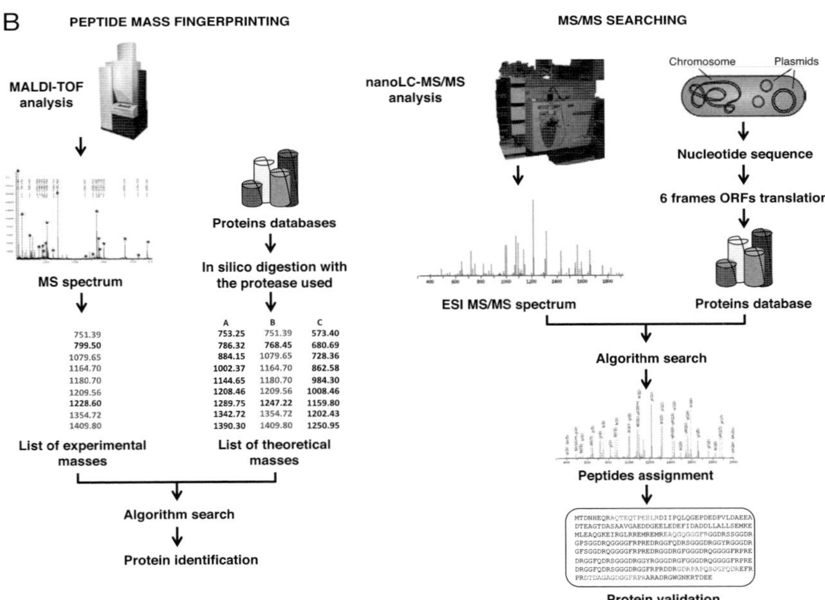

Figure 8.1 Shotgun proteomics: the most common strategies and current tools. Gel-based strategies followed by protein digestion, or protein digestion followed by multidimensional separation techniques, are the most common separation methods used in proteomics (A). These separation approaches are used to further massively analyse mixtures of thousands of different peptides by mass spectrometry and to identify proteins contained in these mixtures. Two approaches are illustrated, namely peptide mass fingerprinting and MS/MS searching (B). Peptide mass fingerprinting (left) is a method where an algorithm searches a protein database for proteins that would produce peptides of these molecular weights. Proteins are then identified by peptide mass mapping. MS/MS searching (right) relies on the multiple MS/MS spectra generated that are assigned to peptide sequences by specific searches against publicly available or homemade databases. Assignments of thousands of MS/MS spectra result in the identification of hundreds of proteins, validated when several peptides (in red) are assigned with confidence. A colour version of this figure is available in the plate section at the back of the book.

weight, provide a low-resolution separation of proteins, but when coupled with tandem mass-spectrometry can be used to identify proteins in moderately complex mixtures. For more complex mixtures, a multidimensional separation may be necessary. With multidimensional separations, two or more independent physical properties of the proteins or peptides are exploited to achieve a higher level of resolution and higher loading capacity than can be achieved in a single dimension. Two-dimensional gel electrophoresis (2DGE) is the most common multidimensional separation technique used to separate a large number of proteins in complex mixtures. Proteins are separated in the first dimension on the basis of their iso-electric point and by their molecular weight in the second dimension. The position occupied by a protein on a 2D gel is a reflection of its approximate p*I* and molecular weight. 2DGE provides separation and visualization of the protein mixture but does not explicitly identify proteins. A second analytical step must be employed. Because of the potential for high-throughput analysis, experimental masses can be determined by a MALDI-TOF mass spectrometer. The protein of interest, 'purified' as a single spot, is excised, digested with a protease and the m/z values of the resulting peptides are recorded. An algorithm searches a protein database for matching proteins that would produce peptides of these observed molecular weights when cleaved with the same protease. A score is generated based upon how many of the observed m/z values in the mass spectrum matched individual sequence entries. This method is known as peptide mass fingerprinting and is used to identify proteins in a simple and straightforward manner (Fig. 8.1). However, the efficiency of the method depends on the purity of the sample, the accuracy of the mass measurement, the number of peptide masses obtained and if the proteins to be identified are from a species that is well represented in sequence databases (Armengaud *et al.*, 2011; Delahunty and Yates, 2005; Johnson *et al.*, 2005; Yates, 2000). Gel-based proteomics has several drawbacks limiting its effectiveness. Manually excising proteins from a gel is time-consuming and when hundreds or more spots need to be processed from a single gel, automation becomes necessary. Most 2D gels can only focus proteins with a p*I* range between 4 and 10, thus excluding highly basic or highly acidic proteins under standard conditions. Proteins with molecular weights below 15 kDa runoff the gel and those above 200 kDa cannot be sufficiently resolved, excluding these proteins from the analysis. Membrane proteins and low-abundance proteins are often not detected on the gel. Thus, gel-based proteomics is rather limited in terms of exhaustiveness compared with gel-free proteomics.

Washburn *et al.* (2001) described a large-scale proteome analysis by means of multidimensional liquid chromatography coupled directly to tandem mass spectrometry. They named this approach MudPIT, standing for multidimensional protein identification technology. This analysis, done on the *Saccharomyces cerevisiae* yeast model grown to mid-log phase, resulted in the identification of 1484 proteins, an impressive record at this time. Such alternative approaches to 2DGE have been developed to directly establish the sequence of each of the peptides in a complex mixture. Usually, the total peptide mixture is loaded onto a reverse-phase nanocolumn. Peptides are then eluted into the ionization source of the mass spectrometer using an HPLC acetonitrile gradient that allows first elution of the polar molecules and then the more hydrophobic entities. As peptides enter the mass spectrometer, a survey scan of the intact peptides is obtained. Using data-dependent acquisition, the instrument can be set to automatically monitor the survey scan and select peptides for fragmentation based on pre-set criteria such as intensity, charge state or m/z. The selected peptides are then fragmented using CID leading to MS/MS spectra specific of individual peptides as output. Complex mixtures of peptides such as those from proteolysis by trypsin of a whole proteome require high-performance separation techniques. In the MudPIT approach, the complex peptide mixture is loaded on a specially packed, biphasic liquid chromatography column using strong cation exchange (SCX) support as the initial phase and reversed-phase (RP) material as the second phase (Delahunty and Yates, 2007; Wolters *et al.*, 2001). The first chromatographic stage will resolve peptides in function of their isoelectric point while the second stage will resolve them upon their hydrophobicity. These two separations are orthogonal leading to a high-resolution separation. This on-line protocol is technically difficult and requires an efficient and stable mass spectrometer connected to the chromatographic system over the whole experiment. Besides the direct on-line protocol of MudPIT, off-line multidimensional separation methods have been used to achieve the same basic aim but in a more robust way. In this case the main advantage is a possible optimization in-between both dimensions for some fractions. However, automation is then decreased and sample loss may occur (See (Delahunty and Yates, 2005; Lin *et al.*, 2003) for a more detailed review on separation techniques). Another means of resolving the peptide complex mixture consists of OFFGel electrophoresis. In this case, the peptides are first separated along their isoelectric point and then, each fraction can be loaded onto the reverse phase column and analysed by nanoLC-MS/MS.

A need for updated bioinformatic tools and statistical assessment of the results

For the assignment of peptide sequences to the MS/MS spectra recorded by nanoLC-MS/MS, bioinformatic tools have been developed. The most widely used software programs are: X!Tandem, SEQUEST, and Mascot (Ansong *et al.*, 2008). They are based on different algorithms and scoring parameters to give the best match for each experimental MS/MS spectrum. The correlation between spectral mass-to-charge values to the best hit among all possible theoretical peptides from a database containing all possible translated amino acid sequences for a specific organism whose complete genome sequence is known (Ansong *et al.*, 2011; Cottrell, 2011) can take a few hours to a few days depending on the size of the database being searched, the complexity of the sample, and the computing resources available. Although assignment algorithms are getting more accurate and reliable, there will always be a large fractions of MS/MS spectra for which no matches or identifications are possible. These arise from poor fragmentation because of the nature of the peptides or from low signal because of the low abundance of the peptides. Results obtained from database searches performed using MS/MS spectra are not necessarily correct. For MS/MS spectra identification, each search engine, depending on the specific scoring scheme used, ranks the sequence candidates according to an assigned score. For each MS/MS spectrum, at least one top-ranked sequence is attributed but is not necessarily correct: it still remains to the users to verify and validate the identification. The individual peptide scores are then merged for the identification and scoring of the corresponding proteins. Because of the correlation of numerous data done at a given probability chosen by the experimenter, specific statistical analysis should be performed such as an evaluation of the number of false-positive hits. This is commonly carried out using a search with the same parameters but with a database comprising protein sequences that do not really exist. The current usage is to search a randomized database of the same size and amino acid composition as the normal database, which is most easily done by reversing the original database.

In summary, current methods and state-of-the-art mass spectrometers such as those based on the Orbitrap analyser are able to list thousands of proteins from the same proteome sample. A recent study reported by Geiger *et al.* (2012) where 11,731 distinct proteins from a human cell line have been identified shows that nowadays, it is relatively easy to catalogue the components of a complex sample. This study shows that in a decade the number of proteins identified was multiplied by eight. Although this fold change does not follow Moore's law, it is an impressive result because of the large dynamic range of protein quantities in such a complex sample. Thus, proteogenomic methodology is now ready to record enough data on a cellular model for getting a real comprehensive view, whatever its origin and nature.

Targeted proteomic analysis for absolute quantification of protein

The complexity and large dynamic range of compounds usually found in proteomic samples challenge the well-established data-dependent workflows requiring very high speed MS/MS acquisition for comprehensive coverage of the analytes. Absolute quantification of a given protein with high reproducibility can be reached using multiple reactions monitoring-mass spectrometry, abbreviated as MRM-MS. This approach consists in defining a given set of proteotypic peptides with known fragmentation pattern and quantifying the intensity of some of the MS/MS fragments called transitions. The set of peptides queried has to be specified prior to MS/MS data acquisition. Triple quadrupole or quadrupole-ion trap hybrid mass spectrometers are usually highly sensitive and selective, isolating the precursor ion in the first quadrupole, fragmenting it within the second quadrupole or in the trap, and finally monitoring the optimum fragment ions using the third quadrupole or the trap (Meng and Veenstra, 2011). However, this method is limited to the measurements of a few hundred transitions per LC-MS/MS run, and thus a few dozen of proteins. Novel strategies have been recently developed to increase the reproducibility and comprehensiveness of data collection. They are based on data-independent MS/MS acquisition. Typically, consecutive survey scans of parent peptides are repeated throughout the LC time range, in which fragment ion spectra for all the precursors contained in a predetermined isolation windows are monitored. The SWATH-MS strategy developed with a Q–TOF hybrid mass spectrometer consists in selecting in the first quadrupole peptides from a wide mass range (25 amu) for collision-induced fragmentation, the resulting fragments being analysed in a TOF analyser at high resolution. This time-and-mass segmented acquisition method generates for a single sample injection fragment ion spectra of all precursor ions within the user-defined precursor RT and m/z space and records a complex fragment ion map which can be interrogated several times (Gillet *et al.*, 2012). Even though a more complex fragmentation spectrum is produced, the high resolution MS/MS enables tighter extraction windows to maintain high specificity. Advantages of this strategy are numerous: (i) no upfront assay development is required on specific targets, (ii) all data are acquired and (iii) targets are mined post-acquisition.

Genome annotation of soil microflora gains in number but not in quality

An avalanche of genome data

Since the first complete genome of *Haemophilus influenzae* Rd KW20 was sequenced in 1995, more than seven thousand complete genomes have been reported. To date, the genomes of 1863 prokaryotes (1742 bacteria and 121 archaea) have been annotated (2012, January 27th NCBI update; see http://www.ncbi.nlm.nih.gov/genomes/lproks.cgi), representing an impressive catalogue of microbial genomes. Besides, an increasing number of large eukaryotic genome sequences and annotations are also available. More than five thousand genomes have also been sequenced and uploaded in official databases but their annotations are currently still in progress. This remarkable evolution in the number of genomes sequenced was made possible because of the advances achieved in sequencing techniques. Genome sequencing was first based on the Sanger sequencing reaction using radioactive compounds. The radiolabelled nucleic acid products were resolved on large electrophoresis gels. The use of fluorescent dyes for labelling nucleotides and the separation of fluorescent products by capillary electrophoresis were two important improvements allowing a drastic reduction of the number of samples to be handled. Moreover, the time-consuming chromosome-walking strategy was replaced by the quick shotgun strategy. This latter approach consists in cloning randomly generated and overlapping fragments into a plasmid and sequencing them with universal oligonucleotides. Today, Whole Genome Shotgun (WGS) sequencing is commonly used. It relies on high-throughput sequencers and computer programs assembling the overlapping ends of millions of short reads. This assembling procedure yields a contiguous sequence or a restricted set of contiguous sequences as closing the circular chromosome may be rather time-consuming and is not always required. With the efficiency of genome sequencing increasing, a significant rise in the release rate of sequenced genomes has been achieved.

Genome annotation relies mainly on automatic procedures

Different levels of genome annotation can be reached with information about the function of cellular components, their spatial location, their interactions and alterations over evolutionary time. Genome annotation is a multilevel process involving the prediction of coding genes, pseudogenes, promoter regions, direct and inverted repeats, untranslated regions and other genome items. It consists of identifying the coding sequences transcribed into functional RNAs or translated into proteins before assigning either predicted or known functionality to the identified gene products. These two levels of annotation are respectively named structural annotation in which the location of each key component is described, and functional annotation in which the function of a gene is assessed. Both are intimately and inherently linked. Genome annotation can be carried out in different ways. Manual curation from experimental data in the literature provides the highest accurate datasets. However, it is a slow process and only a fraction of the data to be annotated can be uncovered by highly trained curators. The information obtained about one sequence can be alternatively transferred to a related sequence considered to be homologous. The evolutionary distance between the related sequences is then crucial for the accuracy of the prediction (Sato and Tajima, 2012). Annotations can also be predicted using *ab initio* methods in which rules established on previous annotations or physico-chemical properties of the molecules are used. Choosing an annotation method is often a compromise in between speed and accuracy. For small datasets, manual curation or experimental methods with higher accuracy are probably more appropriate but they are time-consuming and not adapted to the current avalanche of genome sequences. For larger datasets, methods with higher speed and coverage annotation but lower accuracy may be more relevant for data extraction but should be taken cautiously (Reeves *et al.*, 2009).

Confidence in genome annotation is challenged

Errors in genome annotation can emerge at different stages: during sequencing although the current high sequence redundancy should be further increased in the near future leading to high sequence quality, during the prediction of coding genes or over the process of assigning gene functions. The problems come from information used from previously annotated genomes that have not always been validated by other means than automatic annotation programs. Numerous errors also arise when using automatic systems based on comparative genomics when genome annotation is intended on novel nucleic sequences that do not share any similarities with already known sequences. Functions of the plausible open reading frames are then purely hypothetical, but more importantly what and how they are being translated to proteins is only based on predictions and could produce wrong results (Devos and Valencia, 2001). A recent study on *Deinococcus deserti* genome sequence (de Groot *et al.*, 2009) points out numerous annotation errors in the first two *Deinococcus* species ever sequenced, *Deinococcus radiodurans* BAA-816 (White *et al.*, 1999) and *Deinococcus geothermalis* DSM1130 (Makarova *et al.*, 2007). Annotation errors introduced in the databases are maintained and are accumulating exponentially as they propagate once new, relatively close genomes are being annotated. As a result, current databases are saturated with non-informative data mixed within complete genome sequences. Their use needs specific precautions. The effort of the Swiss Institute of Bioinformatics to specifically, manually validate the genome annotation of each protein from a given set of organisms is worth citing (Jain *et al.*, 2009). Possible reasons for annotation errors, such as overlapping ORFs, determination of start codons or setting the cut-off for filtering short ORFs that might encode small polypeptides, were recently discussed (Poptsova and Gogarten, 2010). Between 30% and 50% of ORFs in complete annotated genomes are identified as hypothetical genes. These predicted ORFs have a much higher probability of being incorrectly annotated start codons. Newly sequenced genomes with the most recent technologies based on ultra-short read data sets (Illumina for example) show some specific trends of wrong base calls. This could be a reason for the increase of sequencing errors currently noted in databases. Second-generation annotation systems, combining multiple gene-calling programs with similarity-base methods, perform in a better way compared with the first annotation tools. Novel improvements of these annotation systems are nevertheless urgently needed. As an example, we recently revealed that approximately 20% of proteins were erroneously annotated in terms of the N-terminus while working on the systematic identification of translation start codons in *Deinococcus deserti* (Baudet *et al.*, 2010). Prediction of start codons estimated by an *in-silico* genome analysis is frequently erroneous (10–20%), reaching up to 60% in some GC-rich prokaryotic genomes (Gao *et al.*, 2010; Nielsen and Krogh, 2005).

Proteogenomics, mapping proteome data onto genome sequence

The alliance between genomics and proteomics

Historically, the genomics and proteomics communities have acted independently. Genomists had to identify genes and the corresponding protein sequences. Large-scale annotation efforts were necessary during and after sequencing of the genomes. The proteins retrieved were considered as a fixed set, although it was admitted that not all proteins are expressed in every cell in a given state. On their side, proteomists had to understand which proteins are expressed under specific conditions or tissues but also to identify the various post-translational modifications and other processing of the genome (Castellana and Bafna, 2010). They were relying on the work of genomists taking as the database for their data interpretation only the list of protein sequences arising from the list of annotated genes. Combining the strengths from both communities was first proposed by Yates and his co-workers (Yates *et al.*, 1995). They proposed the correlation of the MS/MS spectra of peptides with the nucleotide database translated in the six reading frames with the idea of correcting errors of sequencing. Jaffe *et al.* (2004) proposed a decade later a new strategy, called 'Proteogenomic mapping' where peptides are directly mapped onto the genome sequence in order to render easier the validation of genes. This proteogenomic mapping has greatly improved over the last few years because of the recent advances in mass spectrometry (Armengaud *et al.*, 2011). Today, proteogenomics is the method of choice for improving the annotation of really novel genomes by exploiting the mapping of peptides identified by tandem mass spectrometry to the gene loci.

Proteogenomic mapping

Proteogenomics consists of (i) obtaining the largest possible dataset on proteins of a given microorganism and then, (ii) mapping this information onto the genome sequence in order to better annotate genes encoded by this genome. For this, an in-depth analysis of the proteome should be performed listing as many proteins as possible. Two different approaches for mass spectrometry-based protein identification from complex mixtures have been developed; 'top-down' and 'bottom-up' proteomics. The top-down strategy, i.e. the direct analysis of entire proteins, starts with an intact protein which is cleaved in the gas phase rather than in solution (Ferguson *et al.*, 2009). The protein is fragmented inside the mass spectrometer to create a ladder of ions indicative of the sequence. The difference in m/z values of fragment ions defines the position and sequence of the amino acids in the protein. The high complexity of this approach requires a high-resolution mass spectrometer. Furthermore, a specific fragmentation mode such as electron capture dissociation (ECD) is needed (McLafferty and Senko, 1994). The observed molecular weight of a given protein may differ from its predicted molecular weight because of sequence errors, post-translational modifications or proteolytic processing. Therefore, this approach is not yet adapted for high-throughput proteogenomics although some cases of re-annotation have been reported for the *Methanosarcina acetivorans* bacterium (Ferguson *et al.*, 2009). The bottom-up approach identifies proteins by tandem mass spectrometry analysis of peptides derived from digestion of mixtures of intact proteins as explained above (Yates *et al.*, 1995). Typically, the few thousand proteins contained in a proteome sample are fractionated and subjected to trypsin proteolysis. The resulting complex mixture of peptides is analysed by tandem mass spectrometry. Each MS/MS spectrum will be then assigned to a peptide sequence by specific searches against publicly available or homemade protein databases. Proteogenomics requires dealing with the nucleotide sequence of the model organism without any *a priori* on its annotation. The most straightforward procedure is listing all possible open reading frames (ORFs) of a minimum size translated from the six reading frames of the whole genome. An ORF is defined as a stretch of DNA starting just after a translational STOP codon and ending at the next STOP codon in the nucleic acid sequence. The resulting list contains mostly unlikely protein sequences. The recorded MS/MS data will then be used to discriminate amongst all these candidates the truly existing proteins present in the proteomic sample. In parallel, genes can be predicted with specific software or pipelines specialized in the automatic search of the genome for coding domain sequences (CDSs). A simple comparison of both listings will indicate the accuracy of the annotation process. Because extensive coverage of the protein sequences can be reached by means of a shotgun analysis with high-performing tandem mass spectrometers, this approach is commonly used for proteogenomics.

While the approach based on the six reading frame protein database is straightforward for bacterial genomes, proteogenomics appears to be much more complex for archaea, with the existence of inteins, and even more for eukaryotes where the exon/intron structure has to be taken into account (Armengaud, 2009). Inteins are called selfish or parasitic genetic elements and are similar to self-splicing introns. However, in contrast to introns, inteins are transcribed and translated together with their host protein (Gogarten *et al.*, 2002). Only at the protein level, the inteins excise themselves from the host protein. The remaining two portions of the host protein separated by the intein, called exteins, are then joined by a peptide bond. The resulting list of experimental peptides may be lacking all the peptides corresponding to boundaries between inteins and exteins, as they were not present in the theoretical database used for MS/MS assignment. For eukaryotes, the coding regions of the gene are often present in discontinuous regions called exons. Multiple exons are separated by regions that are transcribed but not translated, the introns. They are spliced out of the mRNA prior to translation by an RNA–protein complex called the spliceosome, producing the mature mRNA. Because of these specific features, the strategy for peptide identification based on sequence prediction using the six reading frames of the genome sequence is somehow very complex for eukaryotes. Specific searches should be done for sequences at the boundaries between introns and exons. However, the strategy relying on a six-frame translation offers the most convenient path to directly measure peptides arising from expressed proteins. These peptides are then used to validate gene annotations, identify novel genes, refine the current annotation by correcting initiation start sites or characterize post-translational modifications (Ansong *et al.*, 2008; Armengaud, 2009).

Proteogenomic parameters to be considered

When using software to automatically identify proteins, a number of search parameters need to be considered. One of these is the number of missed cleavages allowed per peptide, as proteases such as trypsin do not cleave all their substrates to completion. They are usually inactivated before completion because of autoproteolytic cleavage. Depending on the preparation of the peptide mixture and chemical reactions that may have been used prior to mass spectrometric analysis, this parameter has to be adjusted case by case. Possible amino acid modifications need to be included as additional search parameters. For example, cysteine residues are often alkylated to assist in proteolytic digestion and avoid disulphide bridge formation between peptides. They are considered as 'fixed' modifications as all the cysteines will be alkylated. The 'variable' modifications are non-stoichiometric (*e.g.* phosphorylation, oxidation of methionine) as only a fraction of residues may have been modified. Problems arise during the MS/MS assignment process when frequently occurring amino acids are said to be variably modified. In this case, many peptides will be affected, slowing the search time and increasing the number of potential false positives that may have the same mass. The correct definition of the database used to search the MS/MS spectra is also important. One should restrict the database to the species of interest to keep the universe of potential matches small. The peptide mass tolerance, or how accurate the masses are being measured, is another parameter that has a great impact on the search. Tight tolerances, provided by instruments capable of high mass accuracy and resolution will greatly reduce the number of false positives. These parameters have to be considered for both peptide mass fingerprinting and MS/MS database searching. One additional parameter has to be taken into account for MS/MS matching, the fragment ion mass tolerance. It is also necessary to mention that a large fraction, usually a majority, of the acquired MS/MS spectra is not matched to a sequence entry in a normal proteomic analysis. Diverse reasons explain this result, such as: (i) the organism studied is not well represented in any sequence database, (ii) peptide modifications *in vivo* or *ex vivo* occur and the observed peptide mass will be different from what would be calculated from a sequence database, (iii) the precursor ion was not derived from an intact peptide that conforms to the anticipated cleavage specificity of the enzyme that was used, (iv) intact peptide ions fragment in the ion source prior to the first stage of mass analysis, (v) the charge state of the precursor ion has not been determined correctly, (vi) the spectrum is of insufficient quality or the MS/MS spectrum was obtained from non-peptide contaminants. To avoid an important ratio of false-positives, a database search should be performed using the simplest criteria: strict tryptic cleavage rule, a single missed cleavage site, and one variable modification only. The unmatched spectra can further be re-analysed with wider mass tolerances, no enzyme specificity, more variable modifications or a larger database if such a search is of interest. For more detailed considerations on the influence of the search parameters on protein identification (Cottrell, 2001; Johnson *et al.*, 2005).

Comparative proteogenomics

Proteogenomics has been proposed to consider the evolutionary constraints in fully interpreting proteogenomic data and improving genome annotations, as proteogenomic peptides provide invaluable information for gene annotation, which is difficult or impossible to ascertain using standard information for gene annotation. The proteogenomic identification might come from a region of the genome not previously known to code for protein. These peptides are referred to as novel and might be intragenic, i.e. falling within the locus of a known gene model, or intergenic, i.e. falling outside the locus of a known gene structure. For intragenic novel peptides, it is difficult to distinguish if the gene structure needs to be corrected or if it can be explained by a novel splice-form and intergenic peptides, which are not proximal to a known gene, may indicate a novel coding region. While analysing the proteome of a single microorganism, some proteins or some peptides indicating a specific maturation event will not be identified with enough confidence to be taken into consideration for the structural reannotation of the corresponding genes. Therefore, researchers take advantage of mass spectrometry data obtained from multiple genomes belonging to the same genus giving more confidence to the reannotations (Armengaud *et al.*, 2011). Reconstructing gene models using mapped peptides is not trivial, mostly because the peptide information is not sufficient to completely determine the structure due to limited coverage. However, the identification of the most N-terminal peptides can, in many cases, be the determining factor in positioning the translational start codons. The confidence of the proteins identified by a unique peptide, the so-called 'one-hit wonders', is also usually low and these identifications are most of the time discarded because they are ambiguous (Veenstra *et al.*, 2004). If such peptides and respective homologues are seen in the proteomes of several closely related microorganisms, then the confidence for this specific event increases.

This has been demonstrated and used in different studies carried out on bacteria such as *Schewanella* (Gupta *et al.*, 2008), *Mycobacterium* (Gallien *et al.*, 2009), *Yersinia* (Payne *et al.*, 2010), and *Roseobacter* (Christie-Oleza *et al.*, 2012a) groups. More confident exhaustive data will greatly improve the proteogenomic annotation of the different microorganisms under consideration. As already mentioned, prokaryotic genomes are usually smaller and less genetically complex than eukaryotes and since prokaryotic genes do not undergo splicing, all proteins can be captured by translating the genomes in all six frames. Proteogenomic validation of predictions is a pragmatic compromise between computational prediction and full experimental validation. Peptides which also map in close proximity to annotated genes may suggest changes to the gene model in these specific loci. Moreover, the detection of a specific, new gene at a novel locus may question whether new conserved ORFs may be indicated in the close neighbourhood and whether an operonic structure may be detected. Annotating translation start sites is not a trivial task because of the existence of three different initiation codons: ATG (Methionine), GTG (Valine), and TTG (Leucine). While ATG is the most frequent start codon, it is usually favoured by automatic annotation software (Hu *et al.*, 2008). Specific databases such as the ProTISA resource collects correct annotations of translation initiation sites (Hu *et al.*, 2008), however its usage is currently limited. Correct identification of translation initiation sites is a key parameter to test for novel annotation software (Hyatt *et al.*, 2010). Proteogenomic mapping of peptides to regions proximal to the N-terminus of the annotation may correct these errors. Another reannotation case relies on peptides which are mapped near an annotated gene but are in a different frame. They may indicate the rare event of programmed frame shift, which is nearly impossible to predict by automated methods (Castellana and Bafna, 2010).

Due to the current procedure for genome annotation, there is no direct experimental evidence for the large majority of genes annotated that they are really translated into proteins. Multiple unique peptides mapping to a single gene provide compelling evidence for the expression of this gene and the production of the encoded product. Thus, proteomic combined with genomic approaches already at the primary annotation stage of a genome sequence annotation project can certify the accuracy of the annotation. The experimental proof accumulated on thousands of proteins represents above all an invaluable proteogenomic resource for the creation and validation of new, ab initio gene-finding software programs for the functional annotation of new genomes.

N-terminomics, new tools for an avalanche of results.

A need for better assessment of the N-termini of proteins

Determination of the correct N-termini of proteins is an important step in genome annotation. Since wrong prediction of protein initiation codons is one of the major sources of problems, specific efforts should be made to improve this point. Errors, especially important in GC-rich regions, are estimated to be higher than 10% when analysing previous annotations. The incorrect prediction of initiation codons in prokaryotic genomes is particularly widespread, although many genomes are annotated and could be used for consensus prediction (Aivaliotis *et al.*, 2007; Baudet *et al.*, 2010; Sato and Tajima, 2012). The identification of a correct translational start site is crucial in the biochemical analysis of a protein or in the genetic analysis of a gene, since errors can seriously impact subsequent biological studies. If the N-terminus is not correctly identified, the protein sequence itself can be either truncated or extended, leading to errors in estimating parameters such as the molecular weight and the isoelectric point of the product. Moreover, its production in a heterologous system, such as *Escherichia coli* overexpression of the gene from a T7 promoter, will lead to major difficulties, such as an unfolded polypeptide recalcitrant to structure determination, if crucial residues are missing. The problem originates essentially from the current trends in rapid and massive sequencing that do not allow specific experiments for validation, as described above. The identification of translational initiation codons is made difficult because multiple initiation codons may be proposed. While ATG is the most frequent, GTG and TTG are also significantly used. In some specific organisms, such as the hyperthermophilic crenarchaeon *Aeopyrum pernix* K1, a strong bias towards the use of TTG rather than ATG has been reported (Yamazaki *et al.*, 2006). Surprisingly, the translation of 52% of the genes starts with TTG in this case as proved by mass spectrometry measurements, while 28% and 20% start from an ATG and GTG codon, respectively. In addition, in very specific cases, non-conventional start codons may be used to initiate the translation of the mRNA. For example, CTG and ATC codons are used as non-canonical starts in *D. deserti* as proved by chemical labelling followed by mass spectrometry and multiple sequence alignments (Baudet *et al.*, 2010). Several genes have such rare initiation codons: DnaA (the protein involved in the DNA replication initiation process), RpsL (the S12 ribosomal conserved protein involved in translation), InfC, and the Deide_03051 gene with no assigned function. Large-scale experimental determination of N-termini has been proposed to further complement proteogenomics. Mass spectrometry is a really helpful technique to assess multiple initiations, alternative splicing and N-terminal post-translational modifications such as removal of the initial methionine, proteolytic truncation, N-terminal acetylation or peptide signal processing, influencing the localization and activity of many proteins. A low coverage of protein N-terminus peptides has been reported in most proteogenomic studies based on a standard shotgun MS/MS analysis. This is mainly due to the poor peptidic coverage of the less abundant proteins present in the proteome sample. Several analytical strategies have been proposed to selectively increase the detection of these specific N-terminal peptides which have been called by some authors the N-terminome.

A large range of N-terminomic strategies

One of the possible strategies to catalogue N-termini of proteins is based on the properties of N-terminal peptides. As they generally have no positive charge in acidic conditions, these peptides can elute in low-salt fractions when being separated by strong cation exchange chromatography from a peptide mixture (Aivaliotis *et al.*, 2007). More specific strategies have been also described. Gevaert *et al.* (2003) suggested a method to sort N-terminal peptides by a diagonal chromatography of fractions collected from a first chromatography, named COFRADIC (COmbined FRActional DIagonal Chromatography). In this method, free amino groups in proteins are first blocked by acetylation before the proteins are digested with trypsin. The resulting peptides are resolved by reverse-phase chromatography. The different fractions collected are then treated with 2,4,6-trinitrobenzenesulphonic acid (TNBS) which reacts with the N-terminus of internal peptides, changing their hydrophobic character. These peptides will separate from the acetylated or blocked N-terminal peptides in a second reverse-phase chromatography (Gevaert *et al.*, 2003; Staes *et al.*, 2008; Van Damme *et al.*, 2009). Another method consists of acetylation before trypsination, followed by biotinylation of free amino groups and then affinity capture of internal peptides bearing biotin (McDonald and Beynon, 2006; McDonald *et al.*, 2005). Guanidination of lysine lateral chains followed by N-biotinylation of the N-terminal and trypsin digestion is another alternative method. Selective purification of N-terminal peptides is performed with avidin beads (Yamaguchi *et al.*, 2007) or internal peptides can be depleted (Yamaguchi *et al.*, 2008). Dimethyl isotope-coded affinity selection (DICAS) is a method where reductive amination of all free amino groups is achieved with formaldehyde before trypsin digestion. Internal peptides formed with free N-terminal amines after proteolysis are then captured by solid supports with aldehyde functionalities through reductive amination. N-terminal peptides of interest remain uncaptured (Shen *et al.*, 2007). Another method worth mentioning is the selective N-terminus derivatisation with N-Tris(2,4,6-trimethoxyphenyl)phosphonium acetyl (TMPP), a labelling reagent that increases the hydrophobic properties of the N-terminal peptide, improves its ionization ability and modifies its fragmentation pattern due to the positive charge introduced (Baudet *et al.*, 2010; Chen *et al.*, 2007; Gallien *et al.*, 2009). A different technique recently developed and called terminal amine isotopic labelling of substrates (TAILS) is used to distinguish protease-generated neo-N-termini from mature protein N-termini with high confidence (Kleifeld *et al.*, 2010, 2011). This quantitative proteomics approach is based on labelling and isolating N-terminal peptides before and after exposure to a protease of interest. The protease is first inactivated after proteome proteolysis and the sample is denatured and reduced. To quantify the cleavage events specific to the protease of interest, and to distinguish these from proteolytic products present in an untreated sample, stable isotopes are introduced to determine the relative abundances of peptides in the protease-treated and control samples. Primary N-terminal and lysine amines are blocked by dimethylation and isotopically labelled by incorporating heavy and light dimethylation reagents in cell culture labels. After tryptic digestion, N-terminal peptide separation is achieved using a high-molecular-weight dendritic polyglycerol aldehyde polymer that binds internal tryptic and C-terminal peptides that now have N-terminal alpha amines. The unbound, naturally blocked or labelled mature N-terminal and neo-N-terminal peptides are recovered by ultrafiltration and analysed by tandem mass spectrometry (Kleifeld *et al.*, 2010, 2011).

The higher the abundance of the protein in the sample, the more likely it is that it should be identified either as a full protein or as one of its peptide derivatives. Because of the present limitations of the dynamic range of mass spectrometers, a decrease in protein mixture complexity is currently required for identifying low-abundance proteins in proteomic analysis and for increasing the total number of proteins identified in a study (Meinnel and Giglione, 2008). In most N-terminomic methods, specific identification of N-terminal peptides from the digest of a protein extract implies the loss of all internal peptides (Gallien *et al.*, 2009). Consequently, protein sequence coverage is low but sample complexity is also reduced and more proteins are accessible for identification from their N-terminal peptides and for further validation or re-annotation. Importantly, proteins are typically identified by only a single peptide in N-terminomic approaches. Thus, it is really important to carefully examine the data generated by N-terminomics to avoid including new errors. As an example for the refinement of genome annotation, the identification of N-termini of *Deinococcus deserti* was recently carried out on a very large scale (Baudet *et al.*, 2010). The proteome was labelled with TMPP before proteolysis either with trypsin or chymotrypsin. The resulting peptides were analysed by nanoLC-MS/MS with a LTQ-Orbitrap XL high-resolution hybrid mass spectrometer. A set of 664 N-terminal peptidic sequences were listed, leading to the correction of 63 translation initiation codons in the genome of *D. deserti*. Of note, a gene may have more than one translational start site. Such an event, studied in *D. deserti* analysis, is reported in Fig. 8.2. Only the most upstream start site that has supporting peptide evidence will be retained in databases in such cases. However, all possible alternatives should be reported for being considered by experts in the field. Another interesting feature in *D. deserti* was the presence of non-canonical codons which can be used to initiate translation (Baudet *et al.*, 2010).

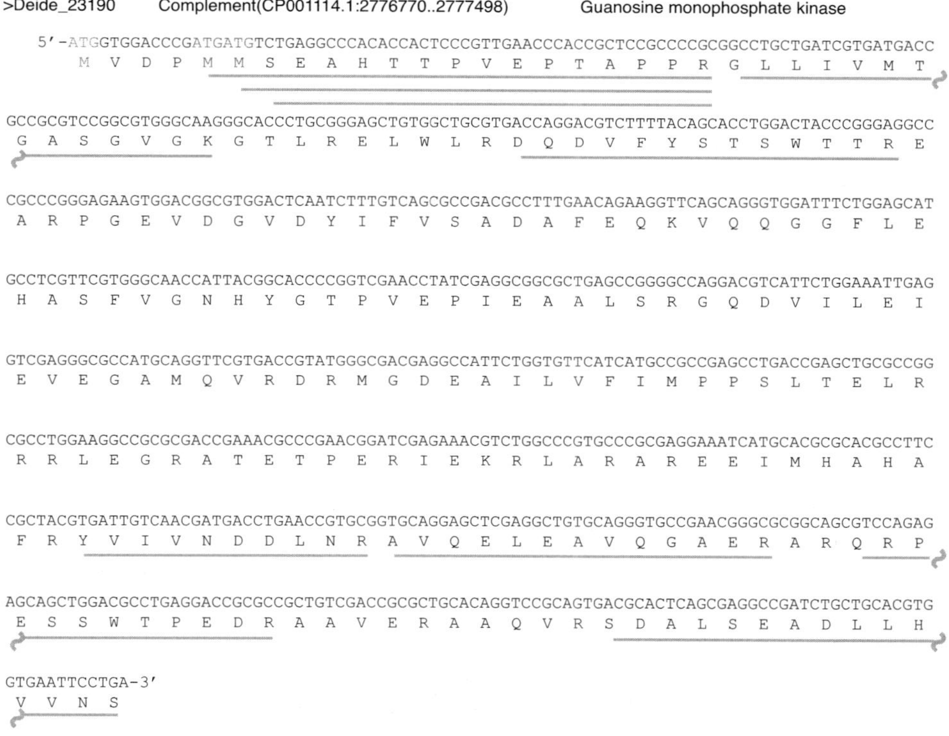

Figure 8.2 Two different initiation codons for the translation of Deide_23190 in *Deinococcus deserti*. The nucleic acid sequence for the locus specifying Deide_23190 is shown. Three different ATG codons found at the 5′-end of the ORF are labelled. Nine different tryptic peptides, detected after mass spectrometric analysis, were assigned to a guanosine monophosphate kinase (de Groot *et al.*, 2009). Three peptides correspond to N-terminal signatures and indicate a misannotation of this predicted gene regarding its translation start codon. The second and third ATG codons, which are contiguous, were used as initiation codons. The cells can then produce two different polypeptides from these two alternatives: one starting with the MMSEAHTTP sequence and the other starting with MSEAHTTP. A maturation of the second polypeptide with the excision of the initial methionine, expected because a serine is found as the second residue, results in the production of a third polypeptide starting with the SEAHTPP sequence. A colour version of this figure is available in the plate section at the back of the book.

Contribution of proteogenomics to a better assessment of soil microflora

Experimental protein-based validation exemplified with *Blumeria graminis*

As explained above, accurate genome annotation is crucial for comprehensive and systematic studies of biological systems. However, the determination of the structure and function of genes encoding proteins is not error-free. In recent years, many examples of proteogenomic-based refinement of genome annotation have been reported for soil micro-organisms. We will develop below two examples that illustrate well the benefits of proteogenomics to better assess soil microflora in terms of protein diversity. *Blumeria graminis* f. sp. *hordei* strain DH14 is an economically important, obligate plant-pathogen fungus, only able to grow on its specific host, barley. The samples were obtained directly from infected plants, given that *Blumeria*, like all powdery mildews, is an obligate biotroph. The proteome from two stages of development representing different functions during the plant-dependent vegetative life cycle of this fungus was analysed and compared by a shotgun proteomic approach carried out to provide protein-level experimental validation, detect incorrectly assigned translational start sites, uncover non-annotated genes and characterize post-translational features (Bindschedler *et al.*, 2009, 2011). Samples from sporulating hyphae and haustoria, the feeding and effector-delivery organs of the pathogen inside barley leaf epidermal cells, were examined. Proteins from each sample were resolved with different strategies before tryptic digestion. The resulting peptides were analysed on a nanoLC-MS/MS with a LTQ orbitrap XL mass spectrometer. The *Blumeria graminis* genome as a draft assembly contained 15,111 contigs of various lengths as well as EST, short sequence tags. MS/MS data were searched against the predicted *Blumeria* ORF sequence database, generated from the annotation of all possible genes. The theoretical protein database contained a rather limited set of protein sequences (5823). A total of 1401 proteins could be identified with at least two significant peptides. A total of 1330 proteins were detected by proteomics in samples corresponding to sporulating hyphae, while 296 were observed in haustoria samples. A total of 71 proteins were only identified in haustoria. The identified peptides could be further aligned to the *Blumeria graminis* genome sequence to visualize their positions relative to each other and to other genome data available. Assembly gaps or mistakes have been highlighted with such approaches, as exemplified with the catalase gene that was encoded within two separated contigs. ORFs truncated in their N-terminal or C-terminal end at the extremity, or misoriented contigs, were also identified with such a proteogenomic approach. The large scale proteogenomic investigation of *Blumeria graminis* allowed an independent and robust validation of gene models and protein existence based on experimental data as opposed to purely bioinformatics-based gene prediction (Bindschedler *et al.*, 2009, 2011).

Genome re-annotation of *Deinococcus* species

We have also carried out an in-depth proteogenomic analysis of the soil bacterium, *Deinococcus deserti* VCD115, which represents a bacterial model of choice. This bacterium was isolated from upper sand layers collected in the Sahara Desert in Morocco and Tunisia (de Groot *et al.*, 2005). *D. deserti* belongs to the *Deinococcaceae*, a family of extremely radiation tolerant bacteria comprising *Deinococcus radiodurans*, known for decades and which has been extensively studied (Blasius *et al.*, 2008; Cox and Battista, 2005; Daly, 2009). To better understand the molecular mechanisms for the radio resistance of *D. deserti* and its adaptation to the harsh conditions encountered in a hot arid desert, its complete genome sequencing and annotation were carried out (de Groot *et al.*, 2009). We determined that its genome comprises a 2.8 Mbp chromosome and three large megaplasmids, totalling 3.8 Mbp of genomic sequence. At the same time as the genome was being sequenced, we performed a large shotgun proteomic analysis using tandem, high-resolution mass spectrometry (de Groot *et al.*, 2009). Fig. 8.3A illustrates the proteogenomic strategy that we developed taking advantage of combining the genome and proteome information. In this study, we built a protein database by listing all possible ORFs with at least a length of 100 nucleotides that could be read with a six reading frame translation of the four nucleic acid molecules. This protein database comprised 65,801 sequences with an average length of 92 amino acids. In parallel, we established a coding domain sequence (CDS) database by means of an automated annotation of the *D. deserti* genome with FrameD, a well-suited software program for gene prediction. The first version of the CDS list comprised 3051 predicted genes. In parallel, the *D. deserti* proteome was fractionated by chromatography and SDS-PAGE in order to get an in-depth coverage of the protein components. Hundreds of fractions were digested with trypsin, and the resulting peptides were analysed by shotgun nanoLC-MS/MS. MS/MS spectra acquired by tandem mass spectrometry were searched against the ORF and the CDS databases. Approximately one-fifth more proteins were identified with the ORF database compared with the search against the CDS database, indicating that settings chosen from FrameD inspection had to be refined. Thus, a more reliable CDS database was created comprising 3455 genes, allowing further confident assignment of 11,129 unique peptides, corresponding to a total of 1348 proteins (Fig. 8.3A). This shows how a back-and-forth strategy in between genome annotation and proteome data are essential to improve protein identification. Furthermore, peptide alignment directly onto the nucleic sequence resulted in an informative proteogenomic map. Such data representation helped in visualizing the real location of the coding domain sequences in the genome. Different cases of annotation errors were highlighted, but they required an in-depth manual sequence analysis to certify the re-annotation of each gene. The mass spectrometry-based data of *D. deserti* blended into the genome annotation at the primary stage revealed different annotation errors. Fig. 8.3B shows a specific locus of the *D. deserti* chromosome where a small ORF, Deide_19965, was not previously annotated despite the use of two annotation software programs (Frame D and Med 2.0). Indeed, in this locus a predicted ORF was annotated incorrectly on one strand whereas the opposite strand was the one that should be read giving the real polypeptide coding sequence. In this case, detection of several peptides by tandem mass spectrometry could be used as multiple proofs. Deide_19965 is a short ORF of 95 amino acids that does not present any detectable similarities to known protein sequences. This ORF is conserved amongst *Deinococcus* species but the wrong annotation of this specific locus, also observed for other species, has been corrected (de Groot *et al.*, 2009).

Concluding remarks

Improvements of the performance of mass spectrometers, with the development of both ionization techniques suitable for analysing biochemical molecules and hybrid mass spectrometry technology including high resolution mass analysers, have significantly helped the identification and characterization of large sets of proteins over the last fifteen years.

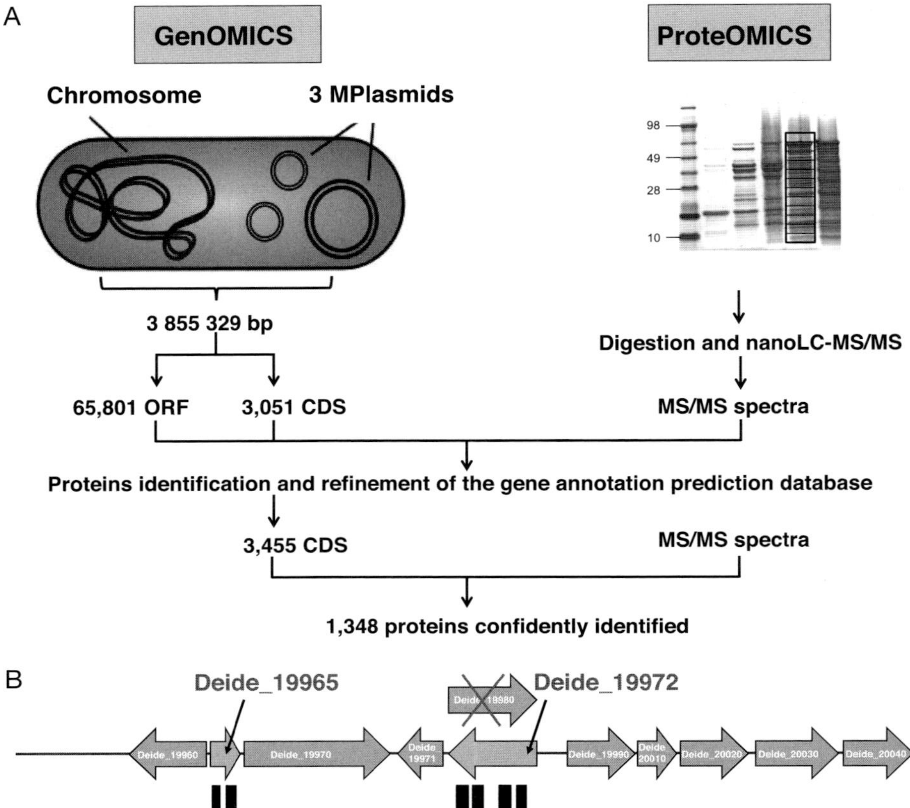

Figure 8.3 Proteogenomic strategy used for *Deinococcus deserti* annotation. (A) Complementary information brought out by genomics and proteomics. Sequencing of the chromosome and the three mega plasmids of *D. deserti* allow the access to the nucleotide sequence of this microorganism and the creation of both the open reading frame (ORF) database, comprising all the possible proteins encoded by the genome, and the predicted coding domain sequence (CDS) database. Proteins are first separated based on an SDS-Page gel, digested with trypsin, extracted and analysed with nanoLC-MS/MS with a high–throughput hybrid mass spectrometer. MS/MS spectra are assigned to peptide sequences and further to proteins by specific searches against the ORF and the CDS databases. However, as more proteins are identified from the ORF than from the CDS database, refinement of the settings to predict the CDS is necessary. As a result, 3455 CDS sequences are listed and a second analysis of the MS/MS data leads to the confident identification of 1348 proteins. (B) use of proteomic data to improve genomic annotation on a specific locus of the *Deinococcus deserti* chromosome. The same locus contained, surprisingly, two important annotations errors. Despite resorting to several annotation software programs, a small ORF was un-predicted and a predicted ORF was incorrectly annotated as the opposite strand was the real polypeptide coding region. Black rectangles represent peptides assigned to MS/MS spectra recorded on the *D. deserti* proteome by tandem mass spectrometry. Alignment of these peptides onto their corresponding reading frame allows the positioning of the genes identified on the forward or reverse strand. Coding sequences previously automatically annotated are indicated with green arrows. Two corrected, miss-annotated proteins are indicated with yellow arrows: a novel identified protein, Deide_19965 (a short polypeptide of 95 residues in the –1 frame), and a protein for which the coding strand has to be re-annotated, Deide_19972 (a polypeptide of 311 residues coded on the –3 frame that differs from the wrongly annotated Deide_19980 protein on the +1 frame). A colour version of this figure is available in the plate section at the back of the book.

In addition, sequencing technologies are now delivering amazingly large sequence data sets leading, currently, to an unprecedented avalanche of genome data. However, efforts to improve genome annotation, relying on software interpretation of the nucleic acid sequence, are urgently needed to avoid errors made in annotation and their propagation throughout newly annotated genomes. Proteogenomics, using direct experimental evidence at the protein level and bioinformatics tools, has proved reliable for better annotating genomes and discovering new genes that were not yet annotated. This strong alliance between genomics and proteomics provides structural information, such as the validation of predicted genes, correction of translational starts, identification of post-translational modifications and constitutive proteolytic processing, as well as functional information on many proteins that can now be quantified in different sets of physiological conditions (Christie-Oleza *et al.*, 2012b). Today, the analysis of a given set of representative proteogenomes that fully cover the tree of life would definitely improve prediction tools. Methods dedicated to metaproteogenomics will probably emerge and will be developed in the following years as a novel generation of mass spectrometers with higher dynamic range will be soon available. Regarding soil environments, metagenomics is currently the most used approach as proteomics of soil is still in its infancy (Chapter 6). Thanks to the new generation of mass spectrometers and improved methodologies, a better knowledge of the protein diversity for different ecosystems such as soil environments is now realizable.

Acknowledgements

We thank the Commissariat à l'Energie Atomique et aux Energies Alternatives, the Agence Nationale de la Recherche (ANR-12-BSV6-0012-01), and the Région Languedoc-Roussillon for financial support.

References

Aivaliotis, M., Gevaert, K., Falb, M., Tebbe, A., Konstantinidis, K., Bisle, B., Klein, C., Martens, L., Staes, A., Timmerman, E., *et al.* (2007). Large-scale identification of N-terminal peptides in the halophilic archaea *Halobacterium salinarum* and *Natronomonas pharaonis*. J. Proteome Res. *6*, 2195–2204.

Ansong, C., Purvine, S.O., Adkins, J.N., Lipton, M.S., and Smith, R.D. (2008). Proteogenomics: needs and roles to be filled by proteomics in genome annotation. Brief Funct. Genomic Proteomic *7*, 50–62.

Ansong, C., Tolic, N., Purvine, S.O., Porwollik, S., Jones, M., Yoon, H., Payne, S.H., Martin, J.L., Burnet, M.C., Monroe, M.E., *et al.* (2011). Experimental annotation of post-translational features and translated coding regions in the pathogen Salmonella Typhimurium. BMC Genomics *12*, 433.

Armengaud, J. (2009). A perfect genome annotation is within reach with the proteomics and genomics alliance. Curr. Opin. Microbiol. *12*, 292–300.

Armengaud, J. (2010). Proteogenomics and systems biology: quest for the ultimate missing parts. Expert Rev. Proteomics *7*, 65–77.

Armengaud, J., Bland, C., Christie-Oleza, J., and Miotello, G. (2011). Microbial Proteogenomics, Gaining Ground with the Avalanche of Genome Sequences. J. Bacteriol. Parasitol. *S3–001*.

Baudet, M., Ortet, P., Gaillard, J.C., Fernandez, B., Guerin, P., Enjalbal, C., Subra, G., de Groot, A., Barakat, M., Dedieu, A., and Armengaud, J. (2010). Proteomics-based refinement of *Deinococcus deserti* genome annotation reveals an unwonted use of non-canonical translation initiation codons. Mol. Cell Proteomics *9*, 415–426.

Beck, D.A., Hendrickson, E.L., Vorobev, A., Wang, T., Lim, S., Kalyuzhnaya, M.G., Lidstrom, M.E., Hackett, M., and Chistoserdova, L. (2011). An integrated proteomics/transcriptomics approach points to oxygen as the main electron sink for methanol metabolism in *Methylotenera mobilis*. J. Bacteriol. *193*, 4758–4765.

Bindschedler, L.V., Burgis, T.A., Mills, D.J., Ho, J.T., Cramer, R., and Spanu, P.D. (2009). In planta proteomics and proteogenomics of the biotrophic barley fungal pathogen *Blumeria graminis f.* sp. *hordei*. Mol. Cell. Proteomics *8*, 2368–2381.

Bindschedler, L.V., McGuffin, L.J., Burgis, T.A., Spanu, P.D., and Cramer, R. (2011). Proteogenomics and in silico structural and functional annotation of the barley powdery mildew *Blumeria graminis f.* sp. *hordei*. Methods *54*, 432–441.

Blasius, M., Sommer, S., and Hubscher, U. (2008). *Deinococcus radiodurans*: what belongs to the survival kit? Crit. Rev. Biochem. Mol. Biol. *43*, 221–238.

Castellana, N., and Bafna, V. (2010). Proteogenomics to discover the full coding content of genomes: a computational perspective. J. Proteomics *73*, 2124–2135.

Chen, W., Lee, P.J., Shion, H., Ellor, N., and Gebler, J.C. (2007). Improving de novo sequencing of peptides using a charged tag and C-terminal digestion. Anal. Chem. *79*, 1583–1590.

Christie-Oleza, J.A., Armengaud, J., and Miotello, G. (2012a). High-throughput proteogenomics of *Rugeria pomeroyi*: seeding a better genomic annotation for the whole marine *Roseobacter* clade. BMC Genomics *13*.

Christie-Oleza, J.A., Fernandez, B., Nogales, B., Bosch, R., and Armengaud, J. (2012b). Proteomic insights into the lifestyle of an environmentally relevant marine bacterium. ISME J. *6*, 124–135.

Cotter, R.J., Woods, A.S., and Cornish, T.J. (1994). Biological applications of time-of-flight mass spectrometry. Biochem. Soc. Trans. *22*, 539–542.

Cottrell, J.S. (2011). Protein identification using MS/MS data. J. Proteomics *74*, 1842–1851.

Cox, M.M., and Battista, J.R. (2005). *Deinococcus radiodurans* – the consummate survivor. Nat. Rev. Microbiol. *3*, 882–892.

Daly, M.J. (2009). A new perspective on radiation resistance based on *Deinococcus radiodurans*. Nat. Rev. Microbiol. *7*, 237–245.

de Groot, A., Chapon, V., Servant, P., Christen, R., Saux, M.F., Sommer, S., and Heulin, T. (2005). *Deinococcus deserti* sp. *nov.*, a gamma-radiation-tolerant bacterium isolated from the Sahara Desert. Int. J. Syst. Evol. Microbiol. *55*, 2441–2446.

de Groot, A., Dulermo, R., Ortet, P., Blanchard, L., Guerin, P., Fernandez, B., Vacherie, B., Dossat, C., Jolivet, E., Siguier, P., *et al.* (2009). Alliance of proteomics and genomics to unravel the specificities of Sahara bacterium *Deinococcus deserti*. PLoS Genet. *5*, e1000434.

Delahunty, C., and Yates, J.R., 3rd (2005). Protein identification using 2D-LC-MS/MS. Methods *35*, 248–255.

Delahunty, C.M., and Yates, J.R., 3rd (2007). MudPIT: multidimensional protein identification technology. Biotechniques *43*, 563, 565, 567 passim.

Devos, D., and Valencia, A. (2001). Intrinsic errors in genome annotation. Trends Genet. *17*, 429–431.

Dongre, A.R., Eng, J.K., and Yates, J.R., 3rd (1997). Emerging tandem-mass-spectrometry techniques for the rapid identification of proteins. Trends Biotechnol. *15*, 418–425.

Ducret, A., Van Oostveen, I., Eng, J.K., Yates, J.R., 3rd, and Aebersold, R. (1998). High throughput protein characterization by automated reverse-phase chromatography/electrospray tandem mass spectrometry. Protein Sci. *7*, 706–719.

Fenn, J.B., Mann, M., Meng, C.K., Wong, S.F., and Whitehouse, C.M. (1989). Electrospray ionization for mass spectrometry of large biomolecules. Science *246*, 64–71.

Ferguson, J.T., Wenger, C.D., Metcalf, W.W., and Kelleher, N.L. (2009). Top-down proteomics reveals novel protein forms expressed in *Methanosarcina acetivorans*. J. Am. Soc. Mass. Spectrom. *20*, 1743–1750.

Gallien, S., Perrodou, E., Carapito, C., Deshayes, C., Reyrat, J.M., Van Dorsselaer, A., Poch, O., Schaeffer, C., and Lecompte, O. (2009). Ortho-proteogenomics: multiple proteomes investigation through orthology and a new MS-based protocol. Genome Res. *19*, 128–135.

Gao, N., Chen, L.L., Ji, H.F., Wang, W., Chang, J.W., Gao, B., Zhang, L., Zhang, S.C., and Zhang, H.Y. (2010). DIGA--a database of improved gene annotation for phytopathogens. BMC Genomics *11*, 54.

Geiger, T., Wehner, A., Schaab, C., Cox, J., and Mann, M. (2012). Comparative proteomic analysis of eleven common cell lines reveals ubiquitous but varying expression of most proteins. Mol. Cell. Proteom. *11*, M111.014050.

Gevaert, K., Goethals, M., Martens, L., Van Damme, J., Staes, A., Thomas, G.R., and Vandekerckhove, J. (2003). Exploring proteomes and analyzing protein processing by mass spectrometric identification of sorted N-terminal peptides. Nat. Biotechnol *21*, 566–569.

Gillet, L.C., Navarro, P., Tate, S., Rost, H., Selevsek, N., Reiter, L., Bonner, R., and Aebersold, R. (2012). Targeted Data Extraction of the MS/MS Spectra Generated by Data-independent Acquisition: A New Concept for Consistent and Accurate Proteome Analysis. Mol. Cell Proteomics *11*, O111 016717.

Gogarten, J.P., Senejani, A.G., Zhaxybayeva, O., Olendzenski, L., and Hilario, E. (2002). Inteins: structure, function, and evolution. Annu. Rev. Microbiol. *56*, 263–287.

Gupta, N., Benhamida, J., Bhargava, V., Goodman, D., Kain, E., Kerman, I., Nguyen, N., Ollikainen, N., Rodriguez, J., Wang, J., *et al.* (2008). Comparative proteogenomics: combining mass spectrometry and comparative genomics to analyze multiple genomes. Genome Res. *18*, 1133–1142.

Hillenkamp, F., Karas, M., Beavis, R.C., and Chait, B.T. (1991). Matrix-assisted laser desorption/ionization mass spectrometry of biopolymers. Anal. Chem. *63*, 1193A–1203A.

Hu, G.Q., Zheng, X., Yang, Y.F., Ortet, P., She, Z.S., and Zhu, H. (2008). ProTISA: a comprehensive resource for translation initiation site annotation in prokaryotic genomes. Nucleic Acids Res. *36*, D114–119.

Hyatt, D., Chen, G.L., Locascio, P.F., Land, M.L., Larimer, F.W., and Hauser, L.J. (2010). Prodigal: prokaryotic gene recognition and translation initiation site identification. BMC Bioinformatics *11*, 119.

Jaffe, J.D., Berg, H.C., and Church, G.M. (2004). Proteogenomic mapping as a complementary method to perform genome annotation. Proteomics *4*, 59–77.

Jain, E., Bairoch, A., Duvaud, S., Phan, I., Redaschi, N., Suzek, B.E., Martin, M.J., McGarvey, P., and Gasteiger, E. (2009). Infrastructure for the life sciences: design and implementation of the UniProt website. BMC Bioinformatics *10*, 136.

Johnson, R.S., Davis, M.T., Taylor, J.A., and Patterson, S.D. (2005). Informatics for protein identification by mass spectrometry. Methods *35*, 223–236.

Karas, M., and Hillenkamp, F. (1988). Laser desorption ionization of proteins with molecular masses exceeding 10,000 daltons. Anal. Chem. *60*, 2299–2301.

Kleifeld, O., Doucet, A., auf dem Keller, U., Prudova, A., Schilling, O., Kainthan, R.K., Starr, A.E., Foster, L.J., Kizhakkedathu, J.N., and Overall, C.M. (2010). Isotopic labeling of terminal amines in complex samples identifies protein N-termini and protease cleavage products. Nat. Biotechnol. *28*, 281–288.

Kleifeld, O., Doucet, A., Prudova, A., auf dem Keller, U., Gioia, M., Kizhakkedathu, J.N., and Overall, C.M. (2011). Identifying and quantifying proteolytic events and the natural N terminome by terminal amine isotopic labeling of substrates. Nat. Protoc. *6*, 1578–1611.

Lin, D., Tabb, D.L., and Yates, J.R., 3rd (2003). Large-scale protein identification using mass spectrometry. Biochim. Biophys. Acta *1646*, 1–10.

Makarova, K.S., Omelchenko, M.V., Gaidamakova, E.K., Matrosova, V.Y., Vasilenko, A., Zhai, M., Lapidus, A., Copeland, A., Kim, E., Land, M., *et al.* (2007). *Deinococcus geothermalis*: the pool of extreme radiation resistance genes shrinks. PLoS One *2*, e955.

McCormack, A.L., Somogyi, A., Dongre, A.R., and Wysocki, V.H. (1993). Fragmentation of protonated peptides: surface-induced dissociation in conjunction with a quantum mechanical approach. Anal. Chem. *65*, 2859–2872.

McDonald, L., and Beynon, R.J. (2006). Positional proteomics: preparation of amino-terminal peptides as a strategy for proteome simplification and characterization. Nat. Protoc. *1*, 1790–1798.

McDonald, L., Robertson, D.H., Hurst, J.L., and Beynon, R.J. (2005). Positional proteomics: selective recovery and analysis of N-terminal proteolytic peptides. Nat. Methods *2*, 955–957.

McLafferty, F.W. (1981). Tandem mass spectrometry. Science *214*, 280–287.

McLafferty, F.W., and Senko, M.W. (1994). Mass spectrometry in the development of drugs from traditional medicines. Stem Cells *12*, 68–73.

McLuckey, S.A., Van Berkel, G.J., Goeringer, D.E., and Glish, G.L. (1994). Ion trap mass spectrometry. Using high-pressure ionization. Anal Chem *66*, 737A-743A.

Meinnel, T., and Giglione, C. (2008). Tools for analyzing and predicting N-terminal protein modifications. Proteomics *8*, 626–649.

Meng, Z., and Veenstra, T.D. (2011). Targeted mass spectrometry approaches for protein biomarker verification. J. Proteomics *74*, 2650–2659.

Nielsen, P., and Krogh, A. (2005). Large-scale prokaryotic gene prediction and comparison to genome annotation. Bioinformatics *21*, 4322–4329.

Payne, S.H., Huang, S.T., and Pieper, R. (2010). A proteogenomic update to Yersinia: enhancing genome annotation. BMC Genomics *11*, 460.

Poptsova, M.S., and Gogarten, J.P. (2010). Using comparative genome analysis to identify problems in annotated microbial genomes. Microbiology *156*, 1909–1917.

Qiu, Y., Cho, B.K., Park, Y.S., Lovley, D., Palsson, B.O., and Zengler, K. (2010). Structural and operational complexity of the *Geobacter sulfurreducens* genome. Genome Res. *20*, 1304–1311.

Reeves, G.A., Talavera, D., and Thornton, J.M. (2009). Genome and proteome annotation: organization, interpretation and integration. J. R. Soc. Interface *6*, 129–147.

Sato, N., and Tajima, N. (2012). Statistics of N-terminal alignment as a guide for refining prokaryotic gene annotation. Genomics *99*, 138–143.

Shen, P.T., Hsu, J.L., and Chen, S.H. (2007). Dimethyl isotope-coded affinity selection for the analysis of free and blocked N-termini of proteins using LC-MS/MS. Anal. Chem. *79*, 9520–9530.

Siggins, A., Gunnigle, E., and Abram, F. (2012). Exploring Mixed Microbial Community Functioning: Recent Advances in Metaproteomics. FEMS Microbiol. Ecol. *80*, 265–280.

Staes, A., Van Damme, P., Helsens, K., Demol, H., Vandekerckhove, J., and Gevaert, K. (2008). Improved recovery of proteome-informative, protein N-terminal peptides by combined fractional diagonal chromatography (COFRADIC). Proteomics *8*, 1362–1370.

Tang, N., Tornatore, P., and Weinberger, S.R. (2004). Current developments in SELDI affinity technology. Mass Spectrom. Rev. *23*, 34–44.

Van Damme, P., Van Damme, J., Demol, H., Staes, A., Vandekerckhove, J., and Gevaert, K. (2009). A review of COFRADIC techniques targeting protein N-terminal acetylation. BMC Proc *3 (Suppl 6)*, S6.

Veenstra, T.D., Conrads, T.P., and Issaq, H.J. (2004). What to do with 'one-hit wonders'? Electrophoresis *25*, 1278–1279.

Washburn, M.P., Wolters, D., and Yates, J.R., 3rd (2001). Large-scale analysis of the yeast proteome by multidimensional protein identification technology. Nat. Biotechnol. *19*, 242–247.

White, O., Eisen, J.A., Heidelberg, J.F., Hickey, E.K., Peterson, J.D., Dodson, R.J., Haft, D.H., Gwinn, M.L., Nelson, W.C., Richardson, D.L., *et al.* (1999). Genome sequence of the radioresistant bacterium Deinococcus radiodurans R1. Science *286*, 1571–1577.

Whitehouse, C.M., Dreyer, R.N., Yamashita, M., and Fenn, J.B. (1985). Electrospray interface for liquid chromatographs and mass spectrometers. Anal. Chem. *57*, 675–679.

Wilmes, P., and Bond, P.L. (2006). Metaproteomics: studying functional gene expression in microbial ecosystems. Trends Microbiol. *14*, 92–97.

Wolters, D.A., Washburn, M.P., and Yates, J.R., 3rd (2001). An automated multidimensional protein identification technology for shotgun proteomics. Anal. Chem. *73*, 5683–5690.

Yamaguchi, M., Obama, T., Kuyama, H., Nakayama, D., Ando, E., Okamura, T.A., Ueyama, N., Nakazawa, T., Norioka, S., Nishimura, O., and Tsunasawa, S. (2007). Specific isolation of N-terminal fragments from proteins and their high-fidelity de novo sequencing. Rapid Commun. Mass Spectrom. *21*, 3329–3336.

Yamaguchi, M., Nakayama, D., Shima, K., Kuyama, H., Ando, E., Okamura, T.A., Ueyama, N., Nakazawa, T., Norioka, S., Nishimura, O., and Tsunasawa, S. (2008). Selective isolation of N-terminal peptides from proteins and their de novo sequencing by matrix-assisted laser desorption/ionization time-of-flight mass spectrometry without regard to unblocking or blocking of N-terminal amino acids. Rapid Commun. Mass Spectrom. *22*, 3313–3319.

Yamazaki, S., Yamazaki, J., Nishijima, K., Otsuka, R., Mise, M., Ishikawa, H., Sasaki, K., Tago, S., and Isono, K. (2006). Proteome analysis of an aerobic hyperthermophilic crenarchaeon, *Aeropyrum pernix K1*. Mol. Cell Proteomics *5*, 811–823.

Yates, J.R., 3rd (1998). Database searching using mass spectrometry data. Electrophoresis *19*, 893–900.

Yates, J.R., 3rd (2000). Mass spectrometry. From genomics to proteomics. Trends Genet. *16*, 5–8.

Yates, J.R., 3rd, Eng, J.K., and McCormack, A.L. (1995). Mining genomes: correlating tandem mass spectra of modified and unmodified peptides to sequences in nucleotide databases. Anal. Chem. *67*, 3202–3210.

Yooseph, S., Sutton, G., Rusch, D.B., Halpern, A.L., Williamson, S.J., Remington, K., Eisen, J.A., Heidelberg, K.B., Manning, G., Li, W., *et al.* (2007). The Sorcerer II Global Ocean Sampling expedition: expanding the universe of protein families. PLoS Biol. *5*, e16, 432–466.

Yooseph, S., Nealson, K.H., Rusch, D.B., McCrow, J.P., Dupont, C.L., Kim, M., Johnson, J., Montgomery, R., Ferriera, S., Beeson, K., *et al.* (2010). Genomic and functional adaptation in surface ocean planktonic prokaryotes. Nature *468*, 60–66.

Analysis of Soil Metagenomes using the MEtaGenome ANalyser (MEGAN)

Daniel H. Huson and Nico Weber

Abstract

Soil is one of the most dense and diverse habitats of microorganisms, and metagenomics is a key technique for studying this environment. Given a dataset of DNA sequencing reads obtained from an environmental sample, the program MEGAN (MEtaGenome ANalyser) can be used to interactively perform an analysis and comparison of the data, both on a taxonomic and functional level. The program attempts to place all reads into the NCBI taxonomy and to map them to SEED functional roles or KEGG orthology groups. To compare various datasets, MEGAN allows the user to load multiple samples at once and combine them into a single comparison document. Multiple views makes it easy to inspect assignments to the different nodes. An alignment viewer is supplied to explore reference-guided multiple alignments of reads. The program supports various input formats for loading data and can export analysis results in different text-based and graphical formats. The program is designed to work with very large datasets. It is written in Java and installers for the three major computer operating systems are available from http://www-ab.informatik.uni-tuebingen.de.

Introduction

Environmental metagenomics is the study of uncultured organisms in their native environment using DNA sequencing (Handelsman *et al.*, 1998). Despite the biological space is very low (Nannipieri *et al.*, 2003) 4×10^7 prokaryotic cells can be found in one gram of forest soil (Richter Jr. and Markewitz, 1995) and thus is a challenge to metagenomics.

Research in this area has benefited from the rise of second generation sequencing technologies and hopes to benefit further from the third generation of sequencing techniques. Sampling and sequencing soil ecosystems can now be done very cheaply and efficiently.

Given a dataset of DNA sequencing reads obtained from an environmental sample, there are three computational challenges to address. The first task is to estimate the taxonomic content. During this step a qualitative and, if possible, a quantitative distribution of source organism is established. The second problem is to determine the functional content of the sample. The third challenge is to compare different samples of interest. In many cases the aim is to detect shifts in taxonomic and or function composition that correlate to external parameters or properties of the samples.

To address these challenges, the first step is usually to align the set of sequencing reads against a database of known reference protein sequences such as NCBI-NR or RefSeq (Wheeler *et al.*, 2008) using a pairwise alignment tool such as BLASTX (Altschul *et al.*,

1990) or RapSearch2 (Zhao *et al.*, 2012). A read is said to *hit* a given reference sequence if a significant alignment is found in this process. The comparison of the sequencing reads against a reference database is usually the computationally most expensive step of the analysis and subsequent steps are based on the obtained alignments.

As an alternative to reference-based methods, one can use alignment-free taxonomic predictors to estimate the taxonomic content of the sample. Such tools often employ machine learning techniques such as SVMs, based on k-mer counts (McHardy *et al.*, 2007). The advantage of machine learning techniques are that they may able to classify reads originating from organisms not represented in the database. A major drawback of such predictors is that most of them are not able to assess the functional content. Moreover, they do not produce pairwise alignments, so comparing reads to sequences of known organisms requires additional work.

Given the result of the alignment step, an analysis program such as MEGAN is then required to explore and analyse the data. MEGAN is a tool for analysing metagenomic sequencing data, allowing the user to interactively explore the taxonomic and functional content of a dataset. It also supports the comparison of multiple datasets on taxonomic and functional levels. The program was originally published by Huson *et al.* (2007) and the most recent version was published by Huson *et al.* (2011). Written in Java, the program runs on all major operating systems. The program can be downloaded from http://www-ab.informatik.uni-tuebingen.de/software/megan.

We would like to emphasize that technical issues such as sample preparation protocol, DNA extraction method and sequencing technology all have a marked effect on the resulting metagenome data.

The basic input to MEGAN is a set of sequencing reads and the result of a pairwise alignment of the reads to a database of appropriate reference sequences. MEGAN supports a number of different input formats, such as BLAST (text, tabular and XML) (Altschul *et al.*, 1990), SAM (Li *et al.*, 2009), RapSearch2 (Zhao *et al.*, 2012), RDP (Wang *et al.*, 2007), NBC (Rosen *et al.*, 2010), QIIME (Caporaso *et al.*, 2010) as well as a number of different CSV (comma-separated value) formats. Processed data are stored in a compressed binary RMA file. The results of analyses produced by MEGAN can be exported in a number of CSV formats and all visualizations provided by the program can be exported in a wide range of graphics formats. The program provides search tools to locate taxa and genes of interest.

In this chapter, we illustrate a hands-on analysis of different soil metagenomic datasets and address typical questions encountered along the way. Sampling, library preparation, sequencing and quality control are beyond the scope of this work and are not discussed further. As sequence comparison software we recommend the afore mentioned programs BLASTX or RapSearch2. We will use an early soil metagenome dataset that was published in Tringe *et al.* (2005). This dataset was sequenced from surface soil sampled from a farm in Waseca County, Minnesota, USA and we will refer to it as the *Minnesota soil dataset*. All discussed files can be downloaded from http://ab.inf.uni-tuebingen.de/software/megan/.

The first step in a MEGAN analysis is to parse the reads and sequence alignment (BLAST) files, using the *Import for Blast* menu item. MEGAN stores the result of parsing a dataset in a so-called RMA (read-match archive) file that contains all reads and matches in a compressed and indexed format. While the initial parsing of a dataset may take a long time (approximately 36 h for 1 terabyte of BLAST input), once generated, an RMA file can be loaded by MEGAN in a few seconds.

Taxonomic analysis

One approach to taxonomic analysis is to focus on specific phylogenetic markers such as 16S rRNA to assess the taxonomic content of a sample. MEGAN allows one to import the result of an analysis of such data obtained, for example, by using the RDP classifier (Wang *et al.*, 2007) or by performing a BLASTN comparison of the reads against the Silva database (Pruesse *et al.*, 2007). Metagenome sequencing proper employs environmental shotgun sequencing of reads from genomic DNA or cDNA from RNA.

To perform a taxonomic analysis of a metagenomic shotgun dataset, MEGAN attempts to place each read onto a node in the NCBI taxonomy, based on an analysis of its hits against a reference database. A key idea is to use all ranks of the taxonomy so as to assign reads specific to a particular species near the leaves of the taxonomy and to map sequences that are conserved across a wider range of organisms to higher-level nodes. For example, a read that comes from a gene than is only present in *Escherichia coli* will be placed on the E. coli node, whereas a read that comes from a gene that is shared widely across different *Proteobacteria* will be assigned to the node labelled *Proteobacteria*.

The input to MEGAN usually consists of a file of DNA reads and a file containing all their hits to a reference database, usually in BLAST or SAM format. In addition, at start-up, MEGAN reads in the whole NCBI taxonomy, which is a rooted tree with over 900 000 nodes. To perform a taxonomic analysis of a metagenome dataset, MEGAN processes each DNA read in turn, assigning each read to the node in the NCBI taxonomy that is the *lowest common ancestor* of the set of species associated with all reference sequences that were hit by the read. This approach is known as the LCA algorithm. In essence, the LCA algorithm places reads by gene content and is thus quite conservative and unlikely to be overly effected by horizontal gene transfer.

The LCA algorithm has a number of parameters. The *minScore* parameter sets a minimum threshold bit score that an alignment must achieve to be considered. For reads of length 100, a value of 35 is appropriate, while for longer reads this threshold should be increased accordingly. The *topPercent* parameter (by default 10%) provides additional filtering of matches. Only those matches are kept whose score lies within the given percentage of the best score for the given read, The *minSupport* parameter specifies the minimum number of reads that a node in the NCBI taxonomy must attract before it is shown in the final output. The reads assigned to a node for which this requirement is not met are pushed up the taxonomy until a node with a sufficient number of assigned reads is reached.

Modern sequencing protocols support sequencing of *paired reads*, that is, pairs of reads that are guaranteed to come from the same DNA insert of a specified length. MEGAN can make use of read pairing information during the LCA assignment. For a given pair of reads, the program determines the set of taxa for which both reads have significant matches and the scores of these matches are boosted by a factor of 20%.

In consequence, the LCA algorithm will base the placement of reads mainly on those taxa, for which both reads of a pair have a good match, often leading to a more specific placement.

In the analysis of 16S rRNA sequences, it is often required that a specific level of DNA identity between two 16S rRNA sequences is achieved, before they can be assigned to the same taxon at a specific taxonomic rank, e.g. more than 99% of identity is required to be counted as the same species. To address this, MEGAN provides a *Percent Identity Filter* that can be used to enforce the following levels of percentage sequence identities for an assignment at a given taxonomic level, namely species 99%, genus 97%, family 95%, order 90%, class 85% and phylum 80%.

Reads that have no hits are assigned to a special node labelled *No Hits,* whereas reads that have hits, but cannot be assigned to a taxon are mapped to a special *Unassigned* node. In addition, reads consisting of highly repetitive sequence are assigned to a *Low Complexity* node.

The part of the NCBI taxonomy to which reads are assigned is displayed in the taxonomy viewer of MEGAN and by default, each node is scaled logarithmically to represent the number of reads associated with it (see Fig. 9.1).

The number of reads assigned to a node, as well as the sum of reads up to this node can be easily seen when selected, or displayed after the name if desired. Nodes can be interactively

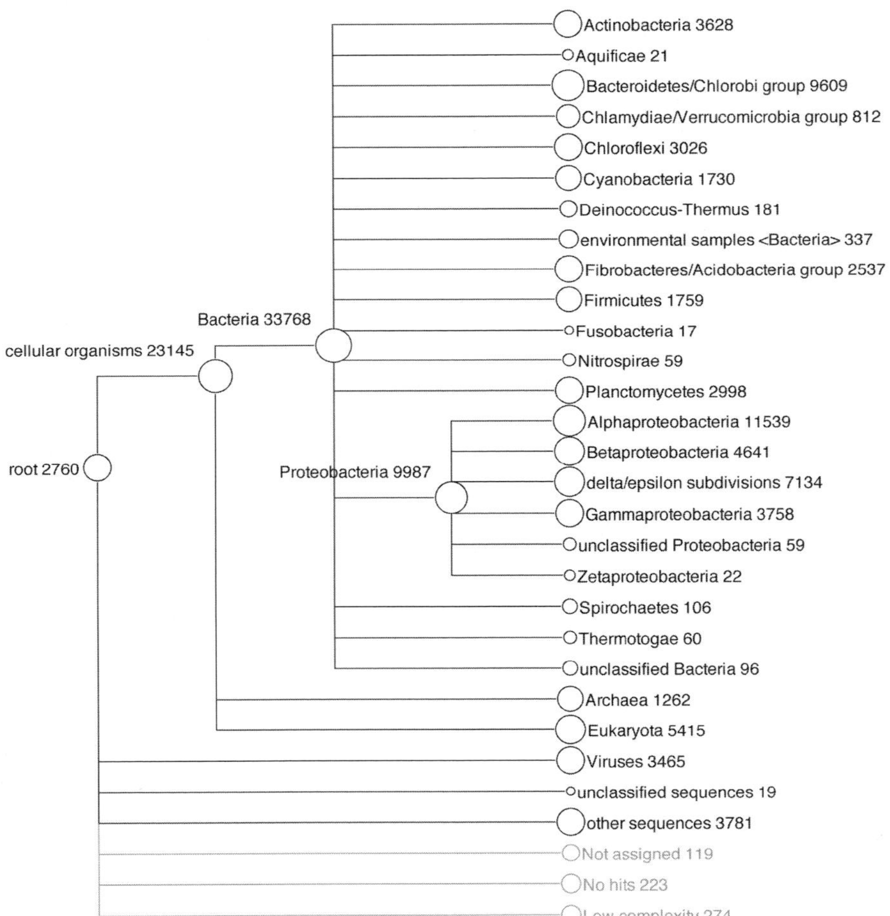

Figure 9.1 Taxonomy analysis of ≈ 140,000 reads originating of a soil sample from a farm in Minnesota (Tringe *et al.*, 2005). Each circle represents a taxon in the NCBI taxonomy and is scaled logarithmically to indicate how many reads have been assigned to it. In addition to the taxon name, each node is also labelled by the cumulative number of reads assigned to, or below, that node.

collapsed or expanded to show more or fewer details of the classification. The user can select nodes of interest, and then either inspect the associated reads and alignments, or save them to a file. The user can also summarize data in a number of ways using different types of charts. Additionally the user can view microbial attributes to get additional information about the selected species (see Fig. 9.2). Nodes can be selected in a number of different ways and many of the menu items provided by the program apply to all selected nodes.

MEGAN allows one to compute a rarefaction curve for a given dataset. The underlying algorithm repeatedly samples replicates of size 10%, 20% ... 100% from the original datasets and then plots the number of leaves produced for each replicate. The steepness of the curve will give an indication how close to saturation the sequencing is. (However, because organisms can be arbitrarily rare in a sample, the rarefaction curve cannot help decide whether all organisms present have been seen.)

Depending on the sampling location, the sequencing technique used and other considerations, the researcher will have to interactively explore the data and adapt the parameters of the LCA algorithm and other options to suit their needs. Typically, after initial parsing of an input file, one will first look at a taxonomic analysis of the data. In a second step, one might then be interested in the functional content of the sample.

Functional analysis

MEGAN allows one to analyse the functional content of a metagenomic dataset using both the SEED classification of subsystems and function roles (Overbeek *et al.*, 2005) or the KEGG classification of KO groups, enzymes and pathways (Kanehisa and Goto, 2000). In essence, the SEED classification maps genes onto functional roles and these appear in different subsystems. Similarly, KEGG maps in both cases, the classification can be represented as a tree with roughly 13,000 nodes. A functional analysis is only possible if reads were compared against a database containing protein information e.g. the NCBI NR database.

To perform a SEED-based analysis, for each read in the input, MEGAN identifies the highest scoring hit to a reference sequence for which the corresponding functional role is known and then maps the read to that functional role. In a KEGG-based analysis, each read is mapped to a KEGG orthology group in a similar fashion. The mapping process is based on NCBI RefSeq accession numbers, which are mapped to SEED functional roles and KEGG KO numbers. The latest mapping files currently contains 1.3 million entries for the SEED mapping and 2.1 million entries for KEGG. With this in mind choosing the right database for comparing reads is essential. Updated SEED mappings will be included on a regular bases in new releases of MEGAN. Due to licensing issues, the version of KEGG shipped with MEGAN has not be updated since July 2011.

Both the SEED and KEGG classifications are displayed as trees in MEGAN and the viewers provide the same interactive features as the taxonomy viewer. In addition to the tree view a separate navigation pane allows the user to browse the structure by subsystem or groups for easy access. Moreover, the number of reads assigned to classes can be displayed as a heat map. The KEGG viewer allows one to see how reads map to different enzymes in a given pathway (see Figs. 9.3 and 9.4). If only interested in one specific function one can select and extract all assigned reads into a new document for further analysis.

As in the taxonomic view the user can calculate a rarefaction curve to estimate whether the number of sequences in a dataset is large enough (see Fig. 9.5 for details). Assume that

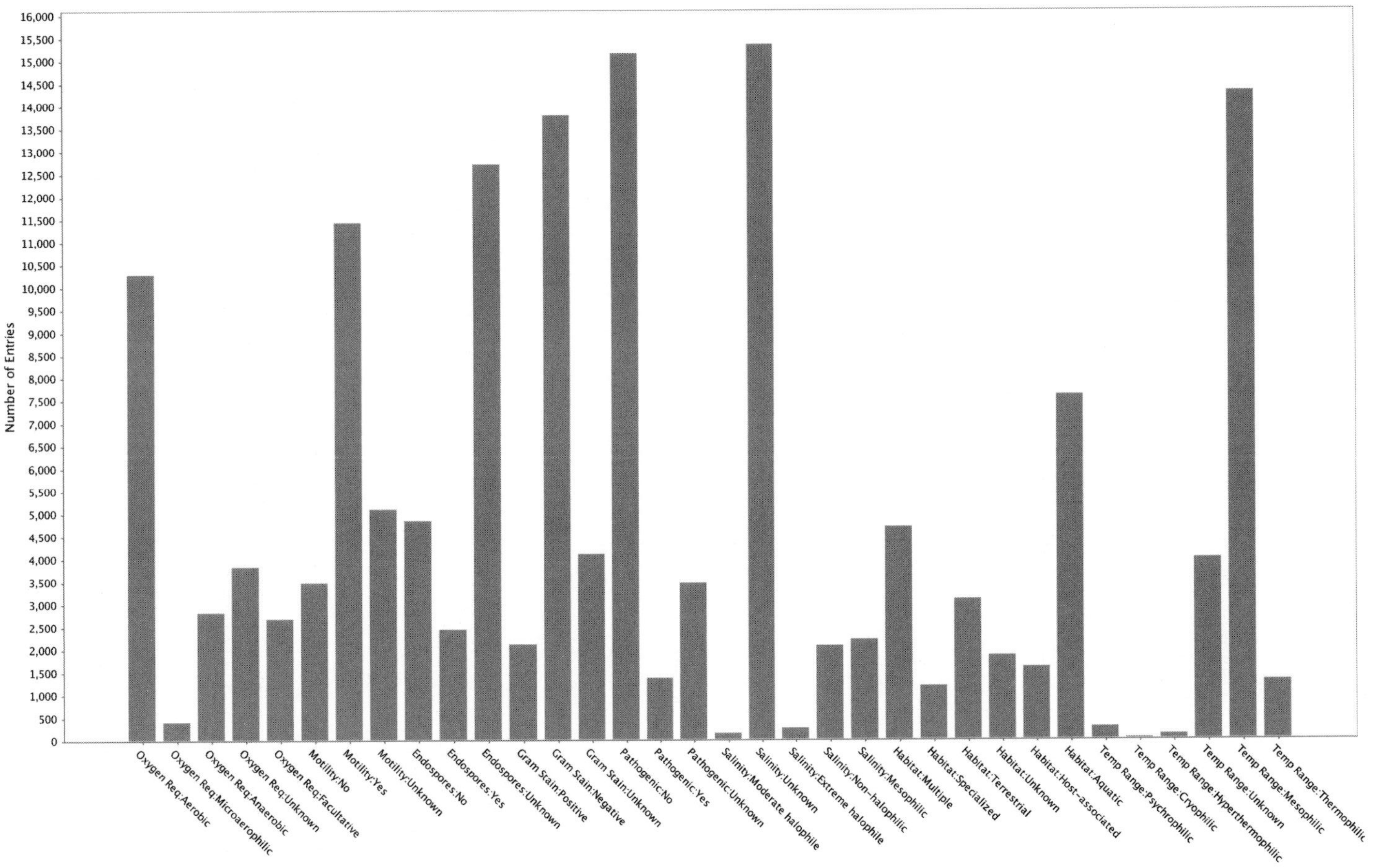

Figure 9.2 Chart of microbial attributes according to NCBI database of the Minnesota farm soil sample.

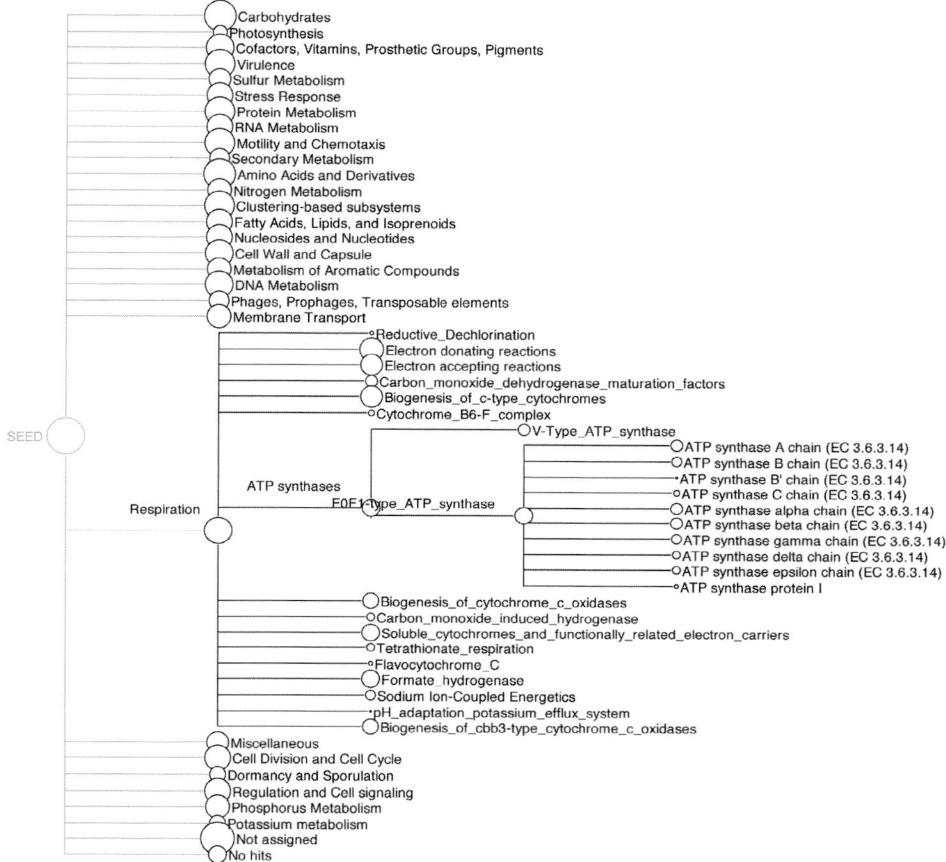

Figure 9.3 SEED-based functional analysis of the Minnesota soil sample. The SEED classification tree has been partially expanded to show details on functional roles involved in ATP synthase. Each circle represents a SEED category and is scaled logarithmically to indicate the cumulative number of reads that have been assigned to it. In addition to the SEED name, each node is also labelled by the number of reads assigned to, or below, that node.

the user is interested photosynthesis and stress response affiliated genes. Those functions can be found in the Seed viewer by selecting the appropriate nodes. In KEGG the user would browse the Metabolism, Energy Metabolism and then the Photosynthesis pathway. The pathway can then be inspected in a graphical representation (Fig. 9.6).

Sequence alignment

As pointed out above, the main computational step is to determine all pairwise alignments between the set of DNA reads and all sequences in an appropriate reference database. Based on this, it is possible to construct a reference-guided multiple sequence alignment between all reads that hit the same reference sequence. This calculation is implemented in a new feature called the alignment viewer (Huson and Xie, unpublished). Once the user has specified

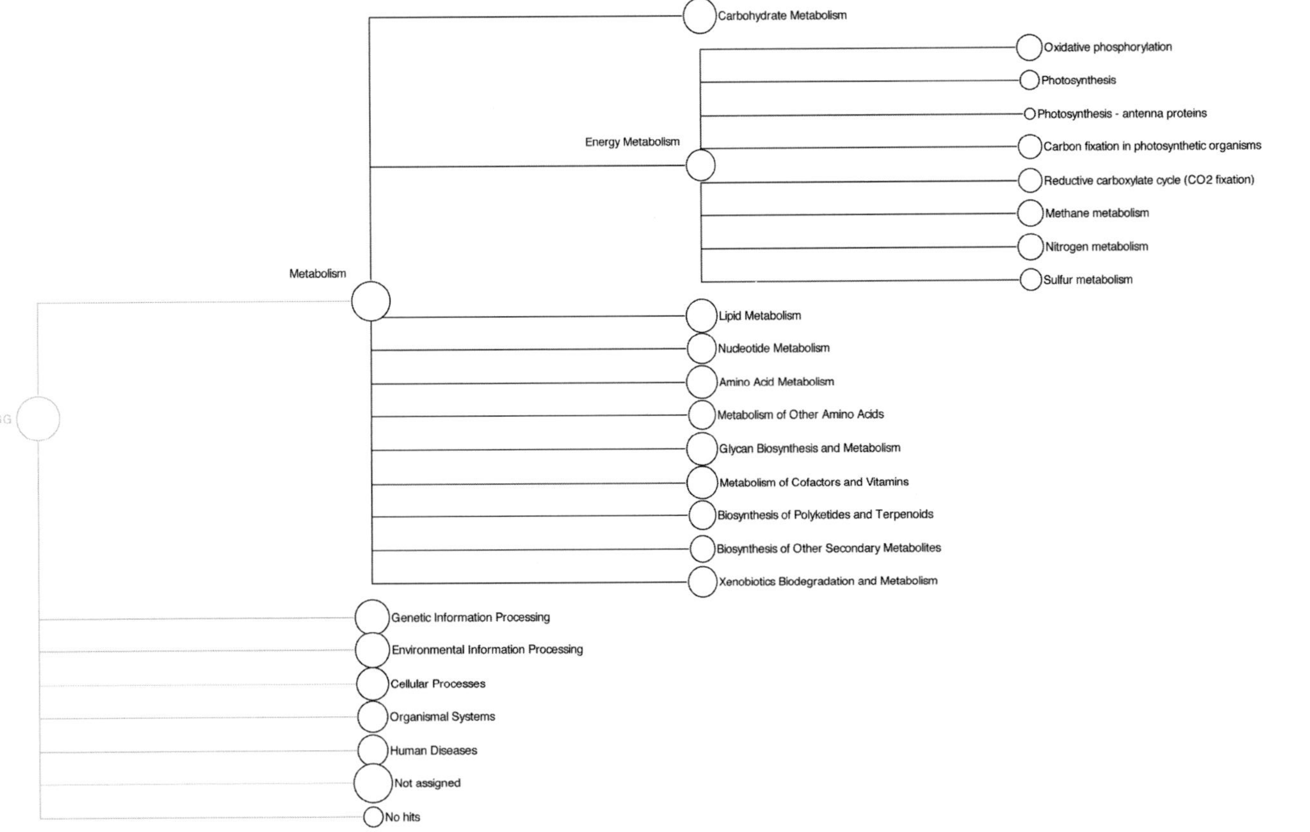

Figure 9.4 KEGG-based functional analysis of the Minnesota soil sample. The KEGG classification tree has been partially expanded to show details on pathways involved in energy metabolism. Each circle represents a KEGG category and is scaled logarithmically to indicate the cumulative number of reads that have been assigned to it. In addition to the KEGG name, each node is also labelled by the number of reads assigned to, or below, that node.

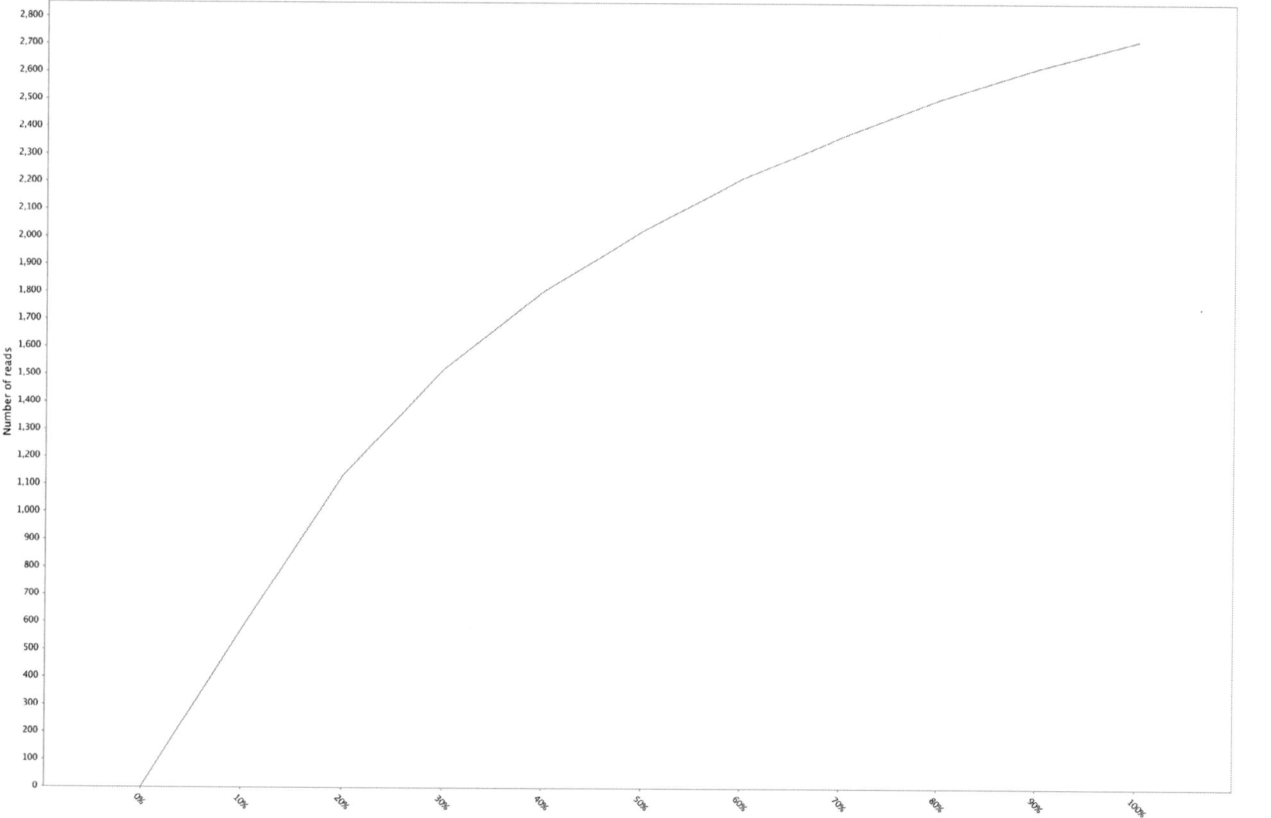

Figure 9.5 Rarefaction curve for SEED functional alignment of one sample dataset. The Plot is generated by randomly and repeatedly sub-sampling 10%, 20%... 100% of the reads and counting the number of leaves in the resulting graph. In an optimal scenario the curve will be asymptotical to 100% of the maximum number of leaves.

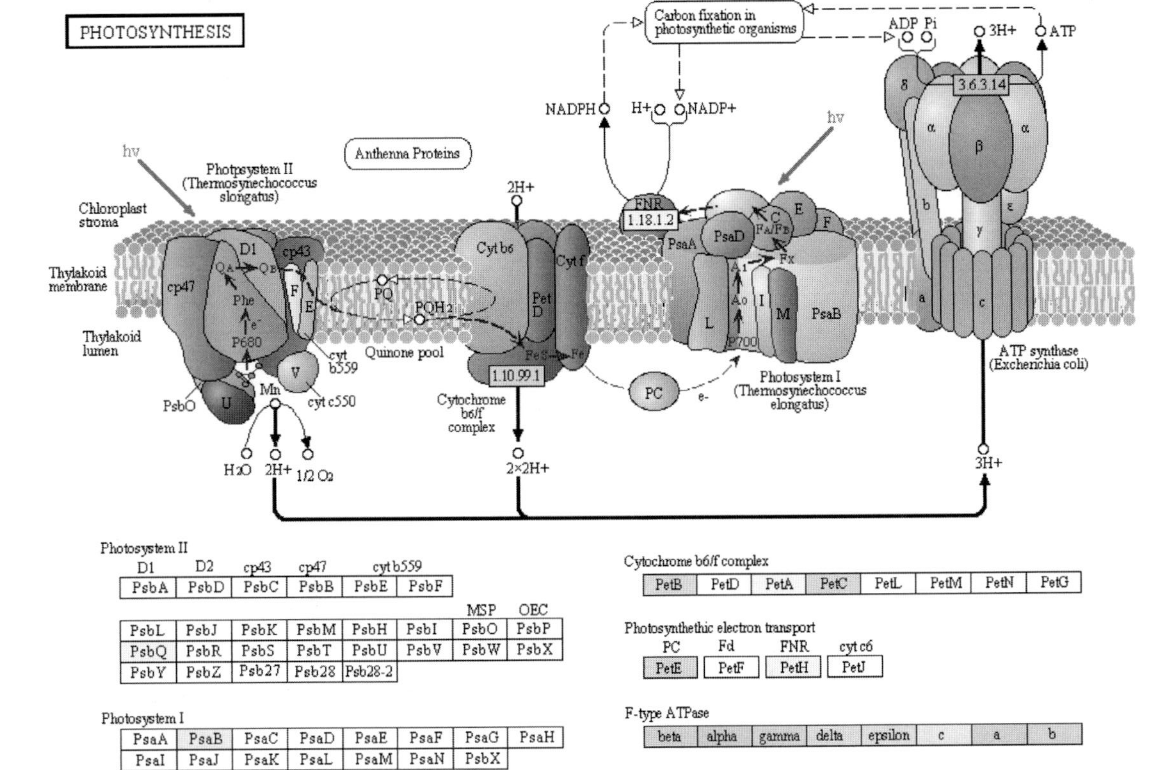

Figure 9.6 Graphical representation of the KEGG pathway for Photosynthesis (Kanehisa and Goto, 2000). Rectangles representing different enzymes are coloured in different shades to indicate the number of reads assigned to them. Tool tips associated with the enzymes provide additional information on the enzymes, such as which KEGG KO groups are associated with the enzyme and a context menu can be used to inspect or save all reads associated with an enzyme.

a node in the taxonomy, SEED or KEGG viewer for which the alignment viewer is to be launched, the program first collects all reference sequences that correspond to the given node and then, for each such reference sequence, the program determines all reads that hit it. The user can then select a reference sequence and the corresponding sequence alignment is subsequently displayed (see Fig. 9.7). The resulting consensus sequence can be exported into a separate file. Depending on the zoom level reads are either displayed as blocks or build out of coloured boxes representing different nucleotides.

This process enables the user to view organism specific alignments without the need of extracting reads and realigning them using an external program. Average coverage of different genes as well as the layout of reads that match to a given reference sequence may help to determine the reliability of the assignment of reads to a specific node. Functional analysis may also benefit from inspecting the alignment. The alignment viewer provides a function for performing a diversity analysis for protein-guided assemblies. The analysis is based on the number of different sequences within a window of the alignment. The analysis estimates how many distinct genomes contributed to this specific alignment.

Comparison of datasets

Most metagenome projects will involve multiple datasets taken from different environments, experimental settings, time points or locations. The comparison of a large number of large datasets is a challenging task. Depending on the project, the goal of a comparison can vary between the detection of simple changes in taxonomic composition to complex functional shifts.

To facilitate the comparison of datasets, MEGAN allows the user to open multiple datasets simultaneously, showing each dataset in a different window. The user can then select a number of open datasets to be combined into a single new comparison document.

To take different sample sizes into account one can select absolute, relative or sub-sampled reads for comparison. When comparing datasets of very different sizes, then the subsample mode is recommended, in which all datasets are randomly and repeatedly subsampled down to the size of the smallest dataset.

The resulting comparison document offers the same taxonomy, SEED and KEGG viewers as in the already familiar single sample view. In each of the three views, the tree indicates how many reads were assigned to each node for each original input document by drawing the node as a pie chart or bar chart, for example (see Fig. 9.8). All the features available for the analysis of a single dataset can also be applied to the comparison of multiple datasets, except for the possibility to save, inspect or align reads assigned to a given node. The number of reads assigned to given nodes can be exported as a CSV file for further statistical analysis with external tools. Attribute charts can also be generated within the comparison view. A general shift or difference in the microbial community can be seen easily without inspecting each node. Examples range from aerobic to anaerobic oxygen requirements or critical changes in distribution of non-pathogen to pathogen microorganisms.

MEGAN also supports the calculation of standard ecological indices for a comparison document (Mitra *et al.*, 2010), such as Goodall, chi-square, Hellinger, Bray–Curtis, Kulczynski as well as normalized functions. Distances between datasets can be calculated from taxonomic or functional data. The calculation only operates on selected nodes, so the user can easily control which parts of the dataset to take into account. If no nodes are

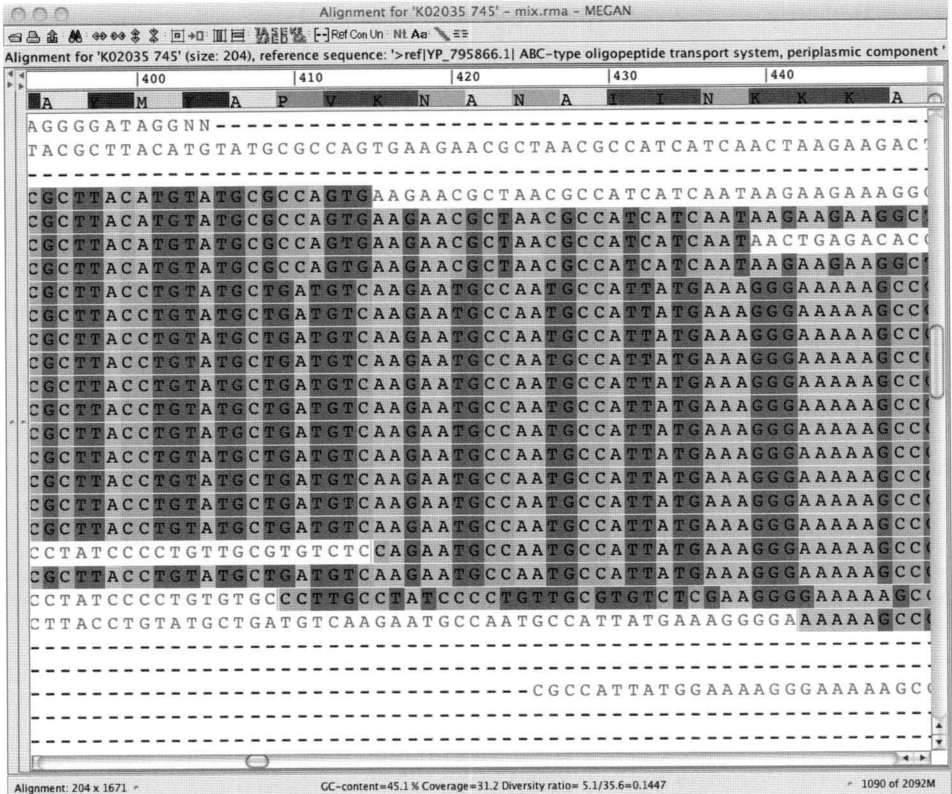

Figure 9.7 MEGAN's alignment viewer constructs and displays a multiple sequence between all reads that map to the same reference sequence. The top track shows the reference sequence and the main panel displays the aligned reads.

selected the whole dataset is used. Resulting distances can be represented as a tree, network or MDS plot Additionally the distance matrix can be exported for later import into other analysis software packages. The tree view represents the relation between different with one possible interpretation of the distances.

For demonstration purposes we created two artificial datasets *MinnesotaSoil-nogeo.rma* and *MinnesotaSoil-nogeo-noacido.rma*. These dataset include all reads from the original dataset, except that the reads assigned to *Geobacter* were removed from the both datasets and all reads assigned to *Acidobacteria* were deleted from the second dataset. Creation was performed using MEGAN's *Extract to new Document* feature. The result of the network comparison is displayed in Fig. 9.9.

As an example the user would inspect the taxonomic contents of a newly created comparison of the original Minnesota soil dataset versus an artificial one. Realizing that the group of Deltaproteobacteria shows a different distribution, the user than plots all leaves of taxa below the specific node. Fig. 9.10 illustrates the results. It can be clearly seen that Geobacter is responsible for the change.

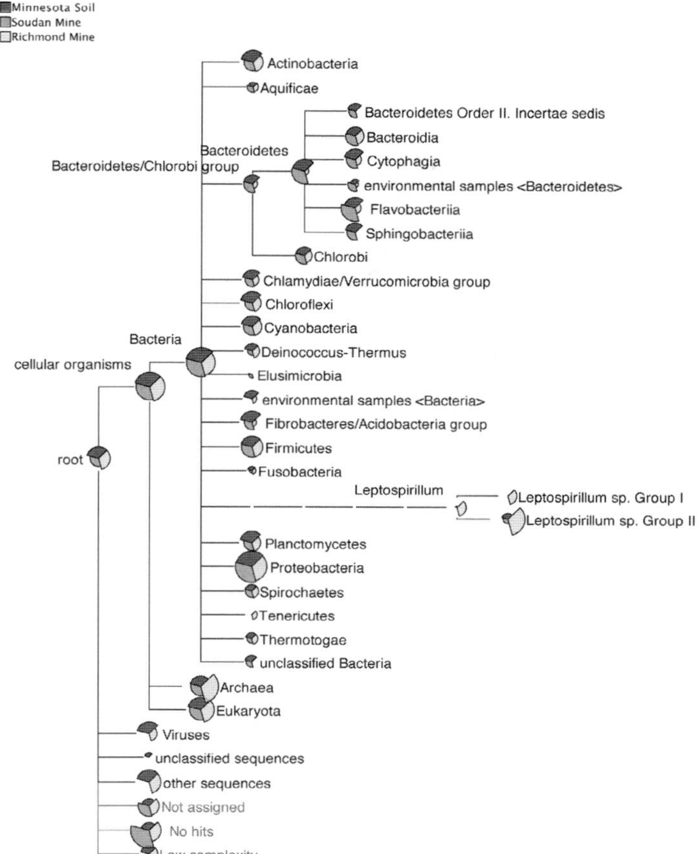

Figure 9.8 High-level comparison of taxonomic content of three different DNA datasets labelled Minnesota Soil (Tringe *et al.*, 2005), Soudan Mine (Edwards *et al.*, 2006) and Richmond Mine (Lo *et al.*, 2007). Part of the tree has been uncollapsed to show differences in underlying taxa. The three different datasets are represented by different colours and each node shows a coxcomb chart that indicates the number of reads assigned to that node, on a logarithmic scale.

Conclusion

MEGAN is an interactive tool for analysing the taxonomic and functional content of metagenomic (and metatranscriptomic) datasets. With MEGAN we hope to provide a versatile tool for analysing single or groups of metagenomes on a desktop computer, aimed at the biologist in the field (or lab) rather than the trained bioinformatician, and thus we try to keep usability as simple as possible. Input is a set of DNA reads and the result of comparing the reads against a reference database.

MEGAN supports various standard file formats to make it easier to import data from different types of alignment or mapping tools. After sequence comparison the taxonomic analysis is performed by placing DNA reads onto nodes of the NCBI taxonomy based on LCA approach. Offering various options the user can tune the result to match specific

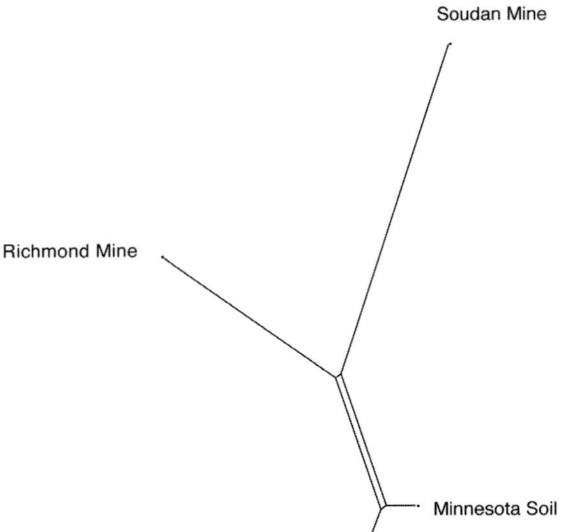

Figure 9.9 Network comparison of the original Minnesota farm dataset (Tringe *et al.*, 2005) and the two artificial soil datasets (created by sub-sampling the Minnesota farm dataset), together with a sample from the Soudan Mine, Minnesota, USA (Edwards *et al.*, 2006) and a sample from the Richmond Mine, Iron Mountain, CA, USA (Lo *et al.*, 2007).

needs of the analysis. Functional analysis is based on mapping reads to SEED and KEGG categories using the NCBI RefSeq identifiers. The program also supports comparative analysis of multiple datasets. As an additional feature an alignment viewer is integrated to inspect and browse the read alignments of each sequence per taxon, which may help explain effects which cannot be explained by the absolute read numbers. Information generated by MEGAN can be exported in standard text and graphical files.

Unfortunately current databases only represent a small percentage of the microbial diversity encountered on earth and are strongly biased towards model organisms. It will be some time before projects such as GEBA (Wu *et al.*, 2009) will have a significant impact on this problem.

MEGAN is written in Java and runs on all major operating systems. When run in command-line mode, the program can also be integrated into larger bioinformatics analysis pipelines. As sequencing technologies continue to improve, the size of analysed datasets continues to increase. MEGAN was reportedly used to perform the taxonomic taxonomic analysis of 124 human gut samples involving around 600 gigabases of sequence (Qin *et al.*, 2010). This work only shows parts of MEGAN capabilities to analyse metagenomic data. The program is under active development and new features are added on a regular basis.

As datasets continue to increase in both size and number, increasingly efficient methods for analysing and comparing data will be needed. We are currently developing a new sequence alignment method called PAUDA that is a companion program to MEGAN and performs sequencing read alignment at a speed of more than 1000 times than of BLASTX.

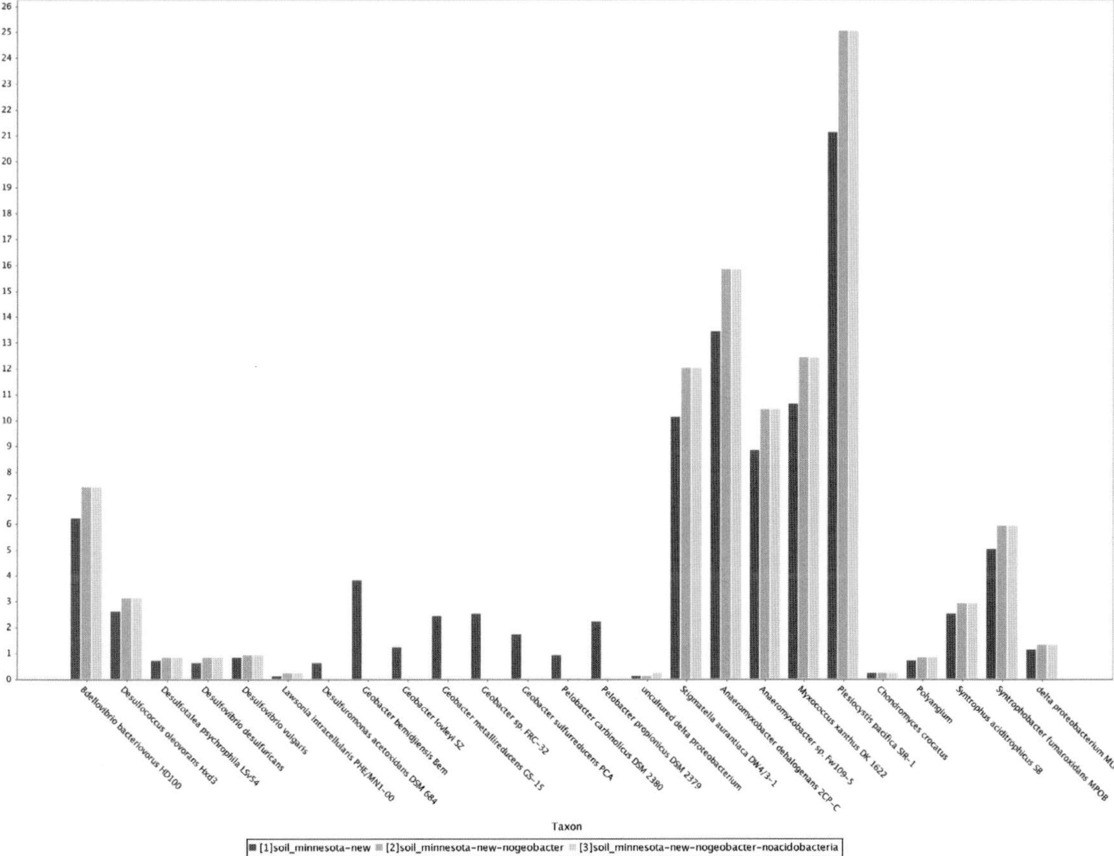

Figure 9.10 Chart of relative occurrences of taxa leaves under the node *Deltaproteobacteria*. The example shows that the group of *Geobacter* is missing in two datasets.

References

Altschul, S.F., Gish, W., Miller, W., Myers, E.W., and Lipman, D.J. (1990). Basic local alignment search tool. J. Mol. Biol. *215*, 403–410.

Caporaso, J.G., Kuczynski, J., Stombaugh, J., Bittinger, K., Bushman, F.D., Costello, E.K., Fierer, N., Pena, A.G., Goodrich, J.K., Gordon, J.I., *et al.* (2010). Qiime allows analysis of high-throughput community sequencing data. Nat. Methods 7, 335–336.

Edwards, R.A., Rodriguez-Brito, B., Wegley, L., Haynes, M., Breitbart, M., Peterson, D.M., Saar, M.O., Alexander, S., Alexander, Jr, E.C., and Rohwer, F. (2006). Using pyrosequencing to shed light on deep mine microbial ecology. BMC Genomics 7, 57.

Handelsman, J., Rondon, M., Brady, S., Clardy, J., and Goodman, R. (1998). Molecular biological access to the chemistry of unknown soil microbes: a new frontier for natural products. Chem. Biol. *5*, 245–249.

Huson, D.H., Auch, A.F., Qi, J., and Schuster, S.C. (2007). MEGAN analysis of metagenomic data. Genome Res. *17*, 377–386.

Huson, D.H., Mitra, S., Weber, N., Ruscheweyh, H., and Schuster, S.C. (2011). Integrative analysis of environmental sequences using megan4. Genome Res. *21*, 1552–1560.

Kanehisa, M., and Goto, S. (2000). KEGG: Kyoto encyclopedia of genes and genomes. Nucleic Acids Res. *28*, 27–30.

Li, H., Handsaker, B., Wysoker, A., Fennell, T., Ruan, J., Homer, N., Marth, G., Abecasis, G., and Durbin, R. (2009). The sequence alignment/map (SAM) format and SAMtool. Bioinformatics *25*, 2078–2079.

Lo, I., Denef, V.J., Verberkmoes, N.C., Shah, M.B., Goltsman, D., DiBartolo, G., Tyson, G.W., Allen, E.E., Ram, R.J., Detter, J.C., Richardson, P., Thelen, M.P., Hettich, R.L., and Banfield, J.F. (2007). Strain-resolved community proteomics reveals recombining genomes of acidophilic bacteria. Nature *446*, 537–541.

McHardy, A.C., Martin, H.G., Tsirigos, A., Hugenholtz, P., and Rigoutsos, I. (2007). Accurate phylogenetic classification of variable-length dna fragments. Nat. Methods *4*, 63–72.

Mitra, S., Gilbert, J., Field, D., and Huson, D. (2010). Comparison of multiple metagenomes using phylogenetic networks based on ecological indices. ISME J. doi:10.1038/ismej.2010.51.

Nannipieri, P., Ascher, J., Ceccherini, M.T., Landi, L., Pietramellara, G., and Renella, G. (2003). Microbial diversity and soil functions. Eur. J. Soil Sci. *54*, 655–670.

Overbeek, R., Begley, T., Butler, R.M., Choudhuri, J.V., Chuang, H.-Y., Cohoon, M., de Crécy-Lagard, V., Diaz, N., Disz, T., Edwards, R., *et al.* (2005). The subsystems approach to genome annotation and its use in the project to annotate 1000 genomes. Nucleic Acids Res. *33*, 5691–5702.

Pruesse, E., Quast, C., Knittel, K., Fuchs, B., Ludwig, W., Peplies, J., and Glöckner, F. (2007). SILVA: a comprehensive online resource for quality checked and aligned ribosomal RNA sequence data compatible with ARB. Nuc. Acids Res. *35*, 7188–7196.

Qin, J., Li, R., Raes, J., Arumugam, M., Burgdorf, K.S., Manichanh, C., Nielsen, T., Pons, N., Levenez, F., Yamada, T., *et al.* (2010). A human gut microbial gene catalogue established by metagenomic sequencing. Nature *464*, 59–65.

Richter Jr., D., and Markewitz, D. (1995). How deep is soil? BioScience *45*, 600–609.

Rosen, G.L., Reichenberger, E., and Rosenfeld, A. (2011). NBC: The naive Bayes classification tool webserver for taxonomic classification of metagenomic reads. Bioinformatics *27*, 127–129.

Tringe, S.G., von Mering, C., Kobayashi, A., Salamov, A.A., Chen, K., Chang, H.W., Podar, M., Short, J.M., Mathur, E.J., Detter, J.C., Bork, P., Hugenholtz, P., and Rubin, E.M. (2005). Comparative metagenomics of microbial communities. Science, 308, 554–557.

Wang, Q., Garrity, G.M., Tiedje, J.M., and Cole, J.R. (2007). Naive bayesian classifier for rapid assignment of rrna sequences into the new bacterial taxonomy. Appl. Environ. Microbiol. *73*, 5261–5267.

Wheeler, D.L., Barrett, T., Benson, D.A., Bryant, S.H., Canese, K., Chetvernin, V., Church, D.M., Dicuccio, M., Edgar, R., Federhen, S., *et al.* (2008). Database resources of the national center for biotechnology information. Nucleic Acids Res. *36*, D13–D21.

Wu, D., Hugenholtz, P., Mavromatis, K., Pukall, R., Dalin, E., Ivanova, N.N., Kunin, V., Goodwin, L., Wu, M., Tindall, B.J., *et al.* (2009). A phylogeny-driven genomic encyclopaedia of bacteria and archaea. Nature *462*, 1056–1060.

Zhao, Y., Tang, H., and Ye, Y. (2012). RAPSearch2: a fast and memory-efficient protein similarity search tool for next- generation sequencing data. Bioinformatics *28*, 125–126.

Classical Techniques Versus Omics Approaches

10

David D. Myrold and Paolo Nannipieri

Abstract

This volume presents the state-of-the-art of omics in soil science, a field that is advancing rapidly on many fronts. The various omics approaches hold much promise but also await further refinement before they are ready for widespread adaptation. One way to judge their readiness is to compare them to methods that have become standards for soil microbiology research. Methods become standards because they provide useful information quickly and inexpensively. There is no question that omics can provide useful information, some of which cannot be obtained with traditional techniques, and integration of omics methods may provide insights into ecosystem functioning. In particular, the potential for omics to provide comprehensive coverage of genes and genes products make them well-suited for the study of general soil microbiological phenomena, such as decomposition, response to water stress, etc.

Introduction

What limits our ability to advance soil microbial ecology? Some suggest that we need to think harder, or better, and focus on identifying underlying principles that are founded, at least in part, on ecological theory (Prosser *et al.*, 2007; Fierer *et al.*, 2009). Others express the long-standing opinion that progress in microbial ecology is intimately linked to the development of new methods (Jansson *et al.*, 2012). Of course, both ingredients – theory and technique – are important drivers of advances in our understanding of the structure and function of microbial communities in soil. Our purpose in this chapter touches on this interplay by contrasting insights that can be gained from classical procedures compared to the newly developed and rapidly evolving omics methods that have been highlighted in this book. We begin by highlighting two, now classic, methods that revolutionized soil microbial ecology.

Microbial biomass by chloroform fumigation

The number, or biomass, of microorganisms is a key property of soil because microbes are catalysts for processing C and nutrients, and key players of several soil functions. The microbial biomass is also an important reservoir of C and nutrients in soil. Until the development and refinement of the chloroform fumigation method (Jenkinson and Powlson, 1976; Brookes *et al.*, 1985), extrapolation of microscopic counts was the only approach available to estimate the content of the main elements (C, N, P and S) present in soil microbiota. The chloroform fumigation method directly determines the content of the main nutrients in soil microbial biomass whereas other methods, such as those based on the components of living cells (e.g. ATP content) or on the response of soil microorganisms to the addition of an exogenous substrate (the substrate-induced respiration method, SIR), permit only the indirect determination of the nutrient content (Anderson and Domsch, 1978; Jenkinson and Ladd, 1981).

The utility of the chloroform fumigation is shown by its rapid and widespread adoption – the articles by Jenkinson and Powlson (1976) and Brookes *et al.* (1985) have each been cited about 1400 times (Web of Science, 23 May 2012). Indeed, prior to 1976 there were no papers in the literature on the topic of 'soil microbial biomass', but thereafter such papers increased exponentially, with 93, 1543 and 5145 articles in the three succeeding decades. Clearly the chloroform fumigation method was instrumental in advancing soil microbial ecology research, particularly in monitoring nutrient fluxes through the soil microbiota. For example, using the chloroform fumigation method it is possible to trace the microbial fate of an isotopically labelled nutrient by monitoring its abundance in the microbial biomass. This in turn has fostered a more holistic approach for studying complex systems, such as soil, which is based on dividing the system into different pools linked by fluxes representing physical or abiotic and/or biotic transformations (Nannipieri *et al.*, 1994). Such current models of nutrient dynamics are based on (i) microbial biomass as a source and sink of nutrients; (ii) simultaneous microbial decomposition of organic matter and microbial synthesis; and (iii) multiple pools of organic matter with different degradation kinetics.

Denitrification rates by acetylene blockage

Quantifying denitrification – the stepwise microbial reduction of NO_3^- to N_2 – has been plagued by several technical challenges, foremost among them was the difficulty of measuring relatively small fluxes of N_2 gas into the large pool of atmospheric N_2 (Groffman *et al.*, 2006). The latter difficulty could be overcome by using ^{15}N-labelled NO_3^- (Hauck *et al.*, 1958); however, this method is also not particularly sensitive and requires access to an isotope ratio mass spectrometer, which greatly limited its adoption. Indeed, it took more than 20 years for this N-isotope distribution approach to be applied (Focht and Stolzy, 1978; Siegel *et al.*, 1982).

In contrast, the discovery by Yoshinari and Knowles (1976) that acetylene can be used to specifically block the final step in the denitrification pathway – the reduction of N_2O to N_2 – was quickly adopted and has been cited about 500 times (Web of Science, 23 May 2012). The so-called acetylene-block method removed the measurement bottleneck because it was relatively quick and inexpensive to measure small changes in N_2O against its ppb background in the atmosphere, and led to an exponential increase in soil denitrification studies. Use of the acetylene-block method spawned the development of additional widely used methods (e.g. Smith and Tiedje, 1979), provided insights into the high temporal and spatial variability of soil denitrification (Parkin, 1987; Sexstone *et al.*, 1988), and generally contributed to a quantifying losses of N by denitrification from soils of diverse ecosystems (Smil, 1999; Janzen *et al.*, 2003; Sietzinger *et al.*, 2006).

The promises of omics methods

Will omics methods – metagenomics, metatranscriptomics, metaproteomics, etc. – have a similar impact on microbial ecology as previous innovations in methodology? It is not possible to predict with certainty at this point, but the rapid and widespread adoption by soil microbiologists of the first generation of molecular methods, such as DNA fingerprinting (e.g. DGGE, T-RFLP) and Q-PCR, and more recently next generation pyrotagged sequencing, suggests that there is a demand for better tools to describe soil microbial communities.

In theory, omics approaches offer two advantages over existing methods: comprehensiveness and complementarity. As their prefix 'meta' implies, omics methods strive to measure all the genes, transcripts, or proteins in a given soil. For example, if a complete catalogue of genes exists, then it may be possible to identify all potential functions of every microorganism in a soil. Included in this complete parts list will be novel genes, whose potential function is not yet known (Vogel *et al.*, 2009). Precisely because they are unknown, their discovery requires a metagenomics approach. The same principle applies for transcripts and proteins.

Because of the canonical relationship of gene to transcript to protein, the three omics methods are inherently complementary. Linking them together provides an integrated picture of the relationship between potential activity – the structure of the microbial community – and realized activity – the function of the microbial community. For example, Jaffe *et al.* (2004) proposed a proteogenomic approach to validate genes and visualize the real location of the coding domain sequences in the genome by directly mapping proteins analysed by the proteomics on genome sequences, as discussed by Bland and Armengaud in Chapter 8. By combining genome and proteome information new genes and new proteins have been detected, giving more useful information than the separate use of metagenomics and proteomics.

Omics data also offer great potential to examine linkages or interactions among organisms or their gene products. For example, network analysis approaches have been used to predict microbial community assemblages and to identify interactions among community members based on gene sequence data (Larsen *et al.*, 2012a; Zhou *et al.*, 2011). These types of analysis may ultimately provide the information needed to develop the framework for ecosystem models (Allison *et al.*, 2010; Treseder *et al.*, 2012).

The pitfalls of omics methods

Will omics methods become widely used? Any method that becomes commonly used meets most, if not all, of the following criteria: uncomplicated and quick to perform, relatively inexpensive, uses readily available instrumentation, and produces clearly interpretable and useful information. Presently, all of the omics methods fall short in several of these criteria, although that is probably the case for most new methods. With greater use, many of the challenges associated with omics methods (e.g. extraction and purification of biomolecules, data handling and analysis) will likely become more tractable.

Challenges typical of working with soils

Some challenges for applying omics methods to soils are also common to any study of the biology of soils (Lombard *et al.*, 2011). These include the collection of a representative soil sample, and the extraction and purification of an adequate amount of the desired biomolecule from the soil matrix.

Soil is a heterogeneous system and the biological space represents a small percentage of the overall soil space, probably because only a small number of soil microenvironments have the right set of conditions (nutrients, protection against predators, growth factors, absence of toxic substances, suitable pH, suitable oxidation-reduction potential, etc.) for the microbial life (Nannipieri *et al.*, 2003). There are some 'hotspots' in soil with increased microbial activity, such as the rhizosphere soil and zones around fresh organic matter. This spatial heterogeneity has led to studies that focus on such microhabitats (Davinic *et al.*, 2012).

Studies at the microscale can help to establish predictive relationships between soil characteristics and microbial processes in soil. Observations by electronic microscopy are useful for visualizing spatial distribution of bacterial and fungi in soil but are limited with respect to their ability to describe microbial community composition or activity. One rare exception is that active acid phosphomonoesterase has been detected in microbial cell membrane fragments in sections without soil particles (Ladd *et al.*, 1996). Consequently, a microsampling strategy has been proposed by Grundmann *et al.* (2001) and used to characterize the spatial distribution of microbial microhabitat and microbial species at the microscale; for example, the spatial distribution of *Agrobacterium* spp. Bioavr 1 in 1 cm^3 of soil was investigated by collecting 865 microsamples, each 500 μm in diameter (Vogel *et al.*, 2003). This sampling strategy is limited in application for measuring the metagenome, metatranscriptome, or proteome at the microscale because the amount of extracted molecules from these microsamples is too small to be analysed by the current techniques. Consequently, larger samples that average across microhabitats are used.

In a study that examined the effects of sample size, Delmont *et al.* (2011) determined that 100 g of soil was sufficient to represent the metagenome of soil microbial community of the Park Grass soil, the reference soil for metagenomic studies (Vogel *et al.*, 2009). As part of this study, a fingerprinting technique, ribosomal intergenic spacer analysis (RISA) of DNA, was used to explore spatial variation at the field scale. That analysis showed that vertical variation in microbial diversity was higher than horizontal variation (Delmont *et al.*, 2011).

This challenge of spatial heterogeneity, which is common to all soils studies, can be further complicated by temporal variability, which is caused by the dynamic nature of the soil microbial community as it responds to changes in environmental conditions or other perturbations. For example, temperature and moisture are important determinants of microbial activity in soil, and these commonly change seasonally. Therefore markedly different results could be obtained when soils are sampled in different seasons, not only for gene expression activities but also for the metagenomic approach. For example, metagenomic approaches usually describe predominant microbial communities (Delmont *et al.*, 2011), which can be different under different soil conditions through the year. For example, and high microbial activity may mean consumption of available P, which represses the synthesis of phosphomonoesterase and thus the expression of the relative genes encoding these enzymes (Nannipieri *et al.*, 2011). The analysis of spatiotemporal data can be difficult; however, methods are being developed to tackle this challenge (Graf *et al.*, 2012; Larsen *et al.*, 2012b).

Regardless of the size of the soil sample used, whenever we extract something from soil, the architecture of the soil is destroyed. According to Kubiena (1938), soil is like a city whose architecture is destroyed after soil sampling, breakdown of soil aggregates, and extraction. Any determination involving soil extracts, such as the characterization of nucleic acids and proteins for evaluating structure and functionality of microbial communities, does not permit the localization of the measured parameter in the soil matrix.

Extractions of nucleic acids and proteins from soil can present several problems affecting metagenomic, transcriptomic, and proteomic analysis, and the relative interpretations of results. It is not only a problem of yields but also completeness: A method should extract these molecules from all members of soil microbial communities. This is not often the case. For example, different methods of DNA extraction from soil give different results about microbial diversity (Bakken and Frostegård, 2006; Delmont *et al.*, 2011). Similarly, Keiblinger *et al.* (2012) found that the numbers and types of proteins varied among four extraction methods. In addition, extraction methods typically release the desired biomolecule from microbial cells but in the process also release other organic compounds present in soil and expose the biomolecule to interactions with other components of the soil, such as clays. Clays can bind the released biomolecule and extracellular enzymes, such as nucleases and proteases, can degrade the biomolecule, reducing yields. The organic 'contaminants' often interfere with downstream analysis of the biomolecule of interest. Procedures have been developed to help alleviate these problems with extraction efficiency and inhibition, but none are universally effective. This has led to the general sentiment that extraction methods should be tested and optimized for individual soils, which is not particularly practical.

Lastly, it is important to remember the challenges inherent in scaling up processes or functions at the field, regional, and landscape level, which is essential for suitable land management (Standing *et al.*, 2007). Scaling from the gene level (10^{-8} m) through plant (10^{-2} m) and field (10^{2} m) up to landscape (10^{5} m) requires an understanding of the connections, linkages, feedbacks, and the scale-related relevance of its constituent parts. In addition, scaling up is made difficult by the use of different techniques at different scales (Standing *et al.*, 2007). For example, soil can be a source and sink of two greenhouse gases, N_2O and CH_4, and the underlying mechanisms of the relative processes as well as the active soil microorganisms are studied by molecular techniques, including the omics approach at the microscale whereas their evolution can be studied with chamber techniques at the plot scale, micro-meteorological measurements at the medium scale, and use of aircraft at the catchment or landscape scale (Standing *et al.*, 2007).

Challenges unique to omics data

Assuming that it is possible to obtain biomolecules of adequate quantity and quality that are representative of the soil being studied, additional hurdles common to meta-omics studies must be overcome. Many of these are related to the quantity of data needed for complete coverage of the genes, transcripts, or proteins in a soil. Although improvements in sequencing technology have enabled ever greater amounts of DNA or RNA sequence data, it is questionable if even terabytes of data will provide complete coverage of the soil metagenome or metatranscriptome (Brown and Tiedje, 2011). Shotgun proteomics are not as readily scalable. In this case further advances are needed to detect more than the most abundant proteins (Siggins *et al.*, 2012).

Analysis of what has become known as the 'data deluge' is another significant challenge. Currently, it is not possible to fully assemble metagenomic and metatranscriptomic data, although some partial assembly is possible and information can be gained from unassembled sequences (Luo *et al.*, 2012; Pell *et al.*, 2012). The small read length generated by the present sequencing systems contributes to this difficulty. Short reads reliably identify sequence homologues only when there is a strong homology to previously described genes (Morales and Holben, 2011). In addition, short sequences reads produce a series of non-contiguous genetic fragments and their assembly into sequences representing genetic organizational units is problematic, particularly for the unknown functions. Advances in bioinformatics algorithms and access to larger computer resources (e.g. cloud computing) will improve this situation, but an additional limitation is the depth and quality of annotated sequences that are queried for gene and protein identification (Raes *et al.*, 2007; Raes and Bork, 2008). A further stumbling block is the need for statistical and modelling tools to interpret the vast quantities of information generated with omics methods. Recent applications of different types of network analysis have shown promise in this area (Zhou *et al.*, 2011; Faust and Raes, 2012; Larsen *et al.*, 2012a).

Successful integration of omics data are still an outstanding challenge for bioinformaticians grappling with the large and complex datasets that are produced, but doing so holds great promise for understanding the connections and interactions that form the basis of microbial processes in soil. In addition to developing bioinformatics tools, it will be necessary to gain a deeper understanding about the time scales and synchronicity of the responses of metagenomes, metatranscriptomes, proteomes, and metabolome, and whether their relationships are linear or nonlinear and how much stochasticity is involved.

Collectively, the challenges that omics data present will make it difficult, and perhaps impractical, to do replicated experiments, which is the hallmark of soil microbial research (Prosser, 2010). Nevertheless, omic-driven research should go beyond being descriptive and be hypothesis-driven, with a greater focus on experimental design rather than on the technological prowess.

Research that is best suited for omics approaches

As noted previously, soils present some common challenges in terms of measuring their properties and processes, such as their variability in space and time, potential interference of methodology with the variable being measured, etc. In most cases, multiple approaches are applied with the hope that the different perspectives will provide insight into the property or processes being measured. An important consideration with omics approaches is to identify the questions that they can answer better than traditional approaches.

Processes that are directly associated with specific enzymes and their affiliated genes and transcripts, such as nitrification with ammonia monooxygenase or denitrification with nitrate, nitrite, nitric oxide, and nitrous oxide reductases, are probably more efficiently studied with a targeted approach. Since the study of Cavigelli and Robertson (1980) first suggested a link between microbial community composition and denitrification, many studies have used molecular methods based on denitrifier genes to explore relationship between denitrifier community structure and function (Wallenstein *et al.*, 2006). More recent studies have also used denitrifier gene transcripts (Pastorelli *et al.*, 2011). For example, Liu *et al.* (2010) found that induction of denitrifier genes was related to denitrification rates; however, the effect of pH on the fraction of denitrification as N_2O appeared to be due to post-translational events.

We are of the opinion that meta-omics data are best suited for obtaining a holistic understanding of generic processes that cannot be easily addressed with a targeted approach, such as soil organic matter mineralization or the response of the entire microbial community to a perturbation, e.g. tillage or irrigation. The mineralization of organic matter involves most of the heterotrophic bacteria and fungi present in soil – making it impossible to focus in on a specific functional taxa, which metabolize a plethora of organic compounds – making it impossible to focus on specific functional genes, and produce a range of metabolites

– making it impossible to follow a single end product. A meta-omics approach has the potential to capture this taxonomic and functional diversity, and explore the interactions among taxa and their metabolic processes assuming, of course, that the challenges to dealing with the large and complex data thus obtained can be overcome.

Acknowledgement

This material is based upon work supported by the National Science Foundation under Grant No. 1051481. Any opinions, findings, and conclusions or recommendations expressed in this material are those of the author(s) and do not necessarily reflect the views of the National Science Foundation.

References

Allison, S.D., Wallenstein, M.D., and Bradford, M.A. (2010). Soil-carbon response to warming dependent on microbial physiology. Nature Geoscience *3*, 336–340.

Anderson, J.P.E., and Domsch, K.H. (1978). A physiological method for quantitative measurements of microbial biomass in soils. Soil Biol. Biochem. *10*, 215–221.

Bakken, L.R., and Frostegård, Å. (2006). Nucleic acid extraction from soil. In Nucleic Acid and Proteins in Soil. P. Nannipieri, and K. Smalla, eds. (Berlin, Germany: Sprimger), pp. 49–73.

Brookes, P.C., Landman, A., Pruden, G., and Jenkinson, D.S. (1985). Chloroform fumigation and the release of soil nitrogen: a rapid direct extraction method to measure microbial biomass nitrogen in soil. Soil Biol. Biochem. *17*, 837–842.

Brown, C.T., and Tiedje, J.M. (2011). Metagenomics: the paths forward. In Handbook of Molecular Microbial Ecology II (John Wiley & Sons, Inc., New Jersey, USA), pp. 579–588.

Cavigelli, M.A., and Robertson, G.P. (2000). The functional significance of denitrifier community composition in a terrestrial ecosystem. Ecology *81*, 1402–1414.

Davinic, M., Fultz, L.M., Acosta-Martinez, V., Calderon, F.J., Cox, S.B., Dowd, S.E., Allen, V.G., Zak, J.C., and Moore-Kucera, J. (2012). Pyrosequencing and mid-infrared spectroscopy reveal distinct aggregate stratification of soil bacterial communities and organic matter composition. Soil Biol. Biochem. *46*, 63–72.

Delmont, T.O., Robe, P., Cecillon, S., Clark, I.M., Constancias, F., Simonet, P., Hirsch, P.R., and Vogel, T.M. (2011). Accessing the soil metagenome for studies of microbial diversity. Appl Environ. Microbiol. *77*, 1315–1324.

Faust, K., and Raes, J. (2012). Microbial interactions: from networks to models. Nature Rev. Microbiol. *10*, 538–550.

Fierer, N., Grandy, A.S., Six, J., and Paul, E.A. (2009). Searching for unifying principles in soil ecology. Soil Biol. Biochem. *41*, 2249–2256.

Focht, D.D., and Stolzy, L.H. (1978). Long-term denitrification studies in soils fertilized with $(^{15}NH_4)_2SO_4$. Soil Sci. Soc. Am. J. *42*, 894–898.

Graf, A., Herbst, M., Weihermuller, L., Huisman, J.A., Prolingheuer, N., Bornemann, L., and Vereecken, H. (2012). Analyzing spatiotemporal variability of heterotrophic soil respiration at the field scale using orthogonal functions. Geoderma. *181*, 91–101.

Groffman, P.M., Altabet, M.A., Bohlke, J.K., Butterbach-Bahl, K., David, M.B., Firestone, M.K., Giblin, A.E., Kana, T.M., Nielsen, L.P., and Voytek, M.A. (2006). Methods for measuring denitrification: diverse approaches to a difficult problem. Ecol. Appl. *16*, 2091–2122.

Grundmann, G.L., Dechesne, A., Bartoli, F., Chassè, J.L., Flandrois, J.P., and Kizungu, R. (2001). Simulation of the spatial distribution of micro-habitat of NH_4^+ and NO_2^- oxidizing bacteria in soil. Soil Sci. Soc. Am. J. *65*, 1709–1716.

Hauck, R.D., Melsted, S.W., and Yankwich, P.E. (1958). Use of N-isotope distribution in nitrogen gas in the study of denitrification. Soil Sci. *86*, 287–291.

Jaffe, J.D., Berg, H.C., and Church, G.M. (2004). Proteogenomic mapping as a complementary method to perform genome annotation. Proteomics *4*, 59–77.

Jansson, J.K., Neufeld, J.D., Moran, M.A., and Gilbert, J.A. (2012). Omics for understanding microbial functional dynamics. Environ. Microbiol. *14*, 1–3.

Janzen, H.H., Beauchemin, K.A., Bruinsma, Y., Campbell, C.A., Desjardins, R.L., Ellert, B.H., and Smith, E.G. (2003). The fate of nitrogen in agroecosystems: An illustration using Canadian estimates. Nutr. Cycl Agroecosys. *67*, 85–102.

Jenkinson, D.S., and Powlson, D.S. (1976). The effects of biocidal treatments on metabolism in soil-V. A method for measuring biomass. Soil Biol. Biochem. *8*, 209–213.

Jenkinson, D.S., and Ladd, J.N. (1981). Microbial biomass in soil, measurement and turnover. In Soil Biochemistry, vol. 5, E.A. Paul, and J.N. Ladd, eds. (New York, USA: Marcel Dekker), pp. 415–471.

Keiblinger, K.M., Wilhartitz, I.C., Schneider, T., Roschitzki, B., Schmid, E., Eberl, L., Riedel, K., and Zechmeister-Boltenstern, S. (2012). Soil meta proteomics – Comparative evaluation of protein extraction protocols. Soil Biol. Biochem. *54*, 14–24.

Kubiena, W.L. (1938). Micropedology. Collegiate Press, Ames, Iowa.

Ladd, J.N., Forster, R.C., Nannipieri, P., and Oades, J.M. (1996). Soil structure and biological activity. In Soil Biochemistry, vol 9, G. Stotzky, and J.M. Bollag, eds. (New York, USA: Marcel Dekker), pp. 23–79.

Larsen, P.E., Field, D., and Gilbert, J.A. (2012a). Predicting bacterial community assemblages using an artificial neural network approach. Nat. Methods *9*, 621–625.

Larsen, P.E., Gibbons, S.M., and Gilbert, J.A. (2012b). Modeling microbial community structure and functional diversity across time and space. FEMS Microbiol. Lett. *332*, 91–98.

Liu, B.B., Morkved, P.T., Frostegard, A., and Bakken, L.R. (2010). Denitrification gene pools, transcription and kinetics of NO, N2O and N-2 production as affected by soil pH. FEMS Microbiol. Ecol. *72*, 407–417.

Lombard, N., Prestat, E., van Elsas, J.D., and Simonet, P. (2011). Soil-specific limitations for access and analysis of soil microbial communities by metagenomics. FEMS Microbiol. Ecol. *78*, 31–49.

Luo, C.W., Tsementzi, D., Kyrpides, N.C., and Konstantinidis, K.T. (2012). Individual genome assembly from complex community short-read metagenomic datasets. ISME J. *6*, 898–901.

Morales, S.E., and Holben, W.E. (2011). Linking bacterial identities and ecosystem processes: can 'omic' analyses be more than the sum of their parts? FEMS Microbiol. Ecol. *75*, 2–16.

Nannipieri, P., Ascher, J., Cecceherini, M.T., Landi, L., Pietramellara, G., and Renella, G. (2003). Microbial diversity and soil functions. Eur. J. Soil Sci. *54*, 655–670.

Nannipieri, P., Badalucco, L., and Landi, L. (1994). Holistic approaches to the study of populations, nutrient pools and fluxes: limits and future research needs. In Beyond the Biomass, K. Ritz, J. Dighton, and K.E. Giller, eds. (Reading, UK: British Society of Soil Science), pp. 231–238.

Nannipieri, P., Giagnoni, L., Landi, L., and Renella, G. (2011). Role of phosphatase enzymes in soil. In Phosphorus in Action: Biological Processes in Soil Phosphorus Cycling, E.K. Bünemann, A. Oberson, and E. Frossard, eds. (Berlin, Germany: Springer Verlag), pp. 215–241.

Parkin, T.B. (1987). Soil microsites as a source of denitrification variability. Soil Sci. Soc. Am. J. *51*, 1194–1199.

Pastorelli, R., Landi, S., Trabelsi, D., Piccolo, R., Mengoni, A., Bazzicalupo, M., and Pagliai, M. (2011). Effects of soil management on structure and activity of denitrifying bacterial communities. Appl. Soil Ecol. *49*, 46–58.

Pell, J., Hintze, A., Canino-Koning, R., Howe, A., Tiedje, J., and Brown, C. (2012). Scaling metagenome sequence assembly with probabilistic De Bruijn graphs. Accepted at PNAS, July 2012; Preprint at http://arxiv.org/abs/1112.4193.

Prosser, J.I. (2010). Replicate or lie. Environ. Microbiol. *12*, 1806–1810.

Prosser, J.I., Bohannan, B.J.M., Curtis, T.P., Ellis, R.J., Firestone, M.K., Freckleton, R.P., Green, J.L., Green, L.E., Killham, K., Lennon, J.J., *et al.* (2007). Essay – The role of ecological theory in microbial ecology. Nat. Rev. Microbiol. *5*, 384–392.

Raes, J., and Bork, P. (2008). Molecular eco-systems biology: towards an understanding of community function. Nat. Rev. Microbiol. *6*, 693–699.

Raes, J., Foerstner, K.U., and Bork, P. (2007). Get the most out of your metagenome: computational analysis of environmental sequence data. Curr. Opin. Microbiol. *10*, 490–498.

Seitzinger, S., Harrison, J.A., Bohlke, J.K., Bouwman, A.F., Lowrance, R., Peterson, B., Tobias, C., and Van Drecht, G. (2006). Denitrification across landscapes and waterscapes: A synthesis. Ecol. Appl. *16*, 2064–2090.

Sexstone, A.J., Parkin, T.B., and Tiedje, J.M. (1988). Denitrification response to soil wetting in aggregated and unaggregated soil. Soil Biol. Biochem. *20*, 767–769.

Siegel, R.S., Hauck, R.D., and Kurtz, L.T. (1982). Determination of $^{30}N_2$ and application to measurement of N_2 evolution during denitrification. Soil Sci. Soc. Am. J. *46*, 68–74.

Siggins, A., Gunnigle, E., and Abram, F. (2012). Exploring mixed microbial community functioning: recent advances in metaproteomics. FEMS Microbiol. Ecol. *80*, 265–280.

Smil, V. (1999). Nitrogen in crop production: An account of global flows. Global Biogeochem. Cy. *13*, 647–662.

Smith, M.S., and Tiedje, J.M. (1979). Phases of denitrification following oxygen depletion in soil. Soil Biol. Biochem. *11*, 261–267.

Standing, D., Baggs, E.M., Wattenbach, M., Smith, P., and Killham, K. (2007). Meeting the challenge of scaling up processes in the plant-soil-microbe system. Biol. Fertil. Soils *44*, 245–257.

Treseder, K.K., Balser, T.C., Bradford, M.A., Brodie, E.L., Dubinsky, E.A., Eviner, V.T., Hofmockel, K.S., Lennon, J.T., Levine, U.Y., MacGregor, B.J., Pett-Ridge, J., and Waldrop, M.P. (2012). Integrating microbial ecology into ecosystem models: challenges and priorities. Biogeochemistry *109*, 7–18.

Vogel, J., Norman, P., Thioulose, J., Nesme, X., and Grundmann, G.L. (2003). Relationship between spatial and genetic distance in *Agrobacterium* spp in 1 cubic centimeter of soil. Appl Environ. Microbiol. *69*, 1482–1487.

Vogel, T.M., Simonet, P., Jansson, J.K., Hirsch, P.R., van Elsas, J.D., Bailey, M.J., Nalin, R., and Philippot, L. (2009). TerraGenome: a consortium for the sequencing of a soil metagenome. Nat. Rev. Microbiol. *7*, 252.

Wallenstein, M.D., Myrold, D.D., Firestone, M., and Voytek, M. (2006). Environmental controls on denitrifying communities and denitrification rates: insights from molecular methods. Ecol. Appl. *16*, 2143–2152.

Yoshinari, T., and Knowles, R. (1976). Acetylene inhibition of nitrous-oxide reduction by denitrifying bacteria. Biochem. Biophys. Res. Commun. *69*, 705–710.

Zhou, J.Z., Deng, Y., Luo, F., He, Z.L., and Yang, Y.F. (2011). Phylogenetic molecular ecological network of soil microbial communities in response to elevated CO_2. Mbio. *2*, e00122–00111.

Index

C

K

L

M

N

Q

R

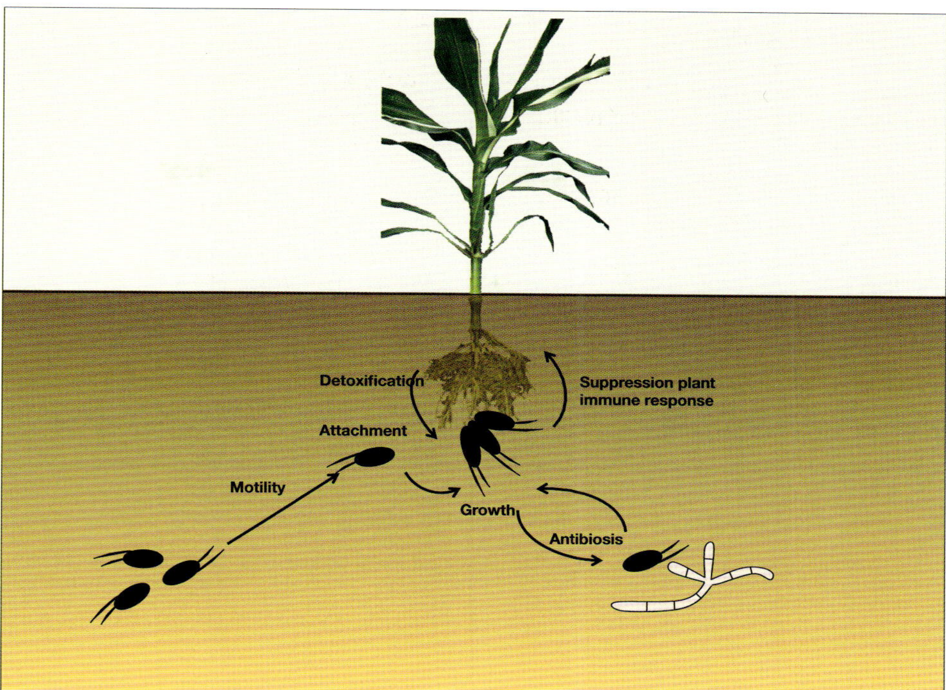

Figure 2.1 Bacterial traits involved in the colonization of the rhizosphere (adapted from Barret *et al.*, 2011). (1) Specific bacterial populations sense and move towards root exudates released in the rhizosphere through chemotaxis and motility. (2) Bacteria adhere on roots or rhizosphere soil surface. (3) Efficient bacterial growth rely on efficient nutrient scavenging systems coupled to synthesis of key components that are not present in the rhizosphere. (4) To survive in the rhizosphere bacteria have to overcome the chemical stress caused by numerous plant-derived toxic compounds through (a) detoxification mechanisms and/or (b) suppression of plant immune response. (5) Finally bacteria have to compete with other microorganisms through antibiosis, competition for nutrient and detoxification mechanisms.

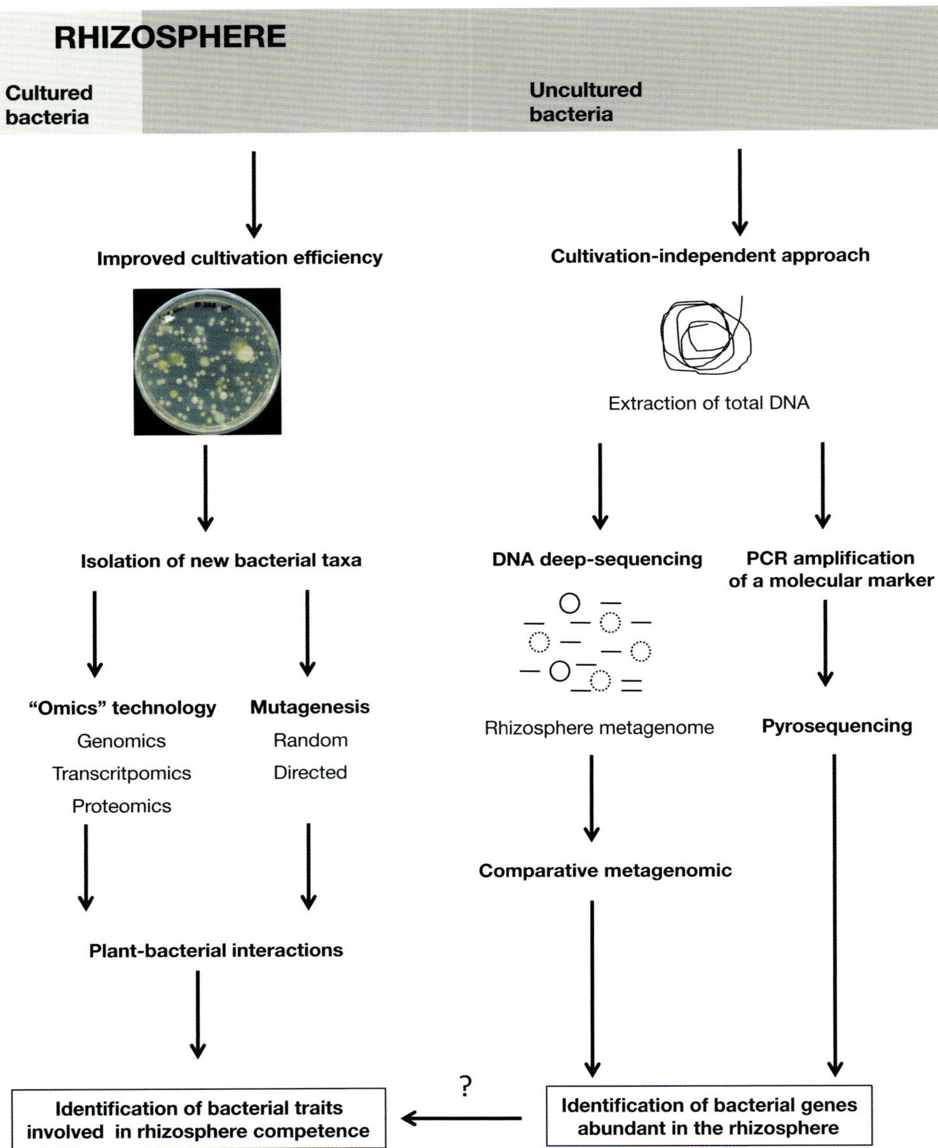

Figure 2.2 Approaches used to decipher the functions involved in the rhizosphere competence of 'uncultured-bacteria'.

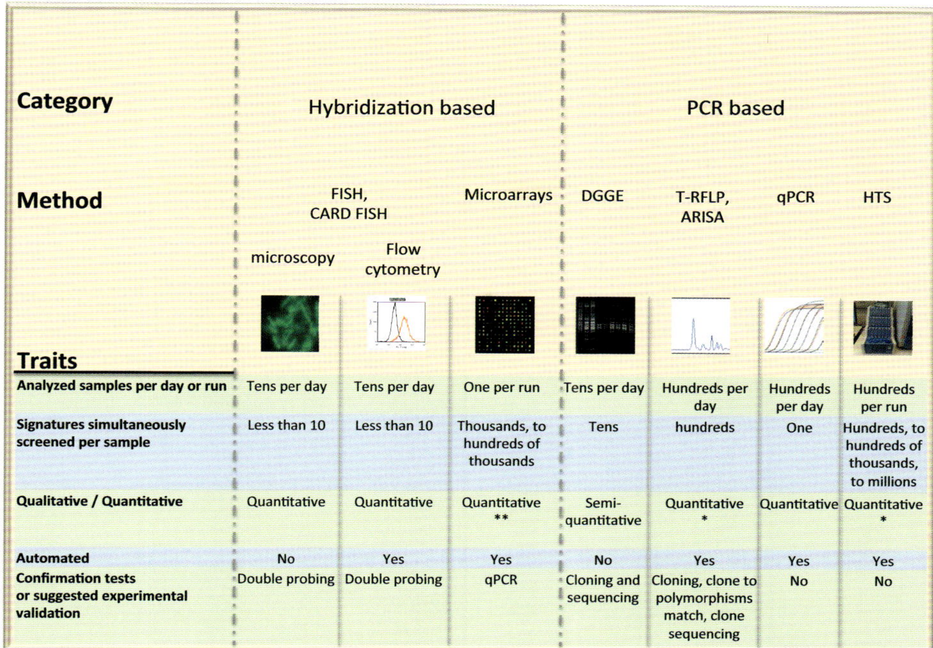

Category	Hybridization based			PCR based			
Method	FISH, CARD FISH		Microarrays	DGGE	T-RFLP, ARISA	qPCR	HTS
	microscopy	Flow cytometry					
Traits							
Analyzed samples per day or run	Tens per day	Tens per day	One per run	Tens per day	Hundreds per day	Hundreds per day	Hundreds per run
Signatures simultaneously screened per sample	Less than 10	Less than 10	Thousands, to hundreds of thousands	Tens	hundreds	One	Hundreds, to hundreds of thousands, to millions
Qualitative / Quantitative	Quantitative	Quantitative	Quantitative **	Semi-quantitative	Quantitative *	Quantitative	Quantitative *
Automated	No	Yes	Yes	No	Yes	Yes	Yes
Confirmation tests or suggested experimental validation	Double probing	Double probing	qPCR	Cloning and sequencing	Cloning, clone to polymorphisms match, clone sequencing	No	No

*: careful preparation (e.g. low number of PCR cycles) is necessary for reducing intensity of the PCR plateau effect on quantification abilities

**: the PCR plateau effect introduced bias is applicable in case a single marker gene like the SSU is screened through multiple taxa after PCR

Figure 4.1 Comparison of examples of hybridization and PCR-based marker screening methods.

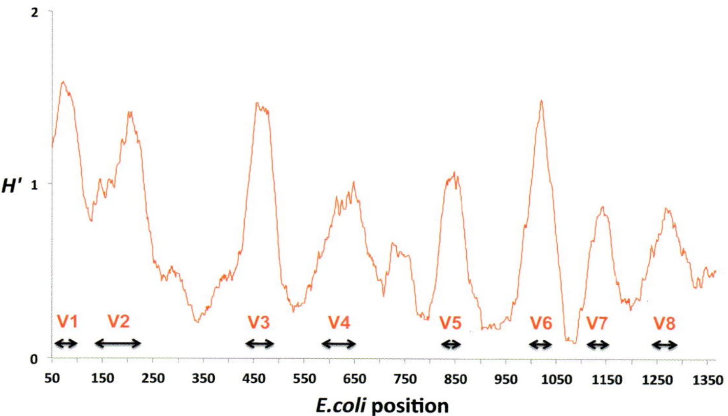

Figure 4.2 Shannon entropy (*H′*) plot of bacterial SSU multiple sequence alignment created using 42,109 soil-derived sequences existing in the RDP database, previously published by Vasileiadis *et al.* (2012) (figure use approved by all co-authors). For the entropy calculation, alignment columns were removed according to reference sequence gaps (sequence derived from *Escherichia coli*, GeneBank accession number 1VS5_A), while the hypervariable regions indicated as designated by Baker *et al.* (2003), according to *E. coli* nucleotide position numbering. Poorly supported areas of the beginning and the end of the sequences were not considered (due to usage of nearly full sequences), excluding this way one more existing V-region the V9.

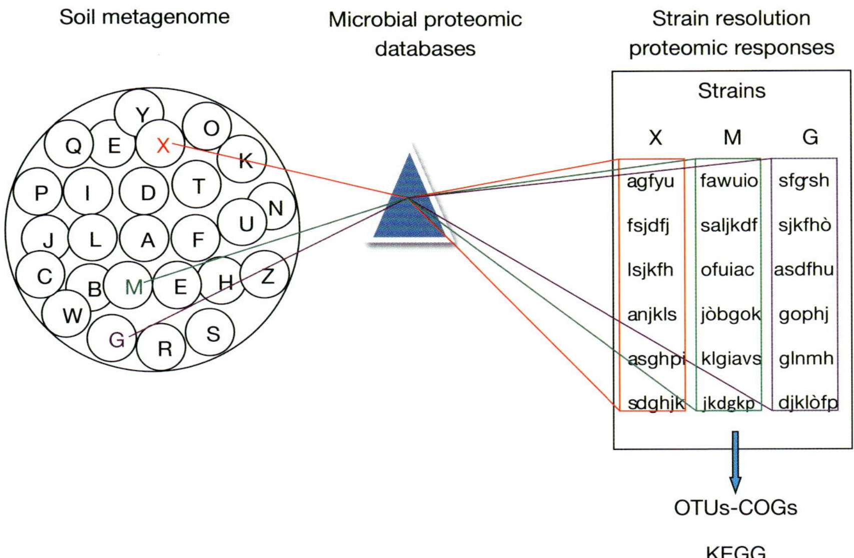

Figure 6.1 Strain resolution approach for soil proteomics.

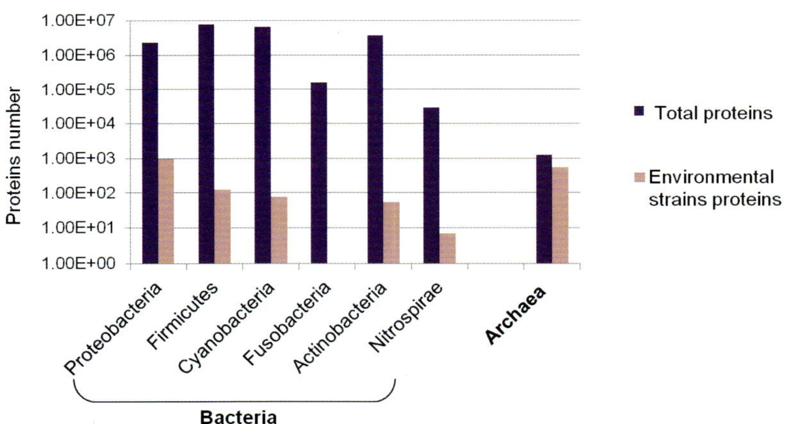

Figure 6.2 Bacterial classes with protein annotation in Expasy SwissProt database (October 2012).

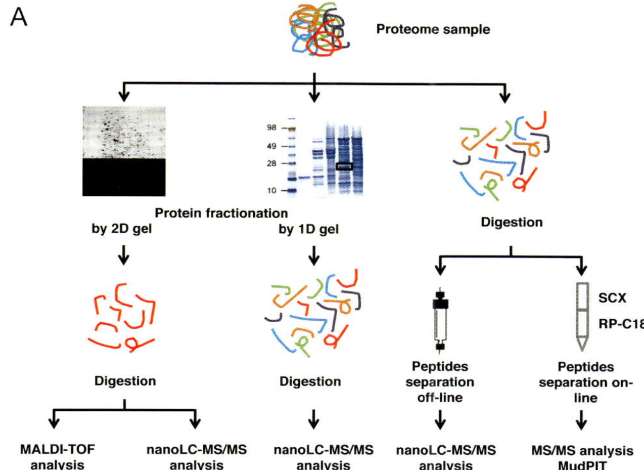

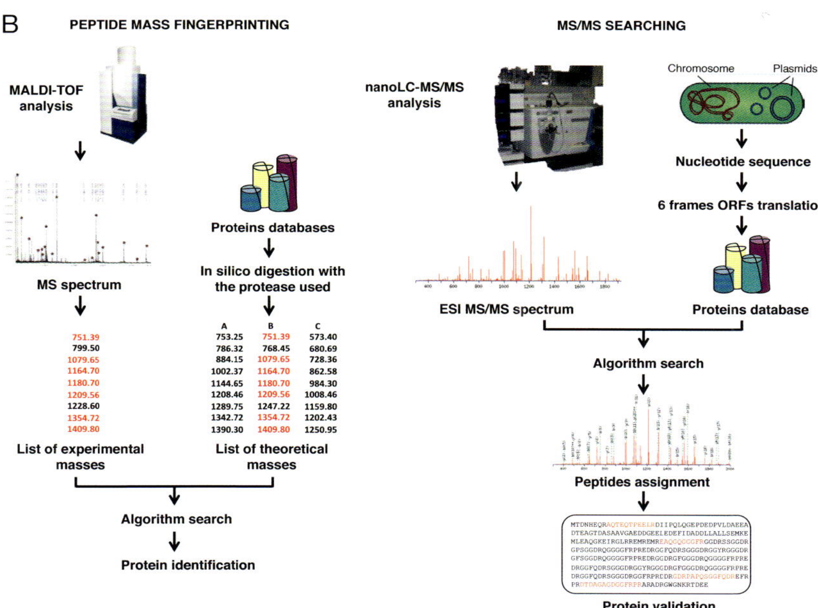

Figure 8.1 Shotgun proteomics: the most common strategies and current tools. Gel-based strategies followed by protein digestion, or protein digestion followed by multidimensional separation techniques, are the most common separation methods used in proteomics (A). These separation approaches are used to further massively analyse mixtures of thousands of different peptides by mass spectrometry and to identify proteins contained in these mixtures. Two approaches are illustrated, namely peptide mass fingerprinting and MS/MS searching (B). Peptide mass fingerprinting (left) is a method where an algorithm searches a protein database for proteins that would produce peptides of these molecular weights. Proteins are then identified by peptide mass mapping. MS/MS searching (right) relies on the multiple MS/MS spectra generated that are assigned to peptide sequences by specific searches against publicly available or homemade databases. Assignments of thousands of MS/MS spectra result in the identification of hundreds of proteins, validated when several peptides (in red) are assigned with confidence.

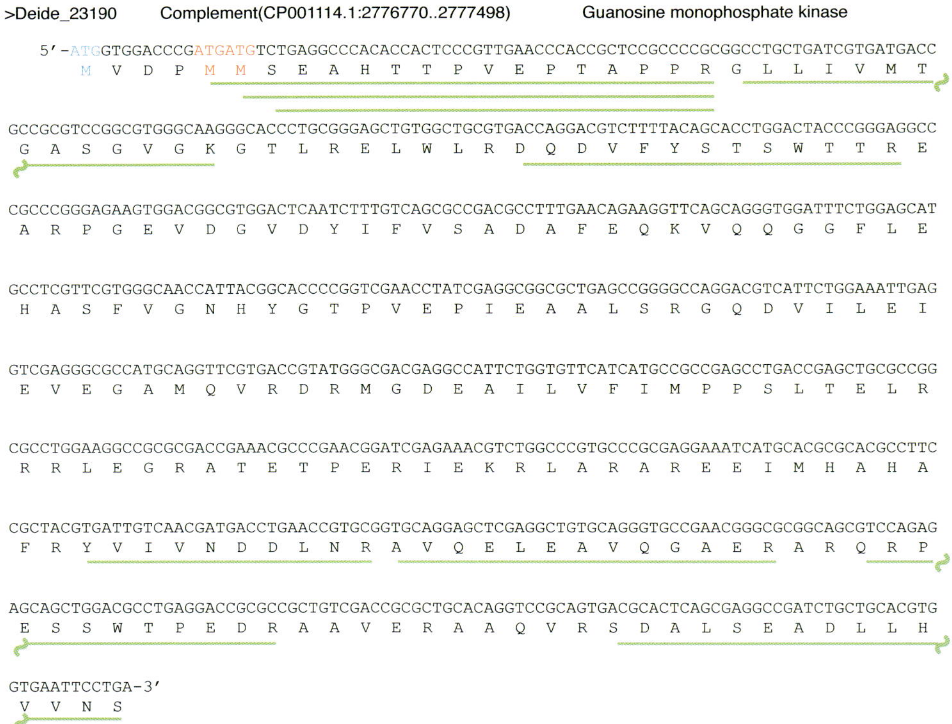

Figure 8.2 Two different initiation codons for the translation of Deide_23190 in *Deinococcus deserti*. The nucleic acid sequence for the locus specifying Deide_23190 is shown. Three different ATG codons found at the 5′-end of the ORF are labelled. Nine different tryptic peptides, detected after mass spectrometric analysis, were assigned to a guanosine monophosphate kinase (de Groot *et al.*, 2009). Three peptides correspond to N-terminal signatures and indicate a misannotation of this predicted gene regarding its translation start codon. The second and third ATG codons, which are contiguous, were used as initiation codons. The cells can then produce two different polypeptides from these two alternatives: one starting with the MMSEAHTTP sequence and the other starting with MSEAHTTP. A maturation of the second polypeptide with the excision of the initial methionine, expected because a serine is found as the second residue, results in the production of a third polypeptide starting with the SEAHTPP sequence.

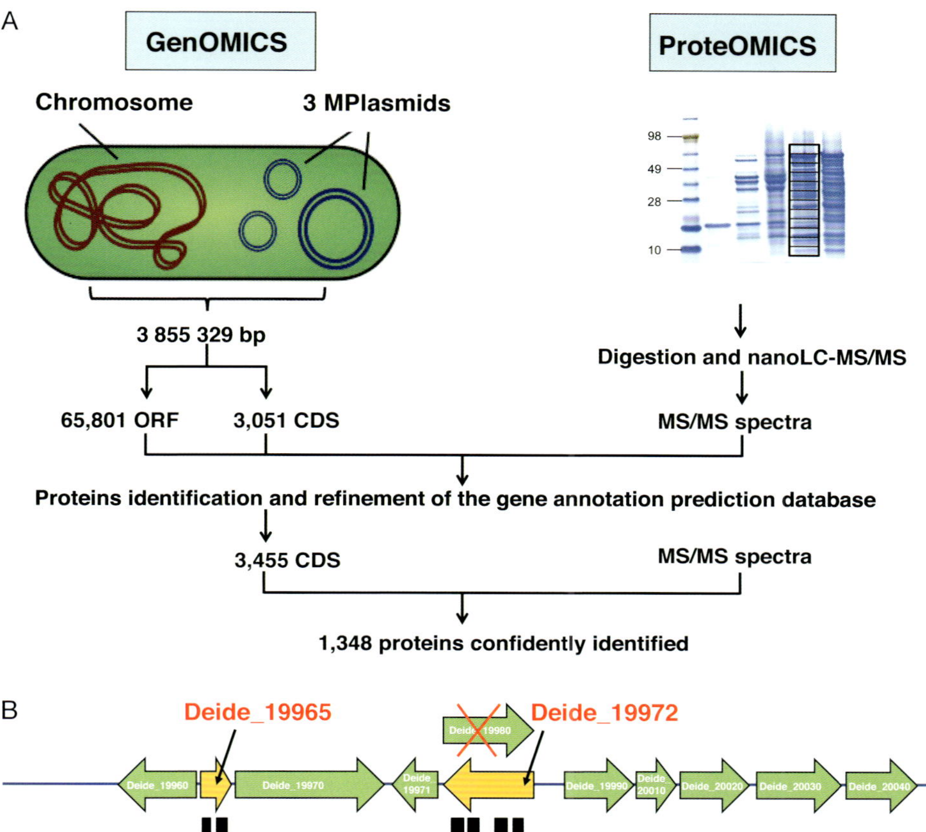

Figure 8.3 Proteogenomic strategy used for *Deinococcus deserti* annotation. (A) Complementary information brought out by genomics and proteomics. Sequencing of the chromosome and the three mega plasmids of *D. deserti* allow the access to the nucleotide sequence of this microorganism and the creation of both the open reading frame (ORF) database, comprising all the possible proteins encoded by the genome, and the predicted coding domain sequence (CDS) database. Proteins are first separated based on an SDS-Page gel, digested with trypsin, extracted and analysed with nanoLC-MS/MS with a high–throughput hybrid mass spectrometer. MS/MS spectra are assigned to peptide sequences and further to proteins by specific searches against the ORF and the CDS databases. However, as more proteins are identified from the ORF than from the CDS database, refinement of the settings to predict the CDS is necessary. As a result, 3455 CDS sequences are listed and a second analysis of the MS/MS data leads to the confident identification of 1348 proteins. (B) use of proteomic data to improve genomic annotation on a specific locus of the *Deinococcus deserti* chromosome. The same locus contained, surprisingly, two important annotations errors. Despite resorting to several annotation software programs, a small ORF was un-predicted and a predicted ORF was incorrectly annotated as the opposite strand was the real polypeptide coding region. Black rectangles represent peptides assigned to MS/MS spectra recorded on the *D. deserti* proteome by tandem mass spectrometry. Alignment of these peptides onto their corresponding reading frame allows the positioning of the genes identified on the forward or reverse strand. Coding sequences previously automatically annotated are indicated with green arrows. Two corrected, miss-annotated proteins are indicated with yellow arrows: a novel identified protein, Deide_19965 (a short polypeptide of 95 residues in the –1 frame), and a protein for which the coding strand has to be re-annotated, Deide_19972 (a polypeptide of 311 residues coded on the –3 frame that differs from the wrongly annotated Deide_19980 protein on the +1 frame).